Contents

Coastal Defences
Processes, problems and solutions

Peter W. French

London and New York

First published 2001 by Routledge
11 New Fetter Lane, London EC4P 4EE

Simultaneously published in the USA and Canada
by Routledge Inc.
29 West 35th Street, New York, NY10001

Routledge is an imprint of the Taylor & Francis Group

Typeset in 9.5 on 12pt Galliard by Wearset, Boldon, Tyne and Wear
Printed and bound in Great Britain by St Edmundsbury Press,
Bury St Edmunds, Suffolk

British Library Cataloguing in Publication Data
A catalogue record for this book is available from the British Library

Library of Congress Cataloging in Publication Data

French, Peter W., 1964–
    Coastal defences: processes, problems and solutions/Peter W.
French.
        p. cm.
    Includes bibliographical references and index.
    1. Shore protection. I. Title.

TC330.F74 2001
627'.58–dc21   00–051834

ISBN 0-415-19844-5 (hbk)
ISBN 0-415-19845-3 (pbk)

PART IV
Coastal defences in a changing environment   301

Figures

Tables

Plates

Boxes

Acknowledgements

No one person can source the amount of information present in a text book. While much additional information comes from published material in journals and other books, there is also considerable material in the grey literature, that vast assortment of unpublished material in the libraries and stores of coastal authorities, planners, non-governmental organisations, etc.

In preparing this text, many of these sources have been approached. Some of these approaches have been unsuccessful, but others have yielded important information and supporting material. Sources of unpublished information are acknowledged in the text as 'personal communication' or 'unpublished report' and all are especially thanked for their time spent in correspondence. The people and organisations listed below warrant specific thanks and gratitude for going out of their way to provide information, which has allowed the inclusion of better photographs, better case studies and better supporting material than might otherwise have been possible.

Chapter 4: Groynes and jetties
Environment Agency, North Wales, for supplying unpublished information on the Dinas Dinlle scheme.

Chapter 5: Cliff stabilisation
Mr. Stevens, Rother District Council for the provision of post-project appraisal and design details of the Fairlight Cove sill scheme. Also for the supply and permission to use Plate 5.4.

Chapter 6: Offshore structures
Mr. Ray Traynor, Arun District Council, for post-project appraisals and design details of the Elmer segmented breakwater scheme.
Cambridge University Air Photograph Unit for permission to use Plate 6.3.

Chapter 7: Beach feeding
Mr. Neal Turner of Bournemouth Borough Council for the supply of unpublished material relating to the Bournemouth recharge schemes, and for several personal communications relating to specific details and questions.

Chapter 8: Dune building
Prof. Ken Pye, Geology Department, Royal Holloway, University of London
for the supply and permission to use Plates 8.3a and b.
Mr. John Houston, Coastal Strategy Office, Sefton Council, for information on
the dune restoration at Formby Point.

Chapter 9: Increasing sedimentation in mudflat environments
Cambridge University Air Photograph Unit for permission to use Plate 9.2.

In addition to the above, my sincere thanks are extended to Justin Jacyno of the
Geography Department, Royal Holloway, University of London for his miracu-
lous conversion of random jottings and scribbles into meaningful diagrams.
Also, to Sue May for photographic advice and support. Finally, thanks go to the
various fellow staff in the Geography Departments of Royal Holloway and Lancaster University for support in writing this book.

Part I

Introduction

1 Introduction to coastal defences

Introduction

In its basic sense, the coastline represents the boundary between marine and terrestrial environments — the place where the land meets the sea. This simplistic statement may describe the coast as a geographic feature, but it does not highlight its complexity; the coast is not a permanent line to be drawn on a map, but a dynamic environment in which land and sea are constantly interacting (eroding and accreting) in response to external factors, both natural and anthropogenic in origin, and acting on a whole range of time scales (Table 1.1). Historically, humans have responded to erosion by building structures to resist it, so ensuring that the boundary between land and sea remains fixed and permanent. This is done under the assumption that the coastline is a feature which has always been in its current position and in its current form, and must never be allowed to change. In more recent times the wisdom of this approach has been questioned and radical changes in how the coastline is protected and managed have occurred.

These changes have been largely driven by new research developments which have provided a greater understanding of how coasts work. A notable example is the work of Per Bruun in the 1960s on the effects of rising sea levels on coastal landforms. More recently, such research efforts have become more focused and coordinated, leading to national and international collaborative programmes. One example is a UK-based programme established to investigate coastal processes and their interaction with engineering structures on a scale reaching beyond the local, scheme-based approach. This project, entitled the Coastal Area Modelling for Engineering in the LOng Term (CAMELOT), was set up to establish and validate coastal models in order to facilitate better and more informed management of long-term coastal change (MAFF 1993). A field-based validation study was also undertaken as part of the wider Land-Ocean Interaction Study (LOIS), funded by a leading UK research council. To date, this work has produced a series of coastal models, such as COSMOS-2D and COSMOS-3D, for the study of beach profile and nearshore bed changes, PISCES for the modelling of coastal behaviour, and BEACHPLAN, which can model variations in coastal position over time. A similar project is the European

Table 1.1 Time scale of coastal changes in relation to both absolute and human time scales (After French, 1997).

Absolute time scale	Human time scale	Coastal process
Millennia	—	Response of sea level to glaciation and tectonics
Centuries	Shifts in settlement	Historic coastal evolution — loss of towns and villages
Decades	Coastal defences	Formation and loss of habitats Coastal response to defences Erosion and accretion cycles
Years	Coastal defences Coastal management Coastal development	Coastal response to defences Longshore drift
Months	Impacts of population — industry and tourism	Natural seasonal adjustment Shore profile adjustment Coastal accretion/erosion
Weeks	Impacts of population Emergency protection works	Shore profile adjustment Tidal cycles Coastal accretion/erosion
Days	Emergency flood defence works	Storm surges, defence breaches Beach scour
Hours	Sewage, litter	Tidal cycles, shore normal sediment movement, storm activity
Minutes	—	Wave and current activities Cliff falls
Seconds	—	Sediment grain movements

Union (EU) funded MArine Science and Technology (MAST) initiative, which has involved close links with US facilities, such as the coastal monitoring facilities at Duck in North Carolina.

By using output from these and other research programmes, we now have a more complete understanding of how coastal processes operate, and how processes operating on one part of the coast may be closely linked to the behaviour of adjacent beaches in respect of sediment supply and sediment movement. A more holistic approach can therefore be adopted towards coastal protection which takes into account the knock-on implications of defending one part of the coast for those adjacent to it. For example, building a sea wall in front of an eroding cliff may well stop that area from eroding, but it will also stop the sediment from the eroding cliff entering the coastal sediment budget. If this sediment were important in supplying beaches down drift with sand, this supply would cease and these beaches would lose their sediment supply and may start to erode. Hence, solving one erosion problem has created another. Figure 1.2 illustrates this process for a hypothetical situation, which will be returned to shortly.

The concept of protecting the coast means different things to different people. These contrasting opinions arise from the plethora of concerns relating to the coast, including industry, tourism and residential sectors. It is obvious to people who live at the coast that their homes and businesses should be protected from flooding and loss of land and, because of this, it is paramount that sea walls, or some other engineered structure, hold the sea in place and do not allow it to encroach onto the land. In contrast, it is equally obvious to many coastal managers that the best way to look after the coastal environment is to leave it in as natural a state as possible, and to allow sediment to build up and protect the coast by natural processes. Immediately, therefore, there is a major difference of opinion between the people who live at the coast, and those who are charged with managing it. The question of which opinion 'wins' will depend on a multitude of factors, ultimately reducing to how much the threatened land is worth. A second issue here concerns the nature of the coast itself. The question of 'naturalness' with respect to the coast will occur time and time again throughout subsequent chapters. Whenever coasts are defended, or when people interact with them in some way, there is a degree of artificiality imparted in them. This may lead to changes in coastal processes, and thus in the behaviour of coastal landforms. The question, therefore, is, just what is a 'natural' coastline or, more philosophically, are there any 'natural' coastlines left, given the amount of interference on the world's coastline?

The coastline is generally a highly populated area, both in respect of high resident numbers, and a large transient population who visit during the summer as tourists. Estimates from Goldberg (1994) state that 50 per cent of the world's population lives within 1 km of the coastal zone, a figure expected to increase by 1.5 per cent by 2010. Given this large population, it is not surprising that many coastal areas are highly developed and, thus, have an associated high land value. In such situations, it is imperative that the development is protected from the sea; some form of defence structure is an obvious solution to the protection of these areas. In contrast, however, smaller areas of development, such as individual hamlets and villages, or individual farmsteads, do not command such high land values and do not justify such expensive defences. As a result, there is an increasing tendency to reject defence construction. This problem touches on another important debate in coastal management, that of whether to construct defences or not. For reasons which will become apparent in subsequent chapters, sparsely developed coasts are increasingly being left undefended in order to maintain sediment supply to the coastal zone, even if this leads to the loss of property.

Historical background to coastal defences

Given that the boundary between land and sea is highly dynamic and constantly changing, one could be forgiven for questioning why some major coastal developments found on today's coasts were ever built. Clearly, however, this is an historical issue and we cannot equate the original reasoning with the relatively well informed decision making of today's planning process.

Initially, sea defences were not built to protect development but were constructed to protect land claim (land taken from the marine environment and converted to agricultural land) from inundation by the sea. Evidence of this process includes Romano-British land claim in many areas such as the Severn Estuary (Allen and Fulford 1990), tenth-century reclamations in the Severn (Allen 1986), twelfth-century reclamations on the Dutch coast (de Mulder *et al.* 1994), thirteenth-century reclamations in Morecambe Bay (Gray and Adam 1973), seventeenth-century reclamations along the north Norfolk coast (Cozens-Hardy 1924), and seventeenth-century reclamations bordering the Dutch Wadden Sea (de Jonge *et al.* 1993). Increasingly, however, as international trade increased and industrialisation occurred, land also started to be claimed for development as a supply of cheap, flat land close to ports and water supplies for industrial use. This development was driven by nations' need to establish a solid industrial and economic base. Much of this industrial expansion necessitated the development of ports and harbours, thus impinging on sheltered areas of coasts, such as lagoons, estuaries and embayments. Similarly, industry requiring water in large quantities found the coast an ideal setting. These developments often involved altering the natural coast by land claim and dredging. Although unaware of this at the time, these developers were initiating many of our current coastal problems as the natural coastal environment became more artificially constrained by man-made structures (see French 1997, Chapter 4, for discussion). Nobody would argue that ports and harbours are unimportant. They play a vital role in a nation's economic and industrial base. Since they are essential facilities, coastal managers need to incorporate them into the overall coastal strategy in as accommodating a way as possible.

Although land claim and port development initiated artificial coastal modification, since the onset of the nineteenth century it also started to become fashionable to 'take the sea air' (Goodhead and Johnson 1996). This trend started a new pressure on the coast, that of amenity and leisure use. During the nineteenth century, resort areas began to develop in Europe and, by the early twentieth century, in the USA as well. By the late twentieth century, increased leisure time and prosperity meant that the tourist market grew; many areas caught the tourism 'bug', and allowed development of coastal areas for tourism. This, again, put new pressures on the coast, requiring further modification of the natural environment (see French 1997, Chapter 5, for discussion).

This drive for economic prosperity has produced many defence problems for coastal areas, ironically costing large amounts of money to put right. Howard *et al.* (1985) illustrates these issues for the American shoreline, and shows how millions of dollars have been spent in response to increased coastal problems due to development. Furthermore, some of this defence work has produced a new set of problems for adjacent coastlines by causing increased erosion which has, in some cases, caused the loss of the very property which schemes originally set out to protect.

However, criticism of historical developers cannot always be justified. All the time that a stretch of coast is accreting and developing seawards, the development of the shoreline or the claiming of part of the intertidal zone for development does not pose too great a problem. In such situations, it is understandable that areas of newly formed land were considered available for development. In reality, however, by claiming this land and constructing a line of coastal defence, we are, in fact, artificially moving the coastline seawards, and by building infrastructure on that land, we are committing ourselves to protecting it from any future changes in coastline position. As the accretion/erosion cycle completes and beaches disappear through erosion, these same coastal developments suffer and demand for protection increases.

Problems of a static coastline

Perhaps the main problem with hard defences is that, once built, they fix the coastline in the position it was in at the time of construction. However, coastlines are not static structures; they migrate landwards and seawards over a variety of time scales in response to forcing factors, such as sea level (SL), wave climate, and seasons. The type of coastline will govern the severity of the restrictions; for example, dune coasts will fluctuate seasonally to changes in sediment transfer to and from the beach without which the complex dynamic stability of the dune/beach system will be lost. On cliffed coasts, the main problem lies with the input of sediment to the coastal sediment budget, the loss of which can have important repercussions for the beaches down drift.

The problems of a static coastline can be summed up as follows:

- inability to respond to sea level changes in the medium and long term (coastal squeeze);
- cessation of beach/dune interactions;
- cessation of sediment inputs to the sediment budget;
- instability in fronting beaches

All of these factors will promote instability in the coastal system and therefore induce erosion as the system attempts to regain a form of dynamic equilibrium. Preventing a coastline being able to respond to sea level rise (SLR) is of paramount importance. In a world where many of our coastlines are experiencing this phenomenon, increased coastal instability will occur as the mean sea level increases. Of critical importance in this is a process referred to as 'coastal squeeze'. This is explained in detail in connection with managed realignment (see Chapter 10); however, it is a problem which can affect all coasts. The Bruun rule of sea level rise (see Chapter 2 for full discussion) states that in order to maintain dynamic equilibrium under conditions of sea level rise, the equilibrium beach and nearshore profile needs to relocate upwards and landwards (Figure 1.1). Clearly this is unproblematic where lack of coastal development and an unprotected coastline occur, but where there are hard defences such landward movement is prohibited and the coastline cannot reach an equilibrium with the new sea level conditions. This is proving a major problem along many shorelines, not least because the inherent instabilities which result from the inability of coastlines to adjust to sea level rise are often manifest in loss of beach sediments, particularly important in tourist areas. Pilkey and Wright (1988) exemplify this process still further, and suggest that not only do sedimentary and floral environments become squeezed, but so do processes. During storms on undefended coasts, the surf zone widens in the landward direction. Where walls have been built, this landward extension may not develop fully, concentrating the greater intensity of surf zone processes into a confined area. This may help explain some of the observations of walled coasts, where sediment losses are greater during storms than on adjacent non-walled coasts. Indeed, Pilkey and Wright develop their argument with studies supporting the intensification of longshore currents, wave reflection, rip-currents, and pressure-gradient related currents; all of these could be responsible for increased sediment losses from beaches fronting walls during storms.

The interaction of dunes and beaches is also important along wind-dominated coasts because it allows the beach to react to summer and winter wave regimes. These ideas are developed fully in Chapter 8 on dune building, but in this context, hard defences will prevent such sediment transfer to and from dunes. This means that during summer (calm) conditions, sediment transfer to dunes is prevented, while in winter, the supply of sand to the beach via draw down from dunes is also prevented; the result is that as beaches change towards their winter profiles, material to support this cannot come from the dunes, but comes from the upper beach instead. This is particularly critical in stormy periods, as there is no available sediment to source beach profile adjustments (Oertel 1974). This may then promote upper beach loss and eventual sediment starvation to upper beach areas. Similar problems also occur in barrier island settings, where over wash is critical in maintaining the stability of the barrier feature. In support of this, Inman and Dolan (1989) identified over wash as

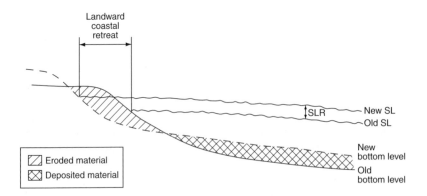

Figure 1.1 Simplified model of landward coastal retreat under sea level rise (Bruun rule). Note landwards and upwards movement of profile.

accounting for 39 per cent of shoreward sediment movement in the Outer Banks of North Carolina, this process being fundamental in maintaining barrier stability.

When dealing with cliffed coasts, the use of hard defences to prevent cliff erosion can be very effective. Although the processes are complex (see Chapter 5), the use of sea walls to stop cliff erosion is common. Regardless of technique, however, the fact that the hard defences are preventing any interaction between waves and cliff base means that erosion will stop — but, more critically, the input of cliff sediment to the coastal sediment budget will also stop, potentially leading to the loss of fronting beaches.

Contemporary coastal managers now have to address a legacy of coastal problems which have arisen following hard defence construction, largely as a result of coastal development which must be protected, even though it may not be built in places which are easily defended. Such problems often restrict best defence practice and prevent the ideal defence scheme being built. French (1997) highlights this problem with respect to nuclear power stations in the UK and shows that of the fourteen localities around the UK coastline where such installations occur, nine may have future defence problems due to low-lying land or coastal erosion. Clearly, such significant developments must be protected from any shifts in coastal position and, no matter what factors come into play with regard to coastal movements, these sites will need to be protected.

Historically then, contemporary coastal managers have been left with a legacy of erosion and defence construction which will be expensive to correct and maintain. The former belief that humans are the masters of the natural environment and that science will provide the solution to all problems was unfounded. The fact that this belief led people to build closer and closer to the coast has left many problems which can only really be solved by building hard structures to keep the sea out and by spending a lot of money. Clearly, this cannot happen indefinitely and so, in response, new ideas are starting to appear.

Recent trends in defence planning

When discussing coastal defence, several distinctions can be made. The term 'coastal defence' is a much used, but also much abused term for any feature along a coast or estuary designed to protect the hinterland. It is more correct, however, to distinguish between two forms of coastal defence. First, *flood* or *sea defences* are structures used to prevent the hinterland from being flooded by the sea, and thus refer purely to the prevention of inundation of the land. Second, *coastal protection* structures are used to prevent the hinterland from being eroded and thus relate to the prevention of land loss to the sea. In some situations, however, a single structure may serve both purposes, leading some commentators to suggest that this distinction becomes confused in places.

While this broad classification is useful in distinguishing the main types of coastal structure, within it there is a wide range of structures, each of which tackles the problem of erosion and flooding in a different way. Each type of structure will form the basis of a subsequent chapter. Before being discussed in detail, however, it is important to understand a more fundamental defence classification which has become the focus of recent coastal management, that of hard and soft defences (Table 1.2). The obvious distinction apparent from Table 1.2 is that hard defences are those which involve the construction of some solid structure within the intertidal zone to reduce wave energy and stop the sea from interacting with the hinterland. Such structures tend to fix and hold the coastline in position and allow no flexibility in response to external variables, such as sea level rise or changes in storminess. In contrast, soft defences generally avoid the solid structure approach, but use natural environments and sediments to reduce wave action. In effect, wave activity can be stopped or reduced by building up beach levels or increasing the width of a salt marsh, thus avoiding the need to construct costly physical barriers to protect the land from the sea.

Prior to World War 2, it was common to protect coastlines with hard structures, typically sea walls and groynes. This trend arose largely due to the relatively poor understanding of coastal processes (Fleming 1992). It was only after the war that a greater understanding of how coasts 'work' began to arise, driven by the development of more informed theoretical models; these increasingly led to the original 'hard' approach being questioned. The models originating out of the CAMELOT and MAST programmes cited earlier are typical examples. In the early 1970s, the National Park Service in the USA adopted a policy of allowing shorelines to remain natural wherever possible; the argument was that, while hard defences undoubtedly protect the hinterland, their impacts on fronting beaches and adjacent parts of the coast could be so severe that the negative impacts outweighed the positive. This could be considered a relatively forward-looking policy and, by the 1980s, traditional methods were being questioned in many countries, resulting in environmentally sensitive coastal planning becoming more common (Pilkey and Wright 1988). This shift in policy was not straightforward, however, as many more 'traditional' coastal researchers continued to argue that despite the clear limitations to hard defence structures, the extent of their negative impacts was inconclusive. However, during the 1980s one of the key research arguments to emerge was that on the seaward side of a sea wall, reflection of wave energy

Table 1.2 Different coastal defence approaches and their respective methodologies.

Hard defences	Soft defences	Indirect solutions
Sea walls	Beach feeding	Building restrictions
Breakwaters	Dune building	True-cost insurance
Revetments	Increasing natural sedimentation	Holistic management
Groynes	Managed realignment	
Gabions	Abandonment	
Offshore breakwaters	Beach drainage	
Hard points	Do nothing	
Armourstone		

would concentrate scour on the fronting beach, leading to the transport of beach sediments away from the beach in either a longshore or offshore direction, and producing a drop in beach elevation. If such a physical barrier was not present, then this beach loss would not occur. Supporting evidence for this was supplied from many sources; one being Porthcawl in south Wales, where beach levels dropped by almost three metres in 75 years following sea wall construction, thus causing the sea wall to be undermined and eventually collapse (Madge 1983). Similarly, work on the Lake Ontario shoreline (Davidson-Arnott and Keier 1982) reported that 70 per cent of structures were damaged due to increased wave action in response to beach lowering within 10 years of construction and that, within 30 years, damage would be expected in 96 per cent. Immediately, there-fore, we have a case against the use of sea walls — although counter to this is the argument that due to the nature of coastal development, a wall may remain the only practical way of protecting some coasts. However, this increased understand-ing, which has led to the advent of soft engineering, allows engineers to make refinements to hard structures, thus negating some of their more detrimental impacts. Such developments are important because, where hard structures are the only solution to an erosion problem, engineers can design them to be increasingly environmentally sympathetic.

Reduced longshore sediment movement in groynes and resulting sediment starvation down drift have also been identified as problems with hard defences. By not allowing the free movement of sediment along the coast, problems of sediment starvation can occur. At this point, it is perhaps useful to return to the earlier example of a cliffed coastline, as it provides a good illustration of what can happen when defences are first instigated. In Figure 1.2a, erosion of the cliffs provides a sediment supply which is moved along the coast by longshore drift to facilitate the build-up of beaches down drift. As a result, the loss of land due to cliff erosion is offset by the build-up of beaches and, in this example, a spit. In economic terms, the healthy beach also provides good protection for settlements down drift, a good tourist amenity, and protection for the port located in the mouth of the estuary. As such, cliff erosion is a critical element in the economic and defence security of the adjacent coastal settlement. After a period of time, a settlement on top of the cliff becomes threatened due to the retreating cliff edge, and it is decided that coastal protection is necessary. As a result, a sea wall is built along the front of the cliff and is successful in prevent-ing further cliff loss. Despite the fact that the cliff is now defended, coastal processes such as longshore drift still operate and remove sediment from the base of the cliff as they did pre-defence. However, this occurs at a higher rate due to the increase in wave energy, as waves are no longer using energy to erode the cliffs. The equation encapsulates this:

Total wave energy = energy for erosion + energy for sediment transport

and shows that, with no change in total wave energy, if erosion of the cliff is prevented, then this part of the equation becomes zero, and the energy available

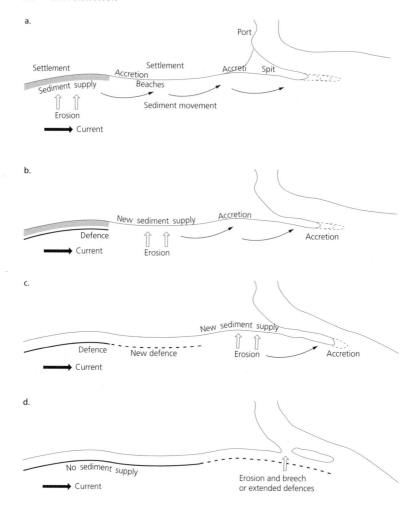

Figure 1.2 Progressive erosion scenario for defended coasts. As more defences are added to prevent erosion, so focus of that erosion is transferred along coast (After French 1997).

for sediment transport is increased. In this case, the nearest available sediment is the fronting beach, and it is this which replaces the cliff as the main sediment supply for longshore drift.

Before the building of the sea wall, beach sediment would still have been moved along the coast by longshore drift, but would have been periodically replaced with material from cliff falls. Now the wall prevents the replacement of beach sediment and so the beaches rapidly lose their store of sediment. With time, all this beach sediment would be removed and transferred down drift. Given the equation above, no transport of sediment will increase the energy for erosion; no more material can be eroded from the beach, so the focus of erosion is shifted down drift to another land-use or settlement. In time therefore, erosion down drift of the cliffs (Figure 1.2b) has started on what was once an accreting beach, which is now the main sediment source for longshore drift. This, therefore, poses the same set of management problems as before, but shifts it to a neighbouring coastal area, possibly even to another defence authority. In time, with ongoing erosion, new defence measures will be required here (Figure 1.2c), setting off the same chain of events as before. This process will continue along the coast until the end of the stretch of coastline which, in this example, is the spit (Figure 1.2d); its erosion may threaten the port, and it therefore also needs to be defended. Ultimately, the decision to build a sea wall along one part of the coast has triggered a chain of events which has resulted in the whole of the coastline needing to be defended, at high cost and with considerably reduced tourist appeal.

This hypothetical scenario represents very much a traditional approach to solving the problem. The chain of events described poses an important question: would it not be possible to tackle the initial problem in a different way, so as to reduce the knock-on impacts of the initial defence technique? If one takes a step back and identifies the original cause of the problem, the whole issue could be looked at differently. Initially, cliff erosion was the trigger for defence construction. Cliff erosion is not a continuous process but occurs intermittently. The reason for this is that when a cliff fall occurs, sediment is piled up at the base of the cliff, thus protecting it from further erosion. With time, waves transport this material along the coast, thus exposing the foot of the cliff again, at which point wave energy is refocused on the cliff base which will eventually lead to a new cliff fall. Herein lies the cause. As an alternative strategy, if the coastal defence policy had targeted its efforts at making sure the foot of the cliff was never exposed to wave activity, new cliff falls would not occur. Given that the reason for this was the longshore transport of beach sediment, artificial replacement of the latter by beach feeding (see Chapter 7) would mean that new sediment supply to the sediment budget occurred without the need for cliff-derived input, thus stabilising its position. The artificial supply of sand would need replacing at intervals but the cost, both in terms of money and aesthetics, would be less than shrouding the whole coast in concrete. A second option would be to have done nothing and to have let the original problem continue and the cliff continue to erode, even though this would result in the cliff-top settlement falling into the sea. Although contentious, such practices do occur in coastal planning, largely in situations where the supply of cliff sediment is so important that to reduce it with sea defences would mean that many other coastal areas would lose critical sediment supplies. A good example of this occurs on the coast of Holderness in the UK, where rapid coastal erosion provides sediment which is of prime importance, not only for UK beaches, but also for other North Sea margin states (de Ruig and Louisse 1991).

It is clear from this example that different methodologies exist to tackle particular problems. The decision about which to adopt will depend largely on economic factors. A further issue, however, lies in how the coastline in question is managed. In many situations, the decision to defend comes as a reaction to the identification of an erosion problem. This reactive management is typical of many hard defence situations and is well exemplified in the progressive erosion scenario detailed in Figure 1.2. Here, the response of the coastal manager is to react to each new defence problem as it arises. An alternative is to anticipate potential problems. The alternative scenario where beach feeding is used will eliminate the further problems highlighted in Figure 1.2 because the reduced sediment supply is anticipated and 'factored' into the beach feeding solution. This more proactive approach typifies many (although not exclusively) softer approaches.

With this increased awareness of the impact of hard defences on the coast-line, the problems of constructing a physical barrier between the land and the sea have become more relevant in coastal policy discussion. Brampton (1992) demonstrates how engineers started to turn to softer methods once they realised the key role which beaches play in coastal protection. By continually adjusting their shape, beaches can absorb much of the incident wave energy which hits the coast, thus reducing the force on the structures behind. This process is typically illustrated in the short term with the formation of winter and summer profiles, where beaches flatten out in winter but steepen in summer. This can be seen in Figure 1.3 for the Holderness coast. Here, the winter profile is flatter than the summer profile. It can also be seen that offshore, the winter profile is actually higher in elevation, representing a transfer of sediment from the upper shore, seawards. This effectively moves the wave base seawards, and increases the distance over which waves are in contact with the sediment surface and, therefore, the area of energy dissipation (see Chapter 7).

By the 1990s, engineers were continuing to question the efficiency of hard defences. Discussions started to focus on the need for defences which interact with, rather than against, natural processes. One aspect of this is the desire to maintain 'natural' coastlines, in which nature and conservation issues are main-tained. A second is the issue of cost. Hard defences are expensive to build and have a finite life span, variable in length according to construction techniques

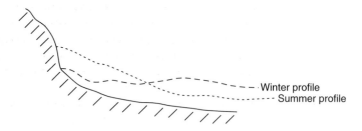

Figure 1.3 Contrast between winter and summer beach profiles. Based on data from Dimlington, Holderness, UK (Modified from Pethick 1992).

and the energy regime in which they are situated. Pearce (1993) highlights this problem of longevity with respect to the coastline of the UK. Repairs alone cost £58 million in 1991, and this programme of repair is not keeping pace with need. Clearly, the financial implications are considerable. This is not to say, however, that soft engineering is cost free. Techniques such as building up beaches or removing sea walls have their own cost implications, but these tend to be less than hard structures, and may also have lower costs associated with maintenance, such as repeat feeding. One way in which coastal authorities have responded to the issue of cost is to proliferate the use of armourstone in coastal defences. While not being either a true 'soft' or 'hard' method, the placing of large rock boulders along coasts represents a shift by coastal authorities to cheaper forms of coastal defence. Many coasts now exhibit armourstone features; they are typically used as a form of additional protection to existing structures, such as by being placed, for example, along the foot of existing walls to reduce toe scour, or marsh edges to reduce lateral erosion.

Given that soft engineering is now becoming more acceptable and popular as a means of defending the coastline, many managers have been asking whether the soft option would be a more effective option for coastal management in their area of jurisdiction. The choice is based on three main factors:

- *Geomorphology* The hinterland erosion which is currently being stopped by hard techniques could be re-activated in places and utilised to provide new sediment to beaches and other areas of coastline, thus providing enhanced sediment supply. As such, hard defences in areas where hinterland value is low could be removed, or not replaced when their natural life span ends, and the land allowed to erode. This would increase the volume of sediment entering the sediment budget and thus the amount available to build up beaches.
- *Financial* The cost of building sea walls lies between £1–£5 million per kilometre, with a design life of around 50 years. Although some coastal areas are clearly of high value and merit such expenditure, many are less valuable and could be as equally well protected by soft techniques, such as building up beaches with sediment.
- *Environmental* The creation of natural environments, such as sand dunes or salt marshes, for coastal defence purposes will also lead to the creation of new habitats for wildlife. Given that throughout the world countries are losing such environments to development, sea level rise and coastal squeeze, any new habitat creation to offset this loss would be welcomed. Furthermore, natural environments can respond to any future changes in sea level, and thus negate many of the additional costs associated with the repair of hard defences.

Thus, during the early 1990s, the emphasis in coastal management switched from hard to soft, although it is clear that hard defences are still required in many situations. This trend continued to dominate coastal defence management throughout the 1990s, supported by an almost continuous flow of scientific publications from around the world relating to the negative impacts of hard coastal structures. As we enter the twenty-first century however new research is now starting to re-examine some of the issues pertaining to soft coastal engineering. Ideas are emerging which claim that techniques, such as managed realignment and enhanced sediment accretion, will also have associated implications, and that these should be considered in planning decision making. Although these research ideas will not be dealt with in detail here, they will be discussed in many of the following chapters; it will suffice to emphasise earlier statements concerning the interference in coastal processes of defence techniques. Whatever the defence methodology employed albeit hard or soft, all will impart some degree of artificiality into the coastal environment and will, therefore, induce some degree of process modification.

Perhaps in concluding this section it should be noted that soft defences do have their detractors. One of the main problems is convincing the public who live in the areas concerned, that effective protection of their homes and businesses is still possible by the placing of sediment on a beach or the building of a salt marsh, instead of building a concrete wall. The public perception of security is critical here and it is true that during the planning process, many people admit to feeling much more secure behind a concrete wall than they do behind a beach. Security is further questioned when decisions are made over their heads by planners and managers who live well away from the area concerned. Clearly, this is partly a problem related to public information and education, and can be best dealt with by the planning and public consultation process; but it also represents a new area of concern for coastal planners and managers.

In summary, a clear trend has emerged, moving from the building of walls and groynes to the increased use of natural processes in defending the coastline. The use of hard structures was fully justified at the time of construction, and arose as a response to the rapid way in which many developed coasts were eroding. Over the years, however, it has emerged that these hard defences do have longer-term problems, the most fundamental of which is that they were constructed at a time when a full understanding of how the coast works was not available. Indeed, because we still do not know the answers to many fundamental questions relating to coastal 'functioning' it is likely that, as with hard defences, current soft defence methods may themselves be subject to re-evaluation and change in the future, as new knowledge emerges. To some extent, this is already happening because practice has shown that certain problems can and do arise in relation to soft defence techniques.

Despite the range of techniques, and the arguments relating to them, it is clear that coastal defence is increasingly being carried out in sympathy with the coast. Even if new hard defences are built, and many are, this is because after full consideration of the local coastline, these structures are considered the best solution to a particular problem. However, because we now have a better understanding of how a coast works, even hard defences can be built so as to reduce their impacts as much as possible. It remains clear, however, that the most effective way to manage the coast is to do it as simply as possible — to leave it alone and let it find a stability of its own. This method is currently being adopted in the UK along large sections of the Holderness coastline. Here soft glacial cliffs are being allowed to erode and fall into the sea, even though this involves the loss of isolated farms and houses. However, also along this coastline are areas of higher value land, where settlements have developed. These are too important to be left to the sea and so hard defences (sea walls and groynes) have been constructed. This provides a good example of the interaction of hard and soft defence techniques along the same stretch of coastline, and it has been argued by some commentators that, along the lines of the progressive erosion scenario detailed in Figure 1.2, these areas of hard defences are leading to accelerated erosion on down drift areas of undefended coastline.

While the policy of doing nothing is not possible in many situations, if we accept that the best way of defending and managing a coast, given local complications, is to do it in a way which most clearly mimics nature, then we will have a coast which produces a minimal number of problems, and the greatest amount of 'naturalness'. If such a strategy is adopted, then coastal defence and management will become more effective with respect to habitat protection, more effective with respect to cost, and, even though it may not appear so to individuals, most effective for the coastal population as a whole. By adopting such an ideology, working with the coast can prove to be a sound long-term investment with reduced repeat and maintenance costs to the protection authority. In this approach, coastal managers and engineers are often faced with a key decision: just what is 'natural' along the coastline? In developed countries, many coasts have a long history of defence construction and it could be argued that any 'naturalness' was lost many years ago. Perhaps, therefore, managing coastal defences in accordance with the 'natural' processes (waves, tide and wind) which operate along any particular piece of coast is the best that can be achieved.

Structure of this book

This book is intended to provide the reader with an understanding of the types of coastal defence strategies which can be used around the world's coastlines, of the problems which can arise from their use, and of the ways in which these may be overcome. Each chapter is presented in the same format, explaining the reasons for using a particular defence type, methodologies, problems, benefits, and illustrated with case studies from around the world. Each chapter also includes a table detailing other relevant case studies which have not been included in detail, and which the reader can use to develop areas of interest further.

The general structure of the book is of an introductory part of two chapters followed by three further parts. This chapter has introduced the reader to the ideas behind coastal defence. Chapter 2 introduces the fundamental natural processes which play a role in governing which type of defence strategy is used.

It is not intended to act as a detailed discussion of coastal processes, for which the reader is referred to such texts as Pethick (1984), Carter (1988), Hansom (1988); nor is it intended to cover all problems associated with coastlines, but will concentrate on those associated with coastal defences. For these, the reader is referred to such texts as Viles and Spencer (1995) and French (1997). By the end of Chapter 2, it is intended that the reader will have enough knowledge of how the coastline works to be able to understand the concepts discussed in the rest of the book.

Within the three main parts, Part II deals with hard approaches to coastal defence and includes chapters on sea walls and revetments, groynes and jetties, cliff stabilisation techniques, and offshore structures for wave reduction. This part deals with the traditional approach to defence but will also stress that there are many situations in which the techniques discussed remain vital for adequate defence and protection of the hinterland. Part III deals with soft engineering solutions to coastal defence problems, and includes chapters on beach feeding, dune building, increasing natural sedimentation and wetland creation and managed realignment. The final part, Part IV, discusses other coastal defence possibilities, such as the potential of using building regulations, insurance, etc., where the problems of defence are removed because the reasons for needing it (i.e. development) are not present, as the costing framework makes development of vulnerable coastlines non-viable. Finally, a concluding chapter draws together key issues from the text and identifies where the topic of coastal defence management is heading.

2 Role of coastal processes in coastal defence management

Introduction

Any person who regularly visits the same stretch of coastline will have realised that it is not a feature which is permanent and fixed in time, but one which continuously changes. These changes occur over a range of time scales (Table 1.1); they may come about in response to large-scale events, such as glaciations or orogenic cycles, that may significantly alter sea level and thus the actual position of the coastline. In such situations, the processes that formed the current coastal features are not operating today in the same way, and can be said to be inherited. Alternatively, this may arise in response to smaller-scale events, such as storms, or just waves and tides which bring in sediment, move the sediment around which is already there, or take sediment away. In this case, observed changes are due to contemporary coastal processes. Clearly, considering which features of the coast are due to inherited processes, and which to contemporary, is fundamental in our understanding of how coasts are functioning at the present time. This book is mainly concerned with these smaller-scale contemporary processes and events. Whenever the tide comes in, waves start to reshape the coast. Meteorologically, a storm will bring higher energy waves and produce correspondingly greater changes than a period of atmospheric calm. The problems for management only occur when these natural processes start to impinge on human activity. It is then that people demand a human response in order to continue their activities, such as tourism, settlement or industry.

The coast is a transient feature, and the natural thing is for it to erode or accrete sediment. There are times when the sea will lay down great sand deposits as sandbanks, or mud as mudflats. It does this when sediment is in surplus and, in effect, just as people put their money in a bank as temporary storage until needed, the sea places its sediment in sediment banks until sediment budgets are in deficit. It is for these reasons that coasts are constantly changing and sediment is constantly being moved from one place to another. Integral in this is the idea of sediment storage. People may describe a coastal feature as a beach, a sand dune, a salt marsh, etc. These descriptions are fine — but an informed coastal manager may well describe them as a 'sediment store'. After all, we can look at a sand dune as a pile of sand, temporarily placed on the upper beach (supra-littoral) area. Similarly, a salt marsh is a store of mud and silt, a beach or spit is a store of sand or shingle, and even a cliff can be regarded as a store of potential beach sediment. There are times when these stores are added to by the sea (e.g. dune development or salt marshes building seawards (Plate 2.1a)), but also periods when the sea wants to 'realise its assets' and starts reworking them (Plate 2.1b).

These natural cycles of accretion and erosion occur all the time and, as has already been stated, may occur without cause for concern. This can be seen in Plates 2.1a and 2.1b. Both show salt marshes in the estuary of the River Kent in Morecambe Bay. In Plate 2.1a, a rapidly prograding marsh is forming, while in Plate 2.1b, the marsh is retreating landwards. Geographically, these two sites are located on opposite sides of the estuary mouth and the contrasting erosion/accretion regime is thought to be driven by a natural cyclicity induced by the lateral movement of the tidal channel within the confines of the estuary. Hence, when the channel is located towards Silverdale (Plate 2.1b), the marsh will be eroded as this will be the main focus for ebb and flood tides. Correspondingly, across the estuary, Kent's Bank receives little tidal energy and sediment can accrete. At intervals, the channel migrates laterally across the estuary and the accretion/erosion situation is reversed (Pringle 1995). In other examples, however, such natural cycles do produce management problems, particularly when humans have built in close proximity to these areas, because then engineers have to interfere with natural processes in order to protect the use of the coast. By such interference natural processes are altered or stopped; immediately this happens, the coast is forced into an unnatural state in which processes no longer behave as they would normally do. At this point engineers are fighting nature because the sea will constantly try to revert to a natural 'equilibrium' state, at the same time that engineers are constantly preventing it from doing so. The art of successful coastal defence is to reduce this conflict as much as possible.

In order to understand how natural processes can affect the management of the coast, it is necessary to consider some fundamental aspects of coastal dynamics, i.e. waves, tides, and wind.

Importance of waves

The most fundamental impact of waves with respect to the coastal zone is that they introduce energy to the coast, and energy is responsible for sediment erosion and sediment transport. In addition, waves can also produce a series of currents, such as longshore drift and shore normal currents, which are themselves important in sediment movement. In this respect, waves can be seen as one of the main formative mechanisms of the coastal environment. The detailed physics of waves are not critical here but are well described by Pethick (1984), Carter (1988) and Komar (1998). Our interest lies in the behaviour of waves as they affect the defence of the coastline. It is important to understand that it is the wave form that moves and not the water, whose movement is restricted to

Plate 2.1 The cyclicity of sediment stores. Salt marsh erosion and accretion in the Kent Estuary, Morecambe Bay, north-west England
    a) Accreting *Spartina* salt marsh at Kents Bank (western shore of Kent estuary)
    b) Eroding salt marsh by lateral marsh cliff erosion at Silverdale (eastern shore of Kent estuary).

circular or elliptical paths (Figure 2.1). In open water, water particle movement is circular although, towards the base where frictional forces increase, these circles become elliptical. Towards the shore, this distortion increases until a point where the ellipses fail to close. At this point, waves start to break and there is a mass transport of water and contained wave energy either into the surf zone or, ultimately, directly onto the beach. When this occurs, waves start to interact with the beach shape, and any coastal defence structures built on it, or landward
of it.

As can be seen from Figure 2.1, in deep water the circular motion of the waves peters out at depth and no motion occurs beyond a certain depth (known as the wave depth and roughly equating to half the wave length (Carter 1988, Hansom 1988)). In terms of sediment movement, sediment which is on the sea bed below this level is largely immobile and unavailable for transfer to the beach area because there is no moving water in contact with the sediment to initiate grain movement. As the sea floor shallows, however, there is a point at which the petering out of the wave form coincides with the sediment surface. At this point there is still some water molecule motion remaining at the interface between sediment and water which, if of sufficient energy, can start to transport sediment. This point is known as the *wave base*, and is of importance because the extent of the sea floor between here and the shore represents that area of sediment which can be transported by the water column and supply the coast with sediment. It can be easily appreciated that the depth at which the wave base occurs will vary according to the size of the waves. Larger waves, as generated during storms, will cause greater disturbance of the water body and greater penetration of wave form to a depth, producing a deeper wave base where more

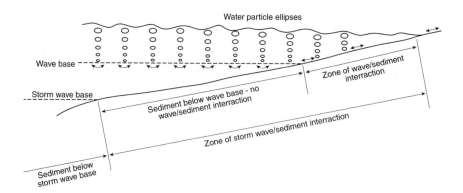

Figure 2.1 Wave form and relationship of wave base to sediment movement. Note amount of sea floor which becomes interactive with wave motion as wave base increases (Modified from various sources).

sediment will lie within the area of reworking. This is known as the *storm wave base*. During calmer conditions, wave base will be much shallower and less sediment will lie within the area of reworking. This is known as the *normal wave base* or simply *wave base*. This is shown in Figure 2.1, with the area of sediment between normal and storm wave base representing sediment which can only be moved under storm conditions.

From the above, it can be seen that whenever there is wave activity on the coastline, there is the potential for sediment movement. One important qualifier to this, however, is that the energy available in the wave has to be sufficient to initiate sediment movement. Clearly, this suggests that waves need more energy to move sediment on shingle beaches than on sand due to the larger, and heavier, grain sizes involved. Similarly, waves need more energy to entrain clay particles than silt and fine sand grains because the sediment is cohesive. These principles were outlined by Hjulström (1935) who assessed the principles of entrainment velocity and sediment grain-size. In support of this early work, Carter (1988) provides a detailed discussion relating to sediment entrainment and the modifications to this process caused by modifications to flow regime and bed shear stress. Clearly, the turbulence of flow can be changed by the roughness of the bed; sandy substrates being generally 'smoother' than shingle. The key factor, however, is the shear stress: when this is sufficient to dislodge grains from the bed, particle motion can begin. In simple terms, this means that sediment can be transported (see Carter 1988 for detailed explanation).

The energy of the waves as they approach the coast, therefore, will govern the area of the sea bed over which sediment can be reworked, and the type of sediment will govern the ease with which particles can be entrained and moved. As the waves move inshore, the sediment surface grades upwards and water depth shallows, thus accounting for another feature often noticed at the coast. When a wave breaks and nears the shore, it becomes higher; this occurs as a result of the physical laws of the conservation of energy, but also because of friction with the bottom sediment. Increased friction is greater at the bottom of the wave than at the top because of contact with sediment, and so the wave base receives greater frictional drag than the surface waters, with the result that the surface waters travel faster. When the surface water 'overtakes' the water at the sediment surface, the wave breaks and collapses. However, this is a very simplistic view. Horn (1997), in her review of developments in coastal research in the 1990s outlines the complexity of surf zone processes and how the interactions between water and bottom topography can produce a range of process modifications. Komar (1998) also provides an extensive review of wave/sediment interactions.

Returning to the breaking of waves on beaches, the style of breaking will also play a part in the nature of beach sediment accumulation. Waves can be classified in several ways (Street and Camfield 1966, Galvin 1968), although the scheme by Galvin is perhaps the most commonly used. In this scheme, waves may be classified as being of one of three main types (Figure 2.2) (Carter 1988): *spilling*, where waves build until the crest becomes unstable before

a) SPILLING WAVES

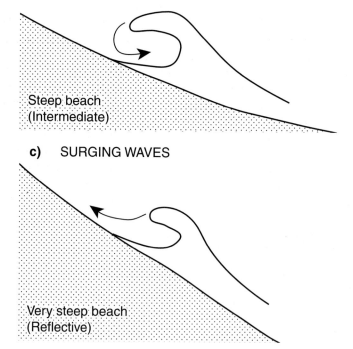

Shallow beach
(Dissipative)

b) PLUNGING WAVES

Steep beach
(Intermediate)

c) SURGING WAVES

Very steep beach
(Reflective)

Figure 2.2 Different wave types and their relationship to beach form
a) Spilling waves on shallow, dissipative beaches
b) Plunging waves on steeper, intermediate beaches
c) Surging waves on steep, reflective beaches
(Modified from various sources).

collapsing as a mass of white water; *plunging*, where the shoreward face of the wave builds until it is vertical, before plunging forward as an in-tact mass of water; and *surging*, which build as if to plunge, but then collapse as a spilling wave. It can be seen that spilling and surging waves possess a strong shoreward movement, while plunging waves have a strong seaward movement. Table 2.1 details the distinctions between these but, generally, a strong shoreward movement as the wave breaks (swash) will move sediment up the beach, while a strong seaward movement (backwash) will move sediment off the beach.

It has already been stated that increased friction at the sediment/water contact slows the wave down above wave base. This is further supported by Komar (1998), who adds that in addition to bottom friction, much energy is also lost through wave breaking, a process which is itself often induced by frictional drag between water and sediment. Thinking laterally suggests that expensive sea walls are partly required because waves reach the coast with too much energy, thus causing damage by flooding and erosion. If, therefore, it is possible to reduce this energy part of the reason for needing sea walls disappears. Clearly, greater amounts of wave energy will be lost if waves flow over greater distances of sediment above wave base. This can be achieved by increasing the beach width: the water depth will be shallower for longer, resulting in greater energy losses through friction and wave breaking, and so lesser need for sea walls. In other words, we immediately have a method by which we can engineer the sediment regime of the beach environment to reduce wave energy received at the coast and remove some of the need for hard defences (see Chapter 7).

The processes described above can determine the degree of onshore/offshore sediment movement (Carter 1988). In addition, given that we know that waves slow down when they enter shallow water, it can been seen that the rate of slow down will depend on the rate of shallowing. If this rate is uniform across a beach, then the waves will slow down uniformly and remain parallel to the coast. In reality, however, sediment surface topography is rarely flat and waves tend to approach the coastline at angles. Thus, as waves approach the shore, parts of the wave could be in deeper water than other parts, so that the

Table 2.1 Wave types and resulting shore morphology and process.

	Spilling	*Plunging*	*Surging*
Wave reflection	low	intermediate	high
Type of beach	dissipative	transitional	reflective
Beach profile	low angle/flat	steep	very steep
Beach width	wide	intermediate	narrow
Sediment transport			
Longshore	low	medium	high
Onshore/offshore	high	medium	low
Transport mode	suspended	mixed	bedload
Texture	silts/sands	sands	sands/shingle
Aeolian processes	common	rare	none

part in deeper water will move faster than the part in shallower water. The resulting differential in wave speed means that waves become refracted towards the shallower water areas. On embayed coasts, water depths approaching the embayments are generally deeper than those approaching the headlands, causing a refraction of wave fronts towards the headlands. This would initially suggest a greater concentration of wave energy, and hence erosion, onto the headlands. In fact, however, we need to remember that the headlands are such because they are more resistant to erosion than the embayments. The important aspect is that along such coasts, and indeed any coasts where sediment surface topography produces focusing of wave fronts, there is a spatial variation in wave foci produced, in which higher levels of wave activity are present in some places than elsewhere. Thus, a series of spatial wave gradients are set up which are themselves important in setting up longshore currents and subsequent shore normal or shore parallel bedload sediment movement. There are many examples of these processes along coastlines. Munk and Traylor (1947) report wave focusing linked to a submerged canyon at La Jolla, California, while Robinson (1980) showed how focusing caused by an offshore sediment bank caused accelerated erosion along the East Anglian coast of the UK.

Before the idea of longshore transport is developed further, it should be mentioned that, while it has been seen how 'primary' waves lose much of their energy as they approach the coast, it is common for a series of 'secondary' wave forms to develop, termed *edge* waves. These edge waves are not fully understood by coastal geomorphologists (see Carter (1988) for a reasoned explanation); they appear to be restricted to the surf zone and are formed as primary waves lose energy, some of this energy being transferred to edge waves. Horn (1997) explains edge waves as forming at 90 degrees to the shoreline on a sloping beach, with a maximum amplitude at the shore, decreasing seawards. Although complex, their significance to coastal morphology, and hence defence strategy, lies with the interference that these edge waves have with other incoming primary waves. Davis (1996) explains how edge waves effectively behave like standing waves and so, according to the physical laws of wave harmonics (see Carter 1988 or Davis 1996), can amplify the incoming primary waves when the crests of the two coincide. Such amplification can increase the impact of the waves on the shoreline, producing greater changes in profile and greater amounts of sediment movement. On open coasts, the interference of edge and primary waves can produce some complex beach topography, including bars, cusps and washovers (Holman and Bowen 1982).

Returning to the main theme, when waves hit the shore obliquely, they will initiate a net longshore movement of beach sediment. Waves will break and wash up the beach according to their angle of approach, but always return to the sea at right angles to the coast (Figure 2.3). As waves wash up the beach (swash), they will move sediment towards the land, and the return to the sea (backwash) will move sediment down the beach. In effect, the swash and backwash directions of the wave will also be the direction of sediment movement and so, with waves having an angular approach, it can be seen how sediment may be moved along the coast by wave activity. If the energy of breaking waves

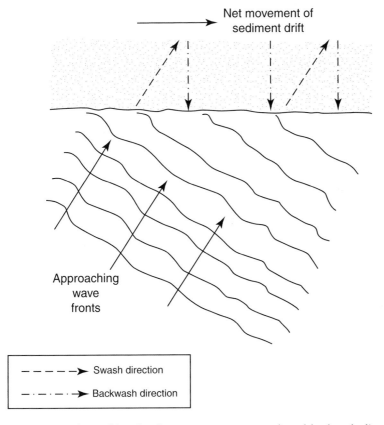

Figure 2.3 Relationship of sediment transport to swash and backwash directions for oblique waves. With backwash always shore normal, oblique waves will initiate sediment movement along the coastline as longshore drift.

is sufficient to move particles in suspension, then the net direction of movement will be the same, although its rate may vary. In some situations, however, such as where there is a strong backwash, suspended sediment may be carried seawards beyond the swash and breaker zone, making it temporarily unavailable for beach accretion. A typical example of this happens during storms, when beach sediment may be transferred offshore beyond normal wave base to a point between this and the storm wave base (refer to Figure 2.1 for terminology).

Many of the problems encountered by coastal defences, however, are due to the interruption of longshore drift and the interruption of bedload sediment transport. It is this sediment which leads to beach formation, and to the accretion behind defence structures. In the case of longshore movement of sediment as bedload, all the time that sediment is being replenished in the source area, there will be plenty of sediment to replace that moved by these processes. In circumstances where defences are built (see Figure 1.1), this source area may become rapidly depleted and a situation of net erosion will occur, because longshore drift is still moving sediment along the coast without any replacement up drift.

This brief overview of wave processes demonstrates how waves govern whether the coasts are erosive or accretional, depending on the type of wave and sediment availability. Earlier, we used a basic equation to describe the action of waves:

Total wave energy = energy for erosion + energy for sediment transport

If waves have little sediment to transport, then the bulk of their energy will be focused on erosion. If there is a lot of sediment is being transported, then energy available for erosion will be small or even zero. This explains the cyclical nature of much of the coastal erosion that occurs. Rates of cliff erosion, for example, are always quoted as $x\,m\,a^{-1}$, whereas in fact cliffs tend to erode at irregular intervals. A cliff fall may happen one winter, but then no more falls may happen for a number of years; the reasons can be traced back to the equation. A cliff fall happens when there is little beach sediment to erode, so that the bulk of the waves' energy is focused on erosion of the cliff. When a fall occurs it results in a large volume of sediment lying at the base of the cliff, with most of the waves' energy now being used to move this sediment along the coast or offshore. When this sediment has been moved, energy will, once again, be used for cliff erosion and a new erosion cycle will begin. Clearly, this only holds true where wave energy is sufficient to erode cliffs. Hard rock cliffs will not provide sediment in this way. Their fronting beaches need to be supplied from other sources, otherwise a barren, wave-cut surface will result. Because hard rock cliffs provide little sediment to coastal sediment budgets, the main sediment supplied from cliffs is derived from soft cliffs.

The basic laws of physics state that wave energy is a function of the square of the wave height; therefore, in storm conditions, wave energy will increase as wave height increases. Hence, more sediment can be eroded and transported, accounting for the dramatic coastal changes which can be observed before and after storms. In calm conditions, the left-hand side of the above equation will decrease and energy available for erosion and transport decrease. In very calm conditions, it is possible for waves to have so little energy that the cohesive strength of the sediment and even very soft cliffs are able to resist erosion.

In summary, it can be seen how waves can shape coasts, and also some of the possibilities that exist to defend and protect them in a soft, 'environmentally friendly' manner. The extent to which waves can reshape coastlines is complicated by a series of factors, such as energy and sediment availability, as discussed here. If waves have insufficient energy, sediments will resist movement and, as waves lose more of their energy, deposition will occur. Conversely, if energy levels are greater than the resistance strength of the sediment, then erosion occurs. The relationship of wave energy to coastal morphology is shown in Figure 2.4. Here, we start with waves approaching the coast; as they do, they

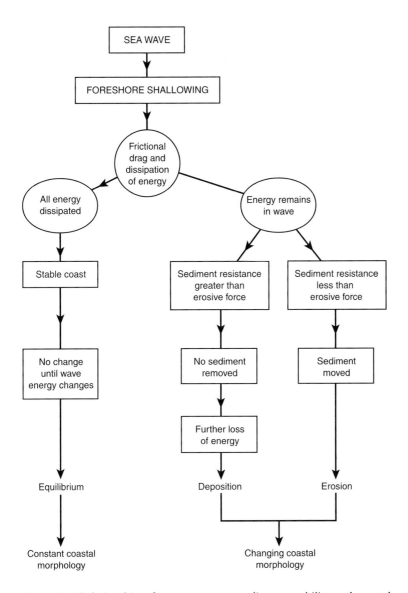

Figure 2.4 Relationship of wave energy to sediment mobility and coastal morphology. Note how the extent of wave energy dissipation will control whether coast becomes erosional, depositional, or 'stable' (After French 1997).

lose energy. If all energy is lost, no change can occur and the coastal form remains constant. If the waves have energy left by the time they hit the beach, one of two possibilities can occur, depending on whether the energy levels are sufficient to produce erosion or not.

Importance of tides

In more sheltered areas, wave action is largely removed either as a result of shallowing or by shelter of an area by some form of sediment deposit, such as a spit or bar. In such cases, wave energy ceases to be a prime factor in coastal formation, although waves may still occur to a lesser degree, and tidal energy becomes more important. As a result, coastal morphology and processes reflect these new environmental conditions. In particular, sediments cease to be dominated by sands and shingle, and become silt and clay dominated, producing such environments as mudflats, salt marshes, mangroves and deltas. Similarly, the exposure of coastal defences to high energy conditions is less than on exposed, wave dominated coasts because peak energy levels in tides occur mid-tide (see French 1997). As a consequence, defences can be 'scaled down', with earth embankments replacing concrete sea walls.

Because of the net finer grain size in tide-dominated areas, suspended sediment transport becomes more important in foreshore accretion. In wave-dominated areas, wave strength and type control sediment input to the beach; in tidal areas, the main factor in controlling the rate at which sediments accrete is the length of time an area is under water, i.e. covered by the tide and/or its proximity to tidal creek networks. A sediment surface which is close to low water will be covered quickly as the tide comes in, and remain covered until the tide is almost out. Hence, it is under water for longer and to greater depth than a sediment surface near to high water, which may only be under water for a short period of time around high water. This factor is reflected in the growth of vegetation, such as salt marshes, and also in sedimentation rates. Hence, the higher a sediment surface gets in the tidal frame, the less sediment it receives and the slower the rate at which it accretes. Studies from the Severn Estuary (French 1996) demonstrate this for three sediment surfaces (in this case distinct salt marsh surfaces); these accrete at rates of $12.1\,\mathrm{mm\,a^{-1}}$ on the lowest surface (which is covered by the tide for the longest period), $6.4\,\mathrm{mm\,a^{-1}}$ on the middle surface, and $2.3\,\mathrm{mm\,a^{-1}}$ on the highest surface (which is covered for the shortest period) (see Figure 2.5). Hence, with higher rates on lower marshes, the elevation difference will decrease with time as lower surfaces in effect catch up with higher ones. While this case is true for marsh surfaces which only receive sediment from tidal inundation, where creeks occur, the nature of marsh surface accretion changes to reflect this new method of delivery — greater volumes of sediment are delivered directly to the high marsh areas. French *et al.* (1995) highlight this with an example from the north Norfolk coast, where high rates of sediment accretion occur not only on the lowest marshes, but also on areas proximal to creeks.

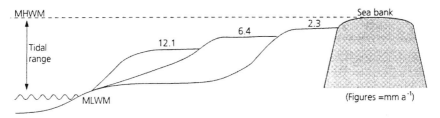

Figure 2.5 Variation in marsh surface accretion rates as function of marsh elevation. Based on data from the Severn Estuary, south-west England (After French 1997).

In mudflat areas, the method of sediment accretion is also not just a simple case of sediment settling from suspension. McCave (1970) details studies from the German Bight in which he identified the problem that deposition rates were too great to be accounted for solely by settling from flood tides. He found there to be a viscous sublayer immediately above the mudflat surface, from which quasi-continuous deposition would occur through the tidal cycle, not just slack water. From this work, accretion rates of between $1.7 \, \text{mm a}^{-1}$ and $2.7 \, \text{mm a}^{-1}$ were calculated, based on a sediment concentration in the sublayer of $2 \, \text{mg l}^{-1}$ and $3.2 \, \text{mg l}^{-1}$ respectively.

It is tempting to think that processes such as those described above are beyond the influence of human activity and, to a certain extent, they are. While humans cannot change the coming in and going out of the tides (construction of tidal barrages can do this, but this is the exception), they can change the rate at which it comes in and goes out, as well as the depth of water. Both these factors can play a key role in accretion rates and also in coastal defence. Tides typically come in and go out twice a day (some areas do receive just one tide a day and, in rare cases, four), meaning that over a 24-hour period, tides come in and go out (flood and ebb) over approximately 12-hour periods. This would suggest a flood period of 5 hours, 1-hour high water stand, ebb of 5 hours and a low water stand of 1 hour, producing a symmetrical tide. Complications to this can arise, however; when tides enter shallow water (nearshore zone or an estuary) they can become distorted by short-period harmonics caused by local sediment surface topography, freshwater flow, and coastal morphology. The result is that the symmetrical tidal cycle may become asymmetrical, such that the flood tide may last 3.5 hours and the ebb 6.5, with 1 hour high and low water stands. This process has important defence implications because irrespective of how long flood and ebb tides last, tides move fixed volumes of water into and out of an estuary on each tidal cycle. When flood and ebb tides are of unequal length, then on the shorter, this fixed volume of water has to move faster than on the longer. In the example above, the estuary would have 3.5 hours to fill up, and 6.5 hours to empty. This can only be achieved if the flow has variable velocity and, therefore, energy. In this example, flood tides occur over shorter periods of time, the estuary has less time to fill and so water must flow faster. If water flows faster, then it has more energy and a greater capacity for transporting sediment, producing a net input of sediment into the estuary. Conversely, the same argument would stand for a shorter period ebb, which will result in a net movement of sediment out of the estuary. In addition to tidal asymmetry, neap/spring tidal cycles can produce a similar effect in estuaries. Spring tides, when tidal amplitude increases, will involve a greater volume of water entering the estuary when compared to neap tides. Hence, tidal velocity will also vary accordingly. In Morecambe Bay, for example, spring tide currents can reach $1.3\,\mathrm{m\,s^{-1}}$ (Aldridge 1997), while in the Mersey and Dee, they can reach 2.5 and $1.5\,\mathrm{m\,s^{-1}}$ respectively (Taylor and Parker 1993).

While issues of flood and ebb dominance are governed by the morphology of the estuarine system, they can be argued to be of natural origin. However, defence activities in an estuary, such as sea wall construction, land claim, dredging, as well as marina development and barrage construction, can change the natural morphology of an estuary and, therefore, its flood/ebb tide relationship, producing a change in sediment flux. Examples of this can be found in southern Ireland where land claim has changed the tidal regime of Rosslare Bay and Malahide (Dublin Bay) (Orford and Carter 1985, Carter 1988, Orford 1988). Alteration in the flood/ebb dominance can cause an estuary to change from being net accretional to net erosional, leading to subsequent defence and management implications.

As well as creating changes in tidal dominance, building sea walls along the banks of an estuary can also restrict the area over which water can flow. Hence, if water is constrained into a narrower channel due to sea walls, then the restrictions in width is compensated for by an increase in depth (French 1997) which, bearing in mind the earlier discussion on salt marsh accretion rates, can also have an effect on sediment build-up in the intertidal zone.

Importance of wind

On some exposed coasts, the wind may also play a part in the function of coastal landforms. Under the right environmental conditions, wind may transfer sediment from the beach environment landward where, if trapped by some obstacle (whether litter, a structure, or vegetation), sediment may accumulate into a dune form. For this to occur, several factors need to be in place: first, a wide, shallow intertidal sand flat; second, a large tidal range; and third, a predominantly onshore wind to dry the sand and initiate landward transfer. The first is perhaps an important limiting factor for dune development along coasts because wind is common on all open coastlines, yet dune formation is not. Fine grade sediments in estuaries are too cohesive to be picked up by the wind and moved.

In order for sand to be transferred into the dune system, the wind has to achieve a sufficient velocity to induce sediment grain movement. This initiation of movement is a function of three things. The first is the velocity of the wind. Wind speeds of around $4–5\,\mathrm{m\,s^{-1}}$ are needed to initiate sand grain movement, although velocities in excess of $20\,\mathrm{m\,s^{-1}}$ are needed for rapid movement

(Bagnold 1954, Pethick 1984). Second, the water content of the sand is important. Wet sand has a greater cohesive strength than dry sand, and so initiating grain movement is considerably harder, if not impossible. This underlies the importance of tidal range, as the sediment needs to be exposed for drying for long periods of time before movement can occur. Jackson and Nordstrom (1998) illustrate this point with their study from Wildwood, New Jersey, USA. They studied the effect of rainfall on wind-blown sediment and found that during rain, the sediment moisture content reached 7 per cent and transport to the dunes was measured as $14.1 \, \text{kg} \, \text{m}^{-1} \text{h}^{-1}$, while 4 hours after rain, following drying, sediment moisture fell to 4 per cent and transport to the dunes was $140.2 \, \text{kg} \, \text{m}^{-1} \text{h}^{-1}$. Although this variation follows rainfall and not tidal inundation, it does provide a clear indication of the impact of increased sediment moisture on the amount of material transported by the wind which, it should be pointed out, did not vary significantly over the study. Kumbein and Slack (1956) indicate that, because of the presence of moisture in the sediments covered by tides, and the effect that this has on sediment transport, up to 80 per cent of the sediment transported into dunes comes from the backshore area, i.e. that area between mean high water and the dunes which is only inundated by storms and exceptionally high tides and as a result, is dry for long periods of time. Hence, the wider this area, the greater surface area from which sand can be derived for dune building. The third control over the initiation of grain movement is the size of the sand particles on the beach, movement being a function of the square root of the grain diameter.

Once movement is initiated, sand grains will be transferred in the direction of the prevailing wind. Movement is generally by a process known as saltation, where particles 'bounce' along the ground, forcing other particles to move as they land. It is only when wind speeds are particularly high that continuous movement occurs in the form of a sand 'cloud' or sand 'storm'. Most sediment moves by saltation, during which particles may move only around 1 m into the air before they return to the sediment surface. Again, therefore, we can identify a process which may influence management strategies. If we need to encourage dune growth by artificially increasing sedimentation, then we must realise that much of the available sediment is only moved within around a metre of the beach surface.

Having investigated the criteria to initiate sediment movement, we need to consider how this transfer landwards can result in the formation of embryo dunes. Clearly, the only way that sediment is going to stop saltating is for the wind to drop. This may occur for meteorological reasons, in which case the whole mass of saltating sand grains will cease movement uniformly, producing a new layer of sand on the beach. This in itself will not cause the initiation of dunes. For this to happen, we need to reduce wind speeds locally, thus allowing sediment to settle in individual areas. To achieve this, we need small obstacles which interrupt wind flow and cause air turbulence, to create pockets of slack air around the obstacle in which sediment will settle. This pattern could emerge in the lee of an obstacle or in the lee of clumps of vegetation. Even at this stage, dune formation is not guaranteed because once the object is buried, sand deposition will cease as there is no longer an obstacle to wind flow. To achieve this, the obstacle needs to keep pace with sediment accumulation. Thus, inanimate objects are not useful alone in initiating dune growth — perhaps a good thing considering the litter on some beaches — but vegetation is, providing its rate of growth is greater than the rate of sediment accumulation around it. Therefore, by putting such observations into the natural beach context, it is possible to develop further dune management tools. The placing of inanimate objects in the wind flow can promote the deposition of sand. Once initiated, the planting of stabilising vegetation which can keep pace with sediment accretion will allow these areas to develop into embryo dunes, and subsequently development into mature dunes.

Concept of the coastal sediment budget

The previous discussion of waves, tides, and wind has shown how sediment and its movement is critical to the stability of a coastline. In order to be able to decide how to manage a length of coast, it is important to know how much sediment is available, where it is coming from, where it is stored, and how it leaves a particular coastal area. The identification of these factors is referred to as the *sediment budget*, and an understanding of it is paramount to any successful coastal defence strategy. As with any 'budget', a sediment budget represents a measure of inputs to the sediment system, areas of storage, and outputs. However, such budgets are very difficult to determine as estimates of sediment inputs and outputs can be difficult, especially when considering sediment movements from and to offshore areas. Nevertheless, attempts have been made, some of which are listed in Table 2.2. Figure 2.6 presents a generalised budget and indicates main inputs and outputs. The significance of each source and output will vary according to different coastlines, and not all will be relevant in all situations. Cliffs, for example, may be locally significant, such as along the Holderness coast of the UK where retreat rates average $1–2\,\mathrm{m\,a^{-1}}$, but in many areas cliffs play only a minor role in sediment supply (Mason and Hansom 1988). Hansom (1988) argues that rivers play the key role in sediment supply to coastal budgets, accounting for around 90 per cent of global marine sediment. While this figure appears huge, fluvial sediment supply varies with latitude, being dominant in low latitudes and less important in mid latitudes, where rivers carry less sediment. What this does illustrate, however, is that not all coastal sediment problems originate at the coast, but may originate inland. In high latitudes, fluvial supply may also vary according to input from glacial outwash. Other important inputs include longshore drift, from offshore, and anthropogenic sources, such as beach feeding.

Outputs to the coastal system include longshore drift, loss to offshore and also transfer to sediment stores. As seen in the discussion of waves, it is possible during storms for sediment to be drawn offshore below normal wave base. This means that large volumes of sediment can be temporarily lost to the sediment

Table 2.2 Examples of sediment budgets from around the world (For full references, see References at the back of the book).

Country	Author	Place of study
Australia	Chapman (1981)	Gold Coast
Brazil	Allison *et al.* (1996)	Amapa Coast
	Barcellos *et al.* (1997)	Sepetiba Bay
Canada	Jordan and Slaymaker (1991)	British Columbia
	MacDonald *et al.* (1998)	Beaufort Shelf
Mexico	Cruz-Colin and Cupul-Magana (1997)	Baja California
Netherlands	van Rijn (1997)	Central coastal zone
New Zealand	Gibb and Adams (1982)	South Island
UK	Bray *et al.* (1995)	South coast
	Clayton (1980)	East Anglia
	Mason and Hansom (1988)	Holderness
	Vincent (1979)	East Anglia
USA	Allen (1981)	Sandy Hook, New Jersey
	Bowen and Inman (1966)	Southern California
	Kana (1995)	Long Island, New York
	Marcus *et al.* (1993)	Chesapeake Bay
	Pierce (1969)	Northern Carolina
	Stapor (1971)	Penhandle, NW Florida
	Stone and Stapor (1996)	Gulf of Mexico

budget. Similarly, aeolian transfer to sand dunes or transfer to mudflats or salt marsh will also produce short-term losses. Anthropogenic losses include beach mining, dredging and land claim, while transfer to stores, as has already been argued, represents temporary places where sediment may be left by the sea.

Being a budget, volumes should balance:

Volume of sediment in = volume of sediment stored + volume of sediment out

When considering defences, these budgets need to be calculated carefully as it is easy to cause an imbalance. For example, given that rivers are regarded as the primary source of sediment to coastal sediment budgets, dams or barrage constructions which prevents downstream sediment movement will have major impacts on the sediment budget. For example, construction of the Aswan Dam reduced the amount of sediment reaching the Nile Delta from 124 million $t a^{-1}$ to 50 million (Stanley and Warne 1983), while completion of the Akosomob Dam stopped 99.5 per cent of river sediment reaching the Volta Delta (Ly 1980); the supply of sediment to the Ebro Delta reduced from 4 million $t a^{-1}$ to 0.4 million $t a^{-1}$ (Marino 1992) and the Rhone Delta lost 90 per cent of its sediment supply (Viles and Spencer 1995). Similarly, on coastlines, any defence scheme that prevents erosion, such as a structure which reduces cliff erosion, will remove a sediment source from the coastal sediment budget, and so if that budget is to remain balanced, a new input has to be found to compensate. An example of this can be found in Fairlight Cove, Sussex (see Box 5.2 for full

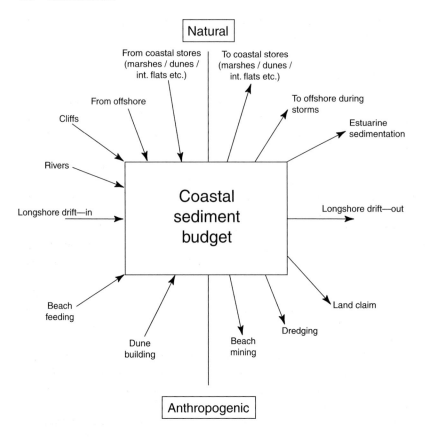

Figure 2.6 Generalised coastal sediment budget. Each component will vary in importance depending on coastal location.

details). In this case, input from the eroding cliffs was estimated at $14\,250\,\mathrm{m^3\,a^{-1}}$, which reduced to $4\,500\,\mathrm{m^3\,a^{-1}}$ after construction of an offshore breakwater: a reduction of 68 per cent.

If a replacement sediment source cannot be found, then the following situation will occur:

Volume of sediment in < volume of sediment stored + volume of sediment out

producing a net loss of sediment to the budget, and the dominance of erosion. This leads us on to a further benefit of soft engineering, in that it works more closely with the budget requirements. Sea walls may remove sediment sources from the budget, and groynes may restrict input down drift, but soft techniques, such as beach feeding, represent a major input to the budget, thus artificially compensating for losses elsewhere.

Problem of changing sea level

The importance of coastal processes in fashioning coastal morphology and moving sediment can be seen from the previous, brief introduction. The link between waves and beach shape or cliff erosion can be clearly seen. However, the medium- to long-term functioning of these processes and, by inference, the management of coastal defences is dependant on one other factor. So far we have only talked about the coast as a zone which is interacting with processes as if they were fixed in time. In reality, however, any defence construction based on estimates of wave, tides, and wind needs to include predictions of what is actually happening to the sea itself. Of prime importance here is the fact that mean global sea level is increasing, giving rise to the problem of sea level rise. Most obviously, as sea levels rise, so water depths increase and wave base will get deeper. This means that waves reaching the coast will have more energy and, therefore, will be able to erode and transport greater quantities of sediment. Similarly, with greater water depth, the volume of water exchanged in estuaries on flood and ebb tides, and therefore ebb/flood current velocities, will also increase.

Bray *et al.* (1997) predict the rate of sea level rise as being between 23 cm by 2050 and 48 cm by 2100, representing a rise of approximately 4 mm a^{-1}. This is a global figure and needs to be adjusted to take into account local and regional differences. Perhaps the greatest local and regional variation will be in relation to isostatic (crustal movement) and eustatic (ocean volume) changes. Clearly, a coast with subsiding land and increasing ocean volume will experience a greater net sea level rise than one with a rising land mass. Areas typically prone to subsiding land are those which experienced coverage by Quaternary ice masses, such as northern Europe, or those areas of thick modern sediment accretion which undergoes compaction and dewatering. Classic examples are delta areas, such as the Mississippi Delta which is experiencing a sea level rise of 10 mm a^{-1} (Day *et al.* 1995). A series of sea level rise estimates from a range of coastal environments are given in Table 2.3. In some locations, isostatic rebound is occurring (i.e. the rising of the land following ice unloading), and this rate of land surface rise exceeds the eustatic rise in sea level, producing a net fall in sea level. Bergen (+0.73 mm a^{-1}) and Stavanger (+0.42 mm a^{-1}) in Norway (Shennan and Wooworth 1992), and Quebec (0.75 mm a^{-1}) in Canada (Shaw *et al.* 1998) are all examples where sea level is falling for this reason.

The issue of sea level rise is multifaceted and produces a range of environmental problems. From our point of view, it means that high tides become higher, wave base increases, and so the energy received at the coast may also increase (see Dolotov 1992 for a detailed assessment). Put another way, the energy received at the defence structure will increase, and many coastal environments may go into disequilibrium if not allowed to adjust to new sea level

Table 2.3 Sea level rise estimates from a range of locations. In considering such rates, it is important to note the time period over which they are calculated, as well as the geographical locality (For full references, see References at the back of the book).

Country	Location	Rate (mm a⁻¹)	Age range	Source
Argentina	Buenos Aires	1.5	1905–1988	Douglas (1997)
	Quequén	0.8	1918–1982	Dennis et al. (1995b)
Bangladesh	Bay of Bengal	3.3	2000–2100	Castro-Ortiz (1994)
Canada	Halifax, NS	3.03	1896–1995	Shaw et al. (1998)
	Quebec	*0.75*	*1940–1989*	*Shaw* et al. *(1998)*
China	General	1.2	1900–2000	Wang (1989)
	North Jiangsu	5.0	until 2050	Chen (1997)
	Pearl River Delta	8.0	until 2050	Chen (1997)
	Yangtze River Delta	9.0	until 2050	Chen (1997)
	Yellow River Delta	6.0	until 2050	Chen (1997)
Denmark	Esbjerg	0.82–1.38	1901–1969	Shennan and Woodworth (1992)
Egypt	Nile Delta	5.0	Current	Day et al. (1995)
France	Brest	1.4	1880–1991	Douglas (1997)
	Ebro Delta	1.0–5.0	Current	Day et al. (1995)
	Marseille	1.2	1885–1991	Douglas (1997)
	Rhone Delta	1.0–5.0	Current	Day et al. (1995)
	Rhone Delta	2.1	since 1905	Suanez and Provansal (1996)
Italy	Venice lagoon	8.0	Current	Day et al. (1995)
Netherlands	North Sea Coast	2.0	1900–1985	den Elzen and Rotmans (1992)
	North Sea Coast	1.5–2.0	20th century	van Malde (1991)
	North Sea coast	0.73–2.53	1901–1987	Shennan and Woodworth (1992)
New Zealand	Auckland	1.3	1904–1989	Douglas (1997)
	Wellington	1.7	1901–1988	Douglas (1997)
Norway	*Bergen*	*0.74–1.18*	*1928–1986*	*Shennan and Woodworth (1992)*
	Starvanger	*0–0.42*	*1928–1986*	*Shennan and Woodworth (1992)*
Senegal	Dakar Harbour	1.4	1943–1965	Dennis et al. (1995a)
UK	Aberdeen	0.54–0.98	1901–1988	Shennan and Woodworth (1992)
	Avonmouth	0.25–1.07	1925–1980	Shennan and Woodworth (1992)
	Dover	1.6–3.0	1961–1987	Shennan and Woodworth (1992)
	East Anglia	3.75–6.25	1990–2030	Clayton (1989b)
	Forth estuary	4.3	1990–2012	Shennan 1993
	Lowestoft	0.45–1.81	1956–1988	Shennan and Woodworth (1992)
	Mersey estuary	5.3	1990–2050	Shennan 1993
	Newlyn	1.7	1915–1991	Douglas (1997)
	Portsmouth	5.0	1962–1987	Woodworth (1987)
	Solent	4.0–5.0	1896–1996	Cundy and Croudace (1996)
	Thames estuary	6.8	1990–2050	Shennan (1993)
	Wash	6.3	1990–2050	Shennan (1993)
USA	Chesapeake Bay	2.7	since c. 1700	Lisle (1986)
	Gulf of Mexico	2.0–2.5	1938–1993	Warren and Niering (1993)
	Key West	2.2	1913–1991	Douglas (1997)
	Mississippi Delta	10.0	Current	Day et al. (1995)
	North Carolina	1.9	1940s–1980s	Hackney and Cleary (1987)
	San Diego	2.1	1906–1991	Douglas (1997)
	San Francisco	1.5	1880–1991	Douglas (1997)

Note. Italic text signifies a falling sea level

conditions. Significantly, the problem is one which is common to low-lying areas the world over, irrespective of politics or wealth. In order to plan ahead and make our defences flexible enough to deal with sea level rise, it is important to understand how the coast will react to sea level changes. To a certain degree, this has been made possible by studying fossil coasts and, from these, gaining an understanding of how the coast has changed in response to Holocene sea level rise, following the last glaciation.

Many countries already have systems in place to defend their coastline and these will also serve to protect against sea level rise. Many of these systems will, however, need to be upgraded or, if this is either not possible or too expensive, more drastic action will need to be taken, such as letting the land concerned flood, with any population or land use having to relocate. The sea level problem is not just a case of the sea becoming deeper. Under conditions of accelerated sea level rise, coastal erosion may become more rapid, removing fronting beaches or salt marshes and exposing the hinterland to increased threats of inundation. Such problems may require either completely new defences to be built, or land to be abandoned. While in some areas such methods will not be a problem, in others, such as the Netherlands, the idea of allowing land to flood is out of the question, as large areas of the country are already below sea level. Many countries which have built extensive defences along their coast now also have to employ barrages across major estuaries to protect the larger cities from flooding; the Thames barrage in the UK was constructed to withstand a 1 000-year flood, but has seen frequent use since its completion in 1982, with 33 closures by the winter of 1998/99 (approximately 2 per year).

Building defences to protect towns and cities is the typical response in countries of the developed world which have the financial resources to do so. Nevertheless, the cost of such defence constructions may still be significant. Burby and Nelson (1991), for example, indicate that the USA would need to defend an additional 1025 km of their coast by 2100 following a 1.0 m rise in sea level. While this will be a large financial burden for a rich nation, countries such as Bangladesh, which do not have the resources to construct hugely expensive coastal defences, are facing predicted sea level rise of up to 345 cm by 2100 (range 56.2–345 cm) (see Castro-Ortiz 1994), and an increasing frequency of storm surges which are a major cause of loss of life, livelihood, and spoiling of land. Much of Bangladesh lies on the delta of the Ganges-Brahmaputra; hence, high land refuges are not common and, as has been seen, isostatic land movements can be considerable. Given that the financial resources are not available to protect large areas of Bangladesh, it is likely that the predicted sea level rise of around 1 m by the middle of the next century will cause the loss of 23 000 km² of land under the 'do nothing' or managed realignment scenarios, coupled with relocation. Some of its administrative districts would disappear altogether, as would 14 per cent of the area of crops, and 29 per cent of its forest (Ince 1990; Viles and Spencer 1995). Much of the population as a result, would need to move, although the possibility of new settlement within Bangladesh would be slight. The rise of 1 m is a mid-range prediction based on current forecasts for the area. What this represents in real terms is the loss of 14 million head of livestock, 100 000 households, 8 000 schools, 1 500 km of railways, 20 000 km of roads, 4 000 km² mangroves. This would result in political and social upheaval on a scale rarely seen in the history of the world.

What is sea level rise and what does it mean?

Sea level rise is, in its simplest terms, the change in sea level relative to the level of the land. The perception of sea level rise has, over the past few decades, developed from alarmist media speculation to a more scientific-based prediction based on real evidence. Since the 1960s, both predictive models and data reliability have improved, with the result that estimates of rise have been gradually revised downwards. Some recent estimates (Wigley and Raper 1990) suggest a rate of 4–5 mm a^{-1} until the year 2100. Other work indicates only 1–2 mm a^{-1}, halving that above. Despite this huge contrast, due in part to the variation in land properties and techniques exemplified in Table 2.3, it is clear that as more data become available, and models more refined, estimates of future rise will become more reliable. More importantly, whatever the rate of sea level rise, that rate is critical in both the response of existing coastal defences and beaches to rising sea levels, and also the planning of new coastal defences.

Predictions of how existing defences and beaches respond to sea level rise are always difficult to make considering the general lack of detailed understanding of the coastal environment. What is certain is that rises in sea level are occurring, and changes in coastal landforms and habitats are being observed. These changes may be as simple as increased flooding due to inadequate defence height and increased elevation of storm surges. Venice, for example, is a city built on a series of islands within a lagoon. This area is already being threatened by increased sea levels and storm surges, exacerbated by increased subsidence due to ground water extraction (Pirazzoli 1991). Considering that today, after a rise of only 10–15 cm over the last century, many more parts of the world now suffer from coastal flooding, and our use of coastal defences in flood prevention have only been partially successful, a further 13 cm predicted in the next 40 years will only make the issue more pressing. The impacts may also be more complex, involving the alteration and erosion of coastal features and the complexities associated with coastal squeeze.

Sea level rise is worthy of extensive discussion, but it is important to keep the issue within the context of this book. From our point of view, the rise in sea level will cause two things to happen:

- an increase in the instability of undefended (soft) coasts with adjustment to new sea levels according to the Bruun rule or other adjustment model (see below). This will involve migration of habitats inland, and the associated loss of hinterland;
- an increase in coastal squeeze and habitat modification on coasts with hard defences, leading to impacts on fronting beaches or marshes, and related

implications for the sediment budget and longshore processes. In addition, increasing water depth along a 'fixed' coastline will cause an increased exposure of defences to wave attack and potential storm damage.

These two issues will be developed in the following chapters.

The Bruun rule

All coastal types respond in different ways to rising sea level. Some physically move inland by processes of roll-over (e.g. Barrier Islands), some revert to different forms of the same morphological type (e.g. salt marshes), some become more active (e.g. cliffs). As we have already seen, all coastlines, regardless of their type, can achieve some degree of dynamic equilibrium profile; changes in sea level are going to alter this state. In order to regain dynamic equilibrium following sea level rise, the profile needs to relocate inland, involving an upward and landward translation (Figure 2.7). In order to do this, the profile will initially erode, the coastline move inland, and the profile re-form in the same position relative to the tidal frame. This process is referred to as the Bruun rule, named after a Norwegian coastal engineer who first defined the process in 1962 (Bruun 1962). Although this rule has been refined by subsequent researchers (see Carter 1988) and criticised by others, the general principles remain the same. The rule (as it appears in Carter 1988) can be stated mathematically:

$$s' = [(zR)/x] . [1 + (r/100)] . [1 + (c/100)]$$

where:

s' is sea level rise
x is width of profile
z is depth of profile
R is landward movement of the shoreline (erosion)
$[1 + (r/100)]$ is constant expressing sediment composition
$[1 + (c/100)]$ is constant expressing losses from the system (i.e. offshore or longshore transport).

The criticisms of the Bruun rule are largely based on its relative simplicity. It does not, for example, allow for the inclusion of longshore processes, thus ruling out important sediment budget and process elements in adjustment estimates. Furthermore, the estimate of closure depth, necessary for sediment budget calculations, is difficult. A further issue is that it also suggests a start and end point which, under the progressive sea level rise experienced by many of the world's coastlines, is not really the case. Despite these reservations, however, the Bruun rule does allow a reasonable understanding of how coasts will respond to sea level rise. Figure 2.7 illustrates the Bruun rule diagrammatically, and shows how the profile moves landwards and upwards in response to rising sea level

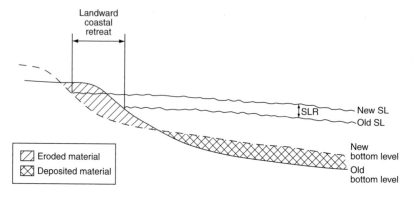

Figure 2.7 The Bruun rule of shoreline adjustment under sea level rise.

(deepening water levels). Clearly, if a new profile is to be achieved, it is not just a question of the profile shifting landwards because this will not take into account the fact that the sea is getting deeper. To compensate for this, the profile also needs to move upwards, so that each point of the profile remains at the same relative water depth. By doing so, normal wave base remains in the same position relative to the morphology, as do mean high and low water marks. To achieve this, eroded material from the profile is transferred onto the sea floor (Figure 2.7). Nicholls (1998) demonstrates the use of the Bruun rule in the assessment of how sandy beaches will react under sea level rise scenarios. From this work, Nicholls was able to estimate the volumes of sand drawn off-shore from the eroding shore face, and thus calculate the amounts of beach nourishment (see Chapter 7) which may be needed. More importantly, however, it is also possible to estimate shoreline change. Work in the USA has predicted that according to the Bruun rule, a sea level rise of 1 metre would produce a coastal retreat of around 75 m in Sea Bight, New Jersey, and Ocean City, Maryland, around 200 m along the barrier islands in Carolina, 300 m in San Francisco, and up to 1 km in Florida (see Titus (1986) for full discussion).

The roll-over model

The Bruun rule relates to coastlines which are dynamic and can migrate into the hinterland area. Offshore bars and barrier coasts are structures which have the sea on both the seaward and landward side. They respond to sea level by moving landwards into the unfilled space behind them, thus 'rolling over' in the landwards direction. Again, by doing this, they maintain their position relative to water depth and tides. Increasing water depths produce greater wave base which leaves the shore face exposed to greater wave energy. This moves sediment up the shore face and over the top of the structure. The process is similar in some respects to wind blowing sand across a dune field, or water formed ripples on a beach or on the base of a river.

As with the Bruun rule, the rate of movement is a function of the rate of sea level rise, and also the angle of the shore face. Carter (1988) also quotes this as an equation:

$$\delta R_1/\delta t = (\delta s'/\delta t)\tan\beta$$

where:

$\delta R_1/\delta t$ is rate of landward migration due to roll-over
s' is sea level rise
t is time
β is shore face angle.

As with the Bruun rule, there are additional factors that need to be borne in mind. First, the landward migration is not a simple case of the structure gradually moving landwards. Its rate of movement can vary, such as when the structure is composed of a range of material of differing resistance to movement (see also Carter 1988). In situations where movement becomes very slow, it is possible that the structure may not be able to keep pace with rising sea levels, and subsequently drown (see below). Another key issue is that not all movement can be directly attributed to sea level rise. Kochel and Dolan (1986), Inman and Dolan (1989) and Fenster and Dolan (1993, 1994) all cite additional factors in shoreward barrier migration, notably storm intensity and severity. This having been said, one main effect of sea level rise is an increase in wave base and a corresponding increase in storm frequency and intensity, thus it could be argued that these controls also have a link to sea level changes.

The drowning model

As has previously been mentioned, structures which are slow to adapt to sea level movements will not be able to keep pace and maintain an equilibrium profile. In this case, they may cease to exist altogether. Where barriers move only very slowly, the rate of increasing water depth will outpace the rate of landward movement and the barrier will become submerged. Similarly in estuaries, the ability of salt marshes to keep pace with rising sea levels is dependant on the supply of sediment to allow them to accrete vertically. If the rate of vertical accretion is less than the rate of sea level rise, then the flora will become unstable due to increased frequency and duration of water coverage, and the marsh will revert to lower vegetation communities and eventually to mudflat. Similar issues pertain to any coastal vegetation community. In Table 2.3, a series of case studies refers to rates of sea level rise in delta areas. Due to their rapid subsidence, the ability of vegetation to avoid drowning and subsequent vegetation community reversion is particularly important. Examples of drowned habitats are common around many of the world's coastlines, both in the geologic and more recent past. Carter (1988) reports on a mid-Holocene forest in western Ireland currently in the intertidal zone; Chesil beach in Dorset is currently rolling over landwards and revealing freshwater peats on the seaward side, previously inundated by the shingle; and Allen (1993) reports on some of the submerged forests on the shores of the Severn Estuary and Bristol Channel (see also Plate 2.2).

*   *   *

The changes in wave, tide and wind processes can, therefore, produce one of three potential response scenarios for coastal habitats under conditions of sea level rise. They could:

a)   cause landforms to erode and reform inland;
b)   cause offshore structures to move landwards; or
c)   if not able to adjust fast enough, could cause inundation and drowning.

A fourth model, not detailed above, concerns 'fossil' habitats which have been isolated from the sea for some time. In these circumstances, increasing sea levels and increased wave attack could cause reactivation, and renewal of

Plate 2.2 Fallen trees in the low intertidal zone of the Severn Estuary, south Wales coast, UK. The peat layers contain pollen reflecting high marsh and transitional, fresh-water communities.

sediment inputs. A typical example would be cliff forms produced at a previous high sea level stand which, during sea level fall, were left above the influence of new sea level.

Implications of sea level rise for sea defences

It is now possible to relate the basic coastal response mechanisms to sea level rise to the response of natural coastlines to changes in wave, tide, and wind conditions by changing shape. Upward trending sea levels are going to induce many such changes in these environments. Increasing water depth and related increases in wave energy are likely to increase beach erosion and sediment instability. The degree of impact from these changes will depend on the coast's ability to relocate inland (Bruun rule). Where coasts are fixed by hard defences, sediments which are not suited to the higher energy environment caused by sea level rise (i.e. fine sediments) may be moved off-shore, longshore, or lost to the system altogether, representing a loss to the sediment budget. The net effect of this loss of finer sediments is that sedi-ment volumes will decrease and so beach levels decline; this will further increase wave base.

The most obvious option on hard coasts is to raise the height of the walls; however, according to Boorman *et al.* (1989), many existing walls do not have sufficient foundations to support any further increase in height. This effectively means that as sea level rises, many walls will not be able to be modified and will meet what is termed the 'end of their natural life span'. They will, as a result, become less effective as defence structures. This does, however, provide the coastal manager with a unique opportunity to change the way coastlines are defended, and perhaps to opt for one of the softer methods of defence discussed in Part III. Using the present situation as a baseline, there are three possible strategies available to provide for adequate coastal defence provision under conditions of sea level rise:

- hold the existing line of defence (rebuild sea walls, increase their height, feed beaches);
- allow the coast to retreat inland (managed realignment and abandon-ment);
- allow the coast to advance seawards (offshore breakwaters, perched beaches).

The impacts of these options will vary according to the type of coastline in question. Many are considered in subsequent chapters but clearly, if the coast-line is to be managed in the most effective manner possible, it is necessary to adopt a solution where the coast can respond to movements in sea level wher-ever possible. Ultimately, this puts a bias towards the movement of defences inland, following the principals of the Bruun rule and roll-over, as outlined above.

Summary

Given that the main aim of this book is to study the impact of different types of defence on coastlines, the understanding of key processes is important. While it has not been possible to discuss all the processes in full, the overview provided gives a sufficient grounding to understand each defence structure discussed in subsequent chapters. For example, by understanding how waves work, it will be possible to study the effects of sea walls on beaches due to wave reflection, and groynes on beach levels due to longshore drift. Similarly, the effect of tides on sediment accretion and marsh formation due to brushwood groyne construction and sea wall removal can also be supported.

Perhaps more importantly, it is important to appreciate that, by necessity, each defence structure is discussed largely as a structure operating under a fixed range of environmental parameters. However, given the consideration of sea level rise issues, it should be remembered how waves, tides, and wind change over this time scale, and so how each defence needs to be constructed with these trends in mind.

In bringing these two introductory chapters to a close, the reader should be able to take key ideas of how coasts work and apply them to the problems which occur in connection with defence construction, and possibly to relate them to their own personal experiences on coastlines around the world.

Part II

Hard approaches to coastal defence

Part II looks at the 'traditional' approaches to coastal defence. These are structures which provide a solid, impenetrable barrier between the land and the sea, and resist the energy of the tides and waves, thus preventing any land/sea interaction from taking place.

In Chapters 1 and 2, it was shown that any structure which prevents natural coastal processes from operating will have a series of impacts on the natural environment. It is reasonable, therefore, to ask why such defences are so common and why the majority of defended coasts have such structures along them. The answer to this is that, historically, when confronted with an erosion or flooding problem, people considered that placing a physical barrier between the cause and problem was an effective solution. However, despite the problems of this approach, which will be seen in this section, hard defences remain the preferred response to coastal problems in many situations. This arises for a variety of reasons:

- tradition — tried and tested methodology and, if sediment supply has been maintained, many of the problems discussed in subsequent chapters may not be too serious;
- perceived security — local people tend to feel safer behind a sea wall than a built-up beach, even though, scientifically, the safety benefits may be the same;
- high value of the hinterland — resort areas or industrial installations need a significant physical barrier between them and the sea;
- politics — electorates like to feel secure and so politicians will often try and placate their concerns;
- inability to adopt a proactive-based defence policy — hard defences are very much reactive, in that they are built once a coastal problem arises.

Accepting that hard defences do have an important role to play in contemporary coastal defence, it is important to understand how they 'work' and how they interact with their local environment. By doing this, amelioration practices can then be incorporated into the design stages so as to prevent problems from happening, or reducing their impacts if they are inevitable. To be successful, however, this does require a detailed knowledge of coastal processes on both an academic and local basis which, unfortunately, may not be available. In the recent past, defence schemes have been built using largely the academic understanding of processes; although these are critical in planning, without an understanding of how these processes interact with local topography and sediments, there will be important gaps in our knowledge of how the local area 'works' and, as such, of how the defence schemes should be planned and operated.

In essence, hard defence structures are used to prevent erosion of the hinterland. Two points are important:

- these structures are designed to protect the hinterland, not the fronting beaches.
- these structures do not provide sediment to maintain a protective beach, they just redistribute it. In fact, because many hard structures have completely the opposite effect, in that they reduce sediment input, additional sediment sources need to be found, if the sediment budget is to be maintained in balance.

The following chapters deal with the four main areas of hard defence structures, linked to the common themes that all prevent the energy of the sea from interacting with the land, and all produce a static coastline which cannot respond to environmental variation. In addition, they may also alter other coastal processes, such as longshore or on/off shore sediment movement, and thus interfere with other areas of the coast down drift. One major criticism of how hard defences have been managed in the past, as outlined by Fleming (1992), is that many of the defence issues have been dealt with parochially, with little allowance made for impacts on adjacent stretches of coast. Any structure which cuts off an area of sediment supply, or which prevents the free movement of sediment along a coast, will have an impact on inputs to the coastal sediment budget and, therefore, may affect the accretion/erosion regime of other parts of the coastline. Such impacts are summarised in the scenario in Figure 1.2 and include:

- excessive draw down of beach material and storage below normal wave base by shore normal currents, leading to excessive beach erosion, caused by high reflectivity from static structures (sea walls) and lack of incoming sediment from up drift due to other defence works;
- loss of beach material as a result of non-replenishment due to retention in shore normal structures (groynes);
- enhanced erosion at the ends of hard structures leading to increased rates of loss and potential future defence problems.

While this Part looks at hard structures, it should be remembered that many of the hard defences built 50–100 years ago are nearing the end of the their useful lives and in some cases the protection which they currently afford the hinterland is minimal. Given that this situation is not going to improve without considerable expenditure from the defence authorities, and that sea levels are continuing to rise in many places, there has been a growing trend of replacing worn-out defences with softer structures (see Part III).

3 Sea walls and revetments

Introduction

Sea walls are the most common form of coastal defence, and that most generally regarded by the public as representing the best form, because by presenting a physical barrier between land and sea, they prevent any erosion of the hinterland and protect it from flooding. In addition, because they are such a solid barrier, the perceived level of protection from the sea is greater than many other forms of defence. However, the main issue regarding sea walls is that, once constructed, they effectively 'fix' the coastline in position, and stop it from responding to environmental changes, such as sea level rise; thus, over time, they start to induce considerable process modification.

The early history of coastal defences has really been that of the development of sea walls, from simple earth structures to the elaborate concrete and block structures seen today. Their use is world wide (see Table 3.1 for examples) and they are found on a range of coastal types. The earliest defences were put in place to stop the sea flooding the land, and so can really be thought of as flood protection, rather than as structures to prevent erosion. However, increased coastal development has necessitated substantial structures to stop land from being washed away and undermining buildings; a vertical wall was originally considered as the best way of doing this. In the UK the Victorians, at the beginning of the twentieth century, started a trend for seaside holidays and a fashion for promenades — walkways along the coastline which combine public amenity with coastal defence structures. These often flanked the shoreline, with vertical walls leading down to the beach; these, as we shall see later, lead to many problems with beach erosion from scour, because waves effectively 'bounce' off them. The trend of Victorians for grandeur meant that some of these structures were visually impressive, if functionally problematical. The tendency to build sea walls vertical lies largely in the engineering experience available at the time. As experience was largely confined to the construction of ports and harbours, where quays were supported by vertical walls, these same practices were adopted along the open coast, (Fleming 1992). This was a mistake, as these two sets of structures were constructed for different purposes — harbour walls are built to provide shelter for ships to dock and unload, while sea walls need to protect the

Table 3.1 Examples of sea wall impact studies from around the world (For full references, see References at the back of the book).

Country	Authors	Location
Bulgaria	Simeonova (1992)	Black Sea coast
Germany	Dette and Gartner (1987)	Isle of Sylt
	Kunz (1987)	Isle of Norderney
Hawaii	Fletcher *et al.* (1997)	Oahu
India	Baba and Thomas (1987)	Trivandrum Coast
Portugal	Granja (1995)	North-west Portugal
	Granja and de Carvalho (1995)	North-west Portugal
UK	Inglis and Kestner (1958a and b)	Wash
	Madge (1983)	Porthcawl, south Wales
	Carter (1988)	Porthcawl, south Wales
	Roberts and van Overeem(1993)	Herne Bay, Kent
	Wood (1978)	Aberystwyth, west Wales
USA	Basco *et al.* (1997)	Sandbridge, Virginia
	Everts (1983)	Cape May City, New Jersey
	Griggs and Fulton-Bennett (1987)	Santa Cruz, California
	Griggs and Tait (1988)	Monterey Bay, California
	Griggs *et al.* (1994)	Monterey Bay, California
	Hall and Pilkey (1991)	New Jersey
	Komar and McDougal (1988)	Oregon
	Morton (1988)	Texas
	Pilkey and Wright (1988)	(various)
	Plant and Griggs (1992)	Monterey Bay, California
	Shepard and Wanless (1971)	Ocean City, Maryland
	Tait and Griggs (1990)	Monterey Bay, California
	Wiegel (1964)	Santa Barbara, California

coast from flooding and erosion. Clearly, both need to resist waves, but the impacts of reflected wave energy on sediment stability, which is not critical in harbour areas, causes problems for sea walls and their fronting beaches. Really, this represents a case of inappropriate technology transfer.

In Chapter 2, the dynamic nature of the coast was seen from the way it reacts and responds to changes in waves, tides, and wind. These processes mean that the coast will change over time, yet because sea walls are solid, static structures, they hold the coastline in the position it occupied when the defences were constructed, and thus remove this ability to change. Where the wall ends, the coast remains free to respond to natural conditions. While part of the coast is fixed, it is clear that the undefended stretch of coastline could move inland under conditions of shoreline retreat. This process will give the coastline a stepped appearance, and also provides the observer with an idea of how much natural adjustment has occurred along a stretch of coastline since defence construction. One example of this phenomenon can be seen along the Fylde coast of north-west England (Plate 3.1). This coast is heavily defended along the frontage of Blackpool with a series of sea walls of varying types. These structures end to the south of the town (coincidentally, the end of the sea walls occurs at the regulatory authority boundary). Beyond them, the coastline is 'natural', with a wide beach backed by sand dunes. What is immediately obvious here is that the dune coast is offset from the protected (walled) coast by around 100 m. This represents the degree to which the dunes have migrated inland since the defences fixed the rest of the coastline, as can be seen in Plate 3.1. It also gives an indication of the problems encountered in coastal management here, because, in effect, the walled coastline at Blackpool is actually 100 m further out to sea than the 'natural', undefended line of the coast (see Box 3.1). Similar offsets at the end of defence structures can be found in many cases. A good example is at the end of the Galveston sea wall, Texas, where rapid landward movement has occurred since its construction in the first years of the twentieth century (see Davis 1996).

Although they create problems of fixing coastlines, it is clear that sea walls have an important role to play in coastal protection, particularly fronting high value land. However, because the way in which this is done prevents land/sea interactions, and thus landward movements of the coastline, the coast is forced to occupy a position further seaward than it otherwise might do. Furthermore, additional problems include loss of beaches, high cost, and impacts on sediment budgets. Despite all of these problems, walls are still built because there remain cases where the hinterland is so valuable that the benefits of using sea walls outweigh the problems. In such cases, it is necessary to reduce the impacts as much as possible. Typically, this will be by design modifications, which can produce walls with less impact on fronting beaches.

Plate 3.1 Typical sea wall end effects, Fylde coast, UK. The wall demonstrates the extent of terminal erosion and landward movement. South Shore, Blackpool, UK.

Box 3.1

Impact of sea walls on beaches at Blackpool, UK

Blackpool, located on the Fylde coast of north-west England, is perhaps best known as one of the UK's top tourist resorts, attracting around seventeen million visitors a year. This importance for tourism makes any threat to its wide, sandy beaches a serious issue, but over recent years, beach levels have been falling in the central part of this coast, particularly around the Bispham area (see Map 3.1).

During the nineteenth and twentieth centuries, the Fylde coast has experienced rapid development, resulting in an almost continuous sprawl of tourism-related and residential development along the whole coastline. This development occurred on a coast with alternating areas of low and higher land, the low land represented by sandy beaches fronting lagoons (hence, 'Black Pool'), and the high land by cliffs of glacial material. Although much of this development was inland, an erosion rate of approximately $2\,\mathrm{m\,a^{-1}}$ (Blackpool Borough Council 1993) meant that coastal retreat was rapid and sea defences became necessary to protect the development. The first sea walls were built in 1868; by 1937, the whole 11 km of the Blackpool frontage was

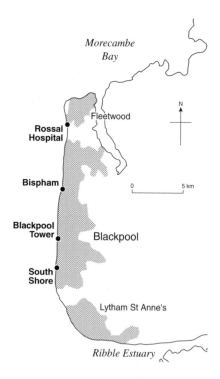

Map 3.1 Fylde coast

protected, largely by vertical concrete or block walls. In 1980, the annual maintenance cost for the shoreline topped £1 million, the highest of any coast in the UK.

Many of these walls are vertical and it is only recently that newer walls are being built to replace these, as they reach the end of their useful life. The new walls have been built with a higher regard for their environmental impacts and incorporate many of the modifications suggested in this chapter. Historically, however, the coastline has witnessed a series of beach impacts.

- The construction of sea walls has removed the cliffs as a source of input to the sediment budget, leading to deficits. With Morecambe Bay to the north acting as a sediment sink, the prime source of sediment to the sediment budget is from either offshore or from the beach itself.
- The vertical or near vertical nature of many of the walls has lead to wave reflection and scour.
- The complex longshore sediment movement (sand waves) appears to drive sediment both northwards and southwards.
- Ongoing sea level and shoreline movements mean that the contemporary shoreline is too far into the surf zone to be stable (see Plate 3.1) due to being fixed in position by the hard defences.

Because sediment inputs from local sources have stopped, and because of the lateral divergence of longshore drift, the central part of the coastline is effectively starved of sediment, and can only obtain renewed sediment supplies by onshore movement, or storm input from other parts of the beach. In addition to this, the older sea walls provide a highly reflective surface for waves which also serve to focus reflected wave scour onto the foreshore. This has produced an interesting situation: because of the limited amount of new sediment being input to the longshore drift circulation, the central part of the beach (around the Bispham) area, is acting as the main source of sediment for this coastline, with the result that levels here have fallen dramatically (see Figure 4.2).

In addition to the loss of beaches due to sediment starvation and scour, the hard nature of the defences has meant that the coastline has not been able to respond to coastal fluctuations, typical of the undefended coastline down-drift. As a result, the current walled coast is fixed around 100 m seaward of its 'natural' position, meaning that it is further out into the surf zone where wave energy is greater. As a result, the scour from wave reflection is enhanced because the energy levels are greater.

Blackpool, therefore, has a highly dynamic coastline, as evidenced by the sand wave structures and wind blown sand (which can cause problems on the coast road), but also one which is starved of sediment, fixed in an unnatural 'position', and backed by highly reflective sea walls. The impacts of these three factors can be clearly seen on the beach.

In overcoming these problems, the local authority is faced with a serious challenge. They need to defend their highly developed coastline, yet are confronted with a rapidly lowering beach. At the present time, the local authority is adopting a policy of holding the existing line while considering softer techniques to reverse the current erosion trend. One potential strategy could:

- accept that holding the existing line is the only viable option, and reduce the causes of the problem. These are basically high wave reflectivity and beach sediment loss at the down-drift end of the system;
- continue the ongoing programme of replacing walls with new designs, such as those recently used to the south of Blackpool (see Figure 3.1). This is already authority policy;
- identify sediment sources which are supplying the current beach. It appears that the main source is in the Bispham area (for a discussion of this, see Chapter 4). This area could be recharged with sediment using material which has already moved southwards and is accumulating around the mouth of the Ribble Estuary. This would represent a source of local, littoral sand, thus fulfilling the main criteria for sourcing recharge sediment (see Chapter 7).

Such a strategy would overcome the main problems of this coastline, yet maintain it as a tourist amenity. The one concern would be the implications for sediment transfer further down the coast. Although this is blocked by the estuary of the Ribble, it is possible under storm conditions for channel changes to occur here, and pulses of sediment to cross the estuary mouth. Such sediment movement may be affected if large-scale removal was to occur.

Types of sea wall and their uses

Sea walls are designed to resist the force of wave attack in order to prevent the erosion of the land. Their design will therefore reflect the amount of wave energy experienced by a particular coastline. In order to accommodate waves into the design of structures, it is important to have a knowledge of the waves occurring in the area in which the structure will be sited. This really necessitates a return to wave theory, but for our purposes, several key things are important. In Chapter 2, it was seen how important wave height can be for structures, given that wave energy is a function of the square of wave height. Clearly, structures have to be designed for the worst case scenario. Komar (1998) refers to this as the *design wave* (often cited in the literature as H_{max}); it is a function of the water depth in front of the structure. Immediately, therefore, it is possible to identify a further concern arising out of beach lowering: as a beach loses sediment and gets lower, the water depth, and thus the size of the design wave, will increase.

For much wave research, it is the *significant wave height* which is important. The value will characterise any sea state chosen, and is taken as an average of the highest one third of all waves occurring over a set time period. This is often cited as $H_{1/3}$ or H_s (Open University 1989). These values are important because they can be used to estimate wave reflection from walls and therefore, scour and wave reflection. Finally, these various wave parameters can be used to estimate the *wave climate*. This value factors in wave height, wave period and direction to give important information on the likely angle of wave approach, and the possible frequencies of waves hitting the coast. With regard to engineering design, this means that it is possible to get an indication of sediment movement direction, such as longshore drift, which is again critical with respect to beach lowering.

Because wave-dominated coastlines can range from low to high energy there needs to be a variety of appropriate defence designs. These range from low earth embankments to high concrete walls. A wall which has to withstand direct, high energy wave attack needs to be a substantial structure, and so is often solid, vertical, and modified with revetments or armourstone. Conversely, a wall which is fronted by a wide beach or salt marsh will receive less wave energy and so can be less substantial.

Perhaps the most fundamental problem with sea walls is that they tend to reflect wave energy and it is this issue which much design research has tried to overcome. In a worst case scenario, they reflected energy may interact with incoming waves to set up a standing wave which can result in scouring of sediment on the beach fronting the wall. This is a major problem in sea wall design; it can even result in wall undermining and eventual collapse. Walls may be modified to include revetments (slopes at the toe of the wall), rip-rap (irregular or interlocking blocks placed at the toe of a sea wall), or armourstone (piles of graded boulders) in order to protect the toe of the wall from sediment loss due to scour, and also to reduce reflected waves by scattering the direction of reflection. An uneven face will reflect wave energy in random directions, but may also act to absorb some of this energy, so that not all will be focused back towards the beach. This latter strategy can be taken further with the use of concrete tetrapods or large stone blocks which are less reflective than solid walls and produce greater dissipation of wave energy. Many walls contain a series of these design modifications, tailored to suit the wave exposure of the area in question. The broad spectrum of sea wall designs is shown in Figure 3.1.

Finally, it should be realised that although wave reflection from walls is often blamed for the loss of fronting beaches, this needs to be put into perspective. While beach losses certainly do occur, in many cases this may not be solely due to the wall itself. After all, building the wall in the first place is often a response to an erosion problem. As such, erosion of the fronting beach may be partly due to this original problem continuing. Pilkey and Wright (1988) have used the term 'passive' erosion to refer to the longer-term erosion processes which were occurring before wall construction, and 'active' erosion, to refer to those new processes which occur in response to the wall. Weggel (1988) argues that many claims of beach loss being a result of 'active' processes are conjecture and unproven by hard scientific data; he also cites an example from Santa Cruz, California, where a beach formed in front of a sea wall where none had been previously. Pilkey and Wright (1988) suggest that the impact of active erosion is the

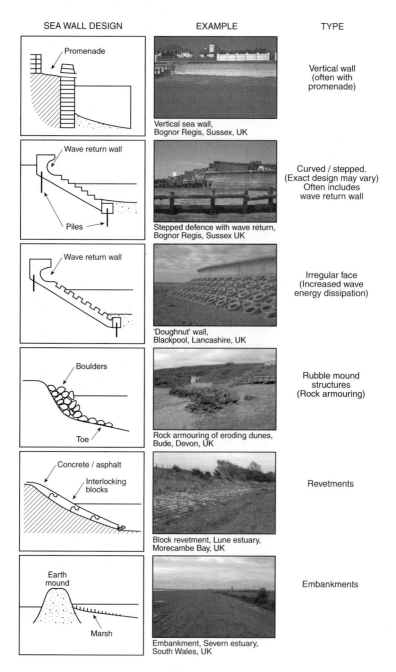

SEA WALL DESIGN	EXAMPLE	TYPE
Promenade	Vertical sea wall, Bognor Regis, Sussex, UK	Vertical wall (often with promenade)
Wave return wall / Piles	Stepped defence with wave return, Bognor Regis, Sussex UK	Curved / stepped. (Exact design may vary) Often includes wave return wall
Wave return wall	'Doughnut' wall, Blackpool, Lancashire, UK	Irregular face (Increased wave energy dissipation)
Boulders / Toe	Rock armouring of eroding dunes, Bude, Devon, UK	Rubble mound structures (Rock armouring)
Concrete / asphalt / Interlocking blocks	Block revetment, Lune estuary, Morecambe Bay, UK	Revetments
Earth mound / Marsh	Embankment, Severn estuary, South Wales, UK	Embankments

Figure 3.1 Variation in design type of sea walls. Hard, concrete structures typical of open coasts are replaced by revetments and embankment-type structures in more sheltered, tide-dominated environments.

area of greatest debate, and name it the *Great Seawall debate*. Other workers (for example, Griggs *et al.* 1994 and Morton 1988) argue that while beaches backed by walls may respond to storm events faster and to greater magnitudes than unwalled beaches (i.e. more rapid and greater erosion), they quickly regain status quo and under non-storm conditions respond in a similar manner to the unwalled areas. It thus appears that some beaches behave differently to walls than others. This would tend to suggest that in cases where no detrimental impacts are experienced, the wall is operating 'in harmony' with the coastal processes, with any scoured sediment being replaced, or where wave reflection is not being focused onto the fronting beach. Walls which initiate large-scale beach losses (see Table 3.2) tend to oppose natural forces and are in conflict

Table 3.2 Summary of the impacts of sea walls on beaches from a series of case studies (For full references, see References at the back of the book).

Author	Place	Impacts
Basco *et al.* (1997)	USA	Increased seasonal variability in beach volume, no increase in erosion rates compared to unwalled beaches, end effects
Everts (1983)	USA	End effects, up-drift impoundment
Fletcher *et al.* (1997)	Hawaii	Beach narrowing, acceleration of natural 'passive' processes, beach loss
Granja and de Carvalho (1995)	Portugal	Increased cliff erosion as end effects, beach lowering
Griggs and Tait (1988)	USA	More rapid berm removal, end effects
Griggs *et al.* (1994)	USA	End effects, otherwise, behaves similarly to unwalled beach
Hall and Pilkey (1991)	USA	Reduction in dry beach width
Madge (1983)	UK	Beach lowering and toe scour
Morton (1988)	USA	Scour in storms, impedance of onshore sediment movement, end effects, increased erosion in storms (normally recover during calm periods), beach lowering
Plant and Griggs (1992)	USA	Increased backwash velocity and duration, reduced infiltration, updrift impoundment
Tait and Griggs (1990)	USA	Scour, end effects, cusp formation, rip currents, beach lowering, updrift impoundment
Wood (1978)	UK	Beach lowering, reduction in width
Wood (1988)	USA/Canada	Beach width reduction, modification of offshore bar behaviour

with them. This further supports ideas mentioned earlier — that a detailed knowledge of coastal processes in the area of construction will allow the most effective wall design to be built, with impacts reduced as much as possible.

When considering the effectiveness of walls, it is necessary to remember their function. Sea walls are built to protect the land behind from erosion and flooding by the sea. Taken like this, it is easy to say that sea walls are very effective. A sea wall is not built to promote sediment accretion and beach development, and so some researchers (see Weggel 1988) argue that these issues are not important. While this may be superficially true, it is important to consider the success of sea walls holistically. Sea walls are good at stopping erosion; this point is not in doubt. However, in the medium to long term, problems of undermining due to scour render walls less efficient at protection; a structure that can initiate its own demise by setting in motion a series of erosive processes which destroy its very foundations may not really be considered so successful and efficient.

Impacts of sea walls on the coast

The construction of a sea wall may solve the problem of erosion, but it will not deal with the causes of erosion. When looking at the impacts of the wall it is important to identify those caused by the processes which continue to operate from preconstruction, and those which have originated as a direct result of the wall being built (i.e. what Pilkey and Wright (1988) refer to as passive and active erosion). Much of the research in this area has focused on the fronting beach, but this is not the entire problem. Sea wall impacts occur in front, at the ends, and behind the wall, and encompass sediments, cliffs, and ecology. All of these areas need to be considered in order to undertake a complete assessment of the role of sea walls in coastal defence.

There are many opinions on the effectiveness and problems of sea walls in many areas of the literature, ranging from scientific papers to more general (grey) literature and the media. All have a role to play in the assessment of environmental problems. In this, local residents should not be dismissed because, although they may have no scientific training, they do know the local area very well, and are probably the best people to comment on how the beach responds under varying conditions (such as seasons or storms). In presenting an account of sea wall impacts, such as that below, it is important to include different points of view and present a balanced argument. It is clear that walls behave differently under different environmental conditions, and that different designs would operate differently under the same environmental conditions. Each case study therefore needs to be considered as relevant to a particular coastal location and care must be taken when making generalisations and extrapolating data to other areas.

Impacts of sea walls on fronting beaches

Of all the impacts commonly cited against the construction of sea walls, the problem of scouring is perhaps the most significant, and is also the one area where research has produced a plethora of published material. Tait and Griggs (1990) claim to have identified a series of commonly observed phenomena on beaches fronting sea walls:

- formation of a scour trough — a linear shore parallel depression fronting a sea wall;
- formation of a deflated profile — the uniform general lowering of the fronting beach;
- formation of beach cusps — semi-circular, seawards opening embayments;
- formation of a rip current trough — a linear shore normal depression;
- terminal scour — accelerated (active) erosion on beaches and coasts immediately down drift;
- up-drift sand accretion, due to impounding at the up-drift end of the wall.

Many of these features will be discussed in the following section, and further detail can be obtained from the review by Kraus (1988). In essence, the first four observations relate to the reflection of waves from the structure, and the interaction of these reflected waves with incoming waves and the beach. Komar and McDougal (1988) also cite the generation of wall-induced rip currents as an important mechanism in the erosion of beaches, and use examples from Oregon, USA to illustrate their argument. The fifth problem, terminal scour, will occur when the reflected energy, in excess of that which moves sediment off the beach, is focused towards the end of the wall, and starts to create an artificial embayment. Finally, where a sea wall sticks out from the coastline due to continued retreat of adjacent undefended coasts (see Plate 3.1), then it can act as a groyne by interrupting longshore sediment movement and causing sediment impoundment, and down-drift starvation.

The problem with scour processes is that by continuing to lower fronting beaches, they can undermine the walls themselves. It has been argued (Hydraulics Research 1992) that the devastating floods and loss of life in both the UK and the Netherlands, as a result of the massive North Sea tidal surge of 1953, owed much to sea wall failure resulting from toe scour. It is further argued that around 11 per cent of all sea wall failures can be directly attributed to toe scour. Carter (1988) cites two examples from Wales demonstrating this problem. In Porthcawl, the level of the beach fell by 3 m in 75 years, necessitating the reconstruction of the whole sea wall in 1934. Similarly, in Aberystwyth, the fronting beach got narrower and lower following promenade construction in 1866–7. Remedial works here, including groyne construction and rock armouring, have not been effective, leaving the wall with an uncertain future. Ongoing research at the Hydraulics Research Institute in the UK is investigating the processes and environmental conditions under which scour occurs, in an attempt to adopt a more proactive approach to hard defence management. By pre-empting the conditions under which scour will become a significant problem, the aim is to produce a series of design guidelines which can predict the behaviour of waves at the foot of sea walls. Initial findings suggest that there is a complex relationship between waves and scour, and that at different water depths, waves with the same properties can have vastly different impacts on the beach sediment. In practical terms different water depths would equate to different states of the tide, beach lowering, or an upward sea level trend. Significantly perhaps, this research is pulling together many of the previous studies which Weggel (1988) argued as being contradictory and conjectural in proving beach loss in front of walls. Emerging research is now showing that fronting beaches can, in some instances, be affected by sea wall design; this provides a framework with which to re-evaluate the many investigations which have occurred over the past few decades.

Other studies have also shown the association of scour and fronting beaches, although these have largely been restricted to laboratory flumes (see Komar 1998), largely because of the difficulties associated with data collection in the field under the range of environmental variables concerned. Despite this, certain common factors tend to emerge from the literature concerning beach scour:

- scour tends to be a greater problem the further the wall is built out into the surf zone (Kraus 1988) — this suggests that as sea level rises, the problem of scour will increase due to the effective landward shift of the surf zone.
- volume of sediment removed from walled and unwalled beaches during a storm tends to be the same, but walled beaches tend to have a restricted width, so this sand volume comes from a narrower beach width and, therefore, a greater depth — the process is most notably from toe scour (Kriebel 1987) and is supported by modelling studies (Rakha and Kamphuis 1997).
- reduction of sediment input to the coastal sediment budget may increase sediment loss if alternative sources are not available. The realisation of such alternatives by coastal processes, however, may increase erosion elsewhere (Walton and Sensabaugh 1978, McDougal *et al.* 1987).
- up-drift end of the wall may cause sediment impoundment and so reduce longshore movement of sediment (compare groynes) (Plant and Griggs 1992, Everts 1983).
- down-drift end of the sea wall is generally marked by increased erosion, causing a 'stepped' appearance to the coastline (Figure 3.2) (see Table 3.2 for references).

The loss of sediment on a beach in front of a sea wall should not be put down solely to wave reflection and scour. Perhaps the most notable feature of walls is that they replace a soft, erodable shoreline with a hard, non-erodable shoreline. This immediately, therefore, changes sediment supply to the beach, both with respect to quantity and source, which could mean that not only do beaches lose sediment due to scour, but also due to non-replacement of sediment because of loss of sediment inputs. In some cases, the beach itself will replace the original undefended coastline as prime sediment source (see also Box 3.1). In many examples, this sediment is moved offshore to form bars (see Kraus 1988), or along shore as part of the longshore drift, adding to the

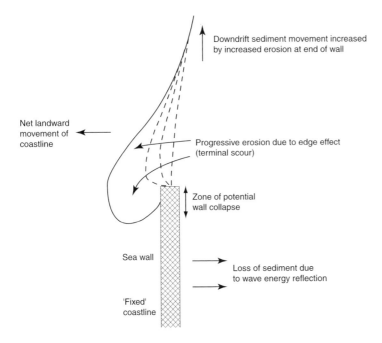

Downdrift sediment movement increased
by increased erosion at end of wall

Net landward
movement of
coastline

Progressive erosion due to edge effect
(terminal scour)

Zone of potential
wall collapse

Sea wall

Loss of sediment due
to wave energy reflection

'Fixed'
coastline

Figure 3.2 Typical end effects (terminal scour) associated with a sea wall. Note the
scour which removes support from end of structure and leads to undermining at end of wall.

volume of sediment lost through wave reflection. Such a loss of sediment produces lowering and width reduction on beaches, an observation supported by
Hall and Pilkey (1991) in their study of New Jersey beaches. In this example,
beaches backed by walls had a mean dry width of 9 m, while those with no walls
had a mean width of 55 m. This study is further supported by Pilkey and Wright
(1988) who also included beaches in North and South Carolina, and still
consistently recorded narrower dry beach widths in those areas backed by walls.

Flanking/terminal scour

It is common to observe at the ends of sea walls an area of increased erosion,
known as flanking or terminal scour. This phenomenon is shown diagrammatically in Figure 3.2, and also in Plate 3.1. Essentially, excess wave energy is being
'transferred' along the coast fronting the wall and is causing the increased
erosion at its down-drift end where it meets erodable sediment. Griggs and Tait
(1988, 1989) have observed reflected waves travelling parallel to sea walls for
up to 30 m; the interference of these reflected waves with incoming waves
results in excessive scour. Perhaps the classic example of this is at Cape May,
New Jersey, where terminal scour has eroded the coastline for over 1 km down-
drift, reaching inland for up to 400 m. In a laboratory study, McDougal *et al.*

(1987) placed a model of a vertical sea wall into an equilibrium beach situation; major erosion was induced, which was at a maximum at the immediate down-drift end of the structure and progressively decreased with distance from the wall in the down-drift direction. This observation is similar to the situation shown in Figure 3.2. Work by Griggs and Tait (1988, 1989) has identified several controlling factors for the magnitude of these end effects, based on their studies in the Monterey Bay area of California, USA. The down-drift extent of the flanking is partly a function of wave height and wave period, but also of the length of the sea wall itself. McDougal *et al.* (1987) estimate that the down-drift extent of the flanking effect will be of the order of 70 per cent of the structure's length. Coupled with this, the extent of the flanking is also dependant on the end geometry of the wall, the permeability of the wall structure, and the angle of wave approach and the stage of the tidal cycle at which it occurs (water depth).

The effects of storm activity should not be underestimated in the formation of flanking effects. Several researchers have identified increased erosion during storm periods at the end of sea walls. While this may be expected due to greater wave energy, Birkemeier (1980) noted a 380 per cent increase in the volume of material eroded down-drift of a sea wall over that predicted for the undefended coastline during a storm event. Even in non-storm conditions, Griggs and Tait (1988) noted a flanking effect penetrating 75 m inland at the end of a wall in Monterey Bay, and 46 m at the end of another. Such observations suggest that there has to be some 'magnification' of the erosion effect of the walls, such as that identified by Griggs and Tait (1988, 1989).

At its most severe, flanking may start to undermine the end of the wall itself. This means that the erosion of the hollow may well increase in width with distance inland (see Figure 3.2), and thus penetrate behind the wall structure, leaving it unsupported from behind and prone to failure in subsequent storm conditions. This happened at Seabroke Island, USA, following Hurricane David, where waves washed out loose sand from behind the wall, leaving it unsupported and inducing its eventual collapse due to continued frontal wave attack (Sexton and Moslow 1981). Regardless of flanking severity, however, the very process will mean that property owners down drift will start calling for extensions of the sea wall because of the accelerated erosion problem, leading to the progressive erosion scenario for defended coasts proposed by French (1997) (see Figure 1.2).

Modification of the tidal regime of estuaries

On the open coast, sea walls impact on the environment largely by wave reflection and lateral wave energy transfer to induce terminal scour. In estuaries, the predominant impact is on the tidal processes of the system. As the tide flows into an estuary, it gradually fills it up until it reaches the top of the salt marsh, then flows over its surface to high land. The *space* which this water fills is known as the tidal volume, while the *volume* of water is known as the tidal prism. When sea walls are built along an estuary (typically for land claim, but also for flood defence), the space for water to occupy is reduced, yet the tidal prism is not. So, with the same volume of water flowing into a smaller space, the level of that water will be higher (Figure 3.3).

French (1997) summarises the main impacts of sea walls in estuaries, and suggests four main ways in which the estuarine environment can be changed:

- the tide will have less area to inundate and so water depth could increase, covering areas for longer and inducing change in vegetation communities, and also increasing the tidal range upstream (see Figure 3.3; French (1997): 66–71).
- as with the open coast, sea defences represent a 'hard' barrier between land and sea which can act as a reflector for waves, causing back scour, and preventing any natural habitat adjustment in response to external factors, such as sea level rise and dredging.

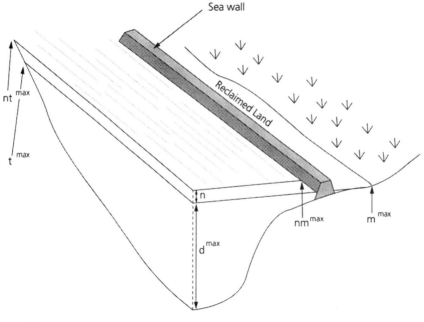

d^{max} = maximum water depth before land claim
t^{max} = maximum upstream limit of tidal range
m^{max} = maximum limit of tidal range on estuary margin
$n+d^{max}$ = new maximum water depth
nt^{max} = new maximun upstream limit of tidal range
nm^{max} = new maximum limit of tidal range on estuary margin

Figure 3.3 Impact of land claim and sea wall construction on intertidal zones of an estuary (After French 1997).

- the cross-sectional area and channel morphology of an estuary will undergo modification through the reduction in intertidal area, thus causing further implications for current activity, wave propagation, sediment movement and ebb/flood dominance.
- removal of part of the temporary sediment store represented by salt marsh sediments from possible reworking and input to the sediment budget.

An additional use of walls in estuaries is to confine and direct flow into certain parts of the channel to maintain stability and access for boats. These structures are termed *training walls*. In the Lune Estuary, Morecambe Bay, for example, training walls altered the duration of ebb and flood tides, concentrating ebb flow into one channel, causing channel enlargement in terms of both width and depth (Inglis and Kestner 1958a). While facilitating channel scour may be considered advantageous to estuary managers, inasmuch as it maintains channel stability, there has to be a corresponding impact elsewhere. In the case of the Lune estuary, this was manifest as increased sedimentation seaward of the estuary. In some situations, this could lead to excessive shoaling in the nearshore zone, with restrictions in longshore drift and possible sediment starvation problems.

The primary impacts in estuaries, therefore, lie in the alteration of the estuarine system which can be fundamental in the functioning of the whole coastline. Clearly, the historic use of sea walls to protect claimed land has forced many of the world's estuaries into an artificial morphology which cannot respond to changes in environmental forcing factors. Under conditions of sea level rise, an estuary needs to respond by lateral shifts of habitats upwards and landwards, following the ideas summarised in the Bruun rule. Clearly, with sea walls, as in the open coast, this will not be possible and coastal squeeze results. The return of the estuary to a natural shape which can respond to sea level changes by the removal of some of these constraining walls forms part of an increasingly popular method of coastal defence, managed realignment (see Chapter 10).

Benefits of sea walls to the coastal environment

It is easy to view the construction of sea walls negatively. However, it should be remembered that they are an effective way of protecting high value land which, in many cases, cannot be effectively protected in other ways. It is also true that some walls do not produce the negative impacts described above; this suggests that under certain design criteria and environmental conditions, sea walls can effectively both protect the hinterland and 'preserve' fronting beaches. If research can answer the question of why, then we will be making a large step forward in developing effective sea wall design. Earlier, we mentioned examples where beaches have not experienced any greater loss than unwalled beaches (Griggs *et al.* 1994, Morton 1988). Weggel (1988) claims that there is a general lack of erosion data to support beach erosion claims. This may reflect the lack of research which focuses on the actual cause of beach loss, i.e. whether it is caused by a continuation of passive processes, or the initiation of new, active processes. Basco *et al.* (1997) in their study in Sandbridge, Virginia, USA, support the statements that there is no significant difference in the loss of sediment from walled and unwalled beaches. However, they do report that seasonal variability of sand fronting walls is greater than unwalled coasts; winter storms draw down greater volumes of sediment, but in summer calm periods, greater beach volumes are built in return. Hence, losses and gains may be greater over a given period of time, but the net outcome is the same. What we have, therefore, is two types of beach (walled and unwalled) which show no net difference between them, but do show differences in the magnitude of their response to storms. It is tempting to suggest that in such cases, storms do lead to greater impacts on walled coasts, but the dynamics of the system and the post-storm availability of sediment is sufficient to rebuild the beaches following these storms. The counter observation, where sea walls lead to more permanent sediment loss, may reflect a system where sediment is lost from the beach during storms and taken offshore below normal wave base, thus making it unavailable to be reworked back onto the beaches under non-storm conditions. This may partly explain the disagreements between researchers as to whether beaches experience adverse impacts from sea walls. It is true, however, that while some walls have provided protection without adverse impacts on the beach, some have been only partly protective, and others have been detrimental (Terchunian 1988).

Increasing the success of sea walls

The position of any undefended coastline is a function of sediment supply, incident wave energy, and sea level. The first complicating issue is that none of these are constant, in that sediment supply varies over time with rainfall, storm erosion, etc.; wave energy is dependant on meteorological conditions; and sea level often reveals upward or downward trends in the medium and longer term. Despite this, however, coastlines can achieve a dynamic equilibrium, meaning that they maintain a net stability in the range of variation offered by the three variables. To gain the optimum result from a sea wall, it needs to be constructed in a way that minimises its impact on these three variables. First, a wall which cuts off an area's only supply of sediment (e.g. a cliff) will have a major impact on the beach, while an area which has many sources of sediment (e.g. from offshore and longshore drift) will have alternate sources and will not have to rely on the fronting beach to source its sediment. Second, wave energy will dissipate naturally over beach profiles (see Chapters 7 and 9) and so the further inland sea walls can be placed, the fewer problems with interference with incident waves, and the less seaward penetration into the surf zone. In effect, this means that a wide beach in front of the wall, supplied from distant sources, will be an ideal situation, with the wall protecting the hinterland from storm erosion or flooding. As an ideal, however, this is often not achievable. In such cases, it is important to minimise the impact of the wall through effective design. The problem of sea level rise (as seen in Chapter 2) is much more complicated, because it is also affects the first two variables.

The careful placement of walls, therefore, can help reduce impacts on beaches. Another method is to alter the design of the wall to suit local conditions. Walls which reflect wave energy are bad and lead to scour. Walls which dissipate energy, either by absorption or by random deflection on an irregular surface, can be effective, while recurves on top of walls can focus wave energy onto a concrete revetment or rock armouring rather than the beach (Figure 3.1). Ahrens and Bender (1992) discuss the merits of recurved walls and their effectiveness in preventing scour and overtopping. Making the surface irregular will mean that waves are not reflected in one direction, but scattered internally. A good example of this is the new wall fronting south Blackpool (Figure 3.1). Here, the 'doughnut' like appearance is effective and the fronting beach has shown some accretion. Care needs to be taken, however: calculations need to be made to ensure that the individual 'nuts' cannot be plucked from the wall during storms.

Another way of overcoming problems of beach loss is to adopt a more proactive approach, to acknowledge its inevitability and to introduce countermeasures for it (Terchunian 1988). If the situation is such that a sea wall needs to be built, but doing so will prevent input from the major sediment source, then it will be necessary to replace the sediment artificially, typically by beach feeding. These techniques are discussed in Chapter 7. In suggesting this Terchunian proposes a mathematical model to calculate the required volumes, taking into account previous, passive erosion rates, post-construction active rates and loss of inputs from the defended source.

Sea walls and sea level rise

We have already seen how wave reflection from sea walls can cause beach scour and loss of beach material, possibly leading to sea wall undermining and collapse. Under a scenario of increasing sea levels but a fixed coastline, the water depth in front of sea walls will increase, as will the wave base, meaning that larger waves, with higher energy, can penetrate further inshore. In areas where reflection from walls is currently a problem, this will exacerbate the rate of beach scour and loss of sediment. Furthermore, the lateral transfer of energy along a sea wall causing terminal scour may also become more acute as these energy levels increase in line with deepening wave base.

In essence, the survival of beaches depends on two things. First, the maintenance of an adequate sediment supply and, second, the ability of the beach to relocate inland (see Bruun rule, Chapter 2). For a beach backed by a sea wall, this landward migration is not possible, forcing the beach to occupy a position further into the surf zone than it would 'naturally' do. As a result, it will occupy a position in the tidal frame at which it will not achieve stability, and will attempt to adjust accordingly. The associated increased erosion and beach lowering caused by rising sea levels will lead to a loss in amenity value as beaches narrow, and also to the potential of sea wall failure due to undermining following wave reflection-induced scour. This impact is particularly acute in areas of amenity usage; in most tourist beach areas the hinterland is often fixed by a sea wall because this area contains the high-value resort infrastructure, such as hotels, promenades and leisure activities. The solution to this problem could take two paths. First, the sea wall should be allowed to retreat inland; this is frequently not a viable proposition in intensive tourism areas, although it may be possible under a scheme of staged retreat (see Chapter 12). Second, along undeveloped beaches where amenity usage is still important, the landward movement of the beach should be able to continue unhindered, and thus beach stability could be maintained. From a socio-economic angle, however, this means that there will be a user shift from areas with degrading beaches backed by walls, to other beaches which are able to maintain some degree of stability through landward migration.

Summary

As far as the general public is concerned, sea walls are the most secure form of defence because they provide a physical, and often substantial, barrier between the land and the sea. This perception of increased security, however, needs to be balanced against the environmental problems which can be caused by their use.

We have seen arguments for and against the use of sea walls, with some authors claiming that there are no detrimental effects on the fronting beach, and others that there is major sediment loss. There is clearly a discrepancy here but, as we have seen, there are more factors involved in the equation than just the presence of a sea wall; and it is likely that different coastal areas respond differently to walls because of the difference in tidal and wave regime as well as the properties and distribution of sediments. What is clear, however, is that coastal managers cannot do without them. Walls are still being built today, and are still producing damaging effects on fronting beaches. It is tempting to ask, therefore, why they are still so popular. In answering this question, we need to consider their role. Basically, a sea wall will stop the sea from eroding the land, thereby protecting coastal development. It is this which provides the answer. Due to the enormous pressure to develop the coastline for residential, industrial and amenity use, there is high investment in coastal property, which needs protecting. Leaving aside the arguments as to whether it was sensible to develop the coastline in some of the ways that we have, we are left with the historic legacy of this development, some of which was built before we had an understanding of sea level rise and coastal erosion. The development needs protecting, and walls are the best method we have for high-value land. In such a situation, we are often left with no other choice than to construct a solid physical barrier, regardless of cost, in order to protect property. To abandon land to the sea would be unthinkable in many coastal areas, because of the sheer scale of development. In some cases, the need to continually upgrade sea walls is critical from a health and safety point of view, particularly considering the location of some nuclear power stations (French 1997).

Building and maintaining sea walls is therefore an ongoing commitment. However, because of their impact on the coastline, we need to either restrict their usage or adapt their design wherever possible in order to avoid the problems of wave reflection, sediment input deficits, terminal scour, etc., discussed here. By studying the range of case studies available (Table 3.1), it is possible to learn a series of lessons relevant to the design and construction of sea walls. In particular, key issues which need to be addressed in the context of improved design are the following:

- reduction in wave reflection and beach scour minimises the focusing of reflected waves on the fronting beach — uneven wall surfaces and armour-stone toe protection are ideal in this context
- detailed research into the nature of the surf zone will help predict beach response to waves under different water depths and storminess — reducing a wall's penetration into the surf zone will help reduce scour and transfer of energy into edge waves
- lateral deflection and transfer of wave energy is a main cause of terminal scour — reduction of these processes will reduce the accelerated rates of erosion in these regions, and help prolong the life of the wall in these terminal regions
- increasing water depth due to scour may be partly offset by the above suggestions, but where local circumstances prevent the construction of a wall fully sympathetic to all potential environmental impacts, proactive measures, such as beach feeding, should be employed to compensate for beach loss.

A further possible measure to include in the above list is the current coastal management trend of avoiding these hard structures, and favouring softer methods which use sediment rather than solid barriers to defend the coast. These methods are addressed at length in Part III of this book; and the reader, bearing in mind the problems caused by the traditional sea wall method of defence, can compare walls with these newer, softer techniques.

Summary of the benefits of sea walls

- Prevention of hinterland erosion
- Increased security for property from flooding
- Physical barrier between land and sea increases perceived security of local people
- Maintenance of hinterland value

Summary of the problems with sea walls

- Potential impacts on fronting beaches
- Interruption of longshore sediment movement

- Increased erosion down drift (terminal scour)
- Fixes coast and prevents its responding to sea level rise
- Stopping of some inputs to sediment budget

Recommended usage

- To protect high-value hinterland development
- To increase and protect amenity usage where other solutions are not suitable

4 Groynes and jetties
Shore normal structures

Introduction

Groynes (US spelling: groins) and jetties may be collectively referred to as shore normal structures, i.e. they are constructed so that they lie at approximately right angles to the coastline. The distinction between the two structures is important. Groynes are smaller scale structures than jetties, tend to occur in groups (groyne fields) and, critically, terminate within the surf zone. In contrast, jetties are often built individually or paired either side of an estuary, and are more substantial structures, penetrating well seaward of the surf zone. As Silvester (1979) points out, because much sediment is held, and much longshore current activity occurs, within the surf zone, then longshore transport of sediment is also concentrated here. Any shore normal structure, therefore, which obstructs all or part of the surf zone will interrupt this sediment transport, causing the retention of sediment and subsequent build up of beaches. Hence, the implications for coastal defence relate to the prevention of the down-drift movement of sediment — an impact associated with both types of structure, although its scale will vary.

The concentration of longshore processes within the surf zone underlies the whole rationale of using groynes to build and protect beaches, in that they are designed to control the movement of sediment along the beach and reduce its loss down drift. In addition to this conventional use of groynes, it is also possible to use them to protect an artificially created beach (see beach feeding, Chapter 7), such as at Capitola in Monterey Bay, California where sand was dumped on the beach and retaining groynes built to prevent down-drift movement (Griggs 1990). In this situation, groynes are not built to interrupt longshore sediment movement in order build up beaches, but to fix that which has been artificially introduced, and prevent it from moving down drift.

Although groynes are a common form of sea defence (see Table 4.1 for examples), a similar methodology can be adapted on a larger scale to prevent the longshore movement of sediment impinging on harbours and estuaries. Jetties tend to be constructed at the mouths of ports and harbours in order to shelter an inlet for shipping, and prevent longshore sediment movement encroaching on the mouth and deep-water shipping channel. The principles

Table 4.1 Examples of groyne and jetty impact studies from a range of locations (For full references, see References at the back of the book).

Country	Authors	Location
Belgium	Charlier and de Meyer (1995a)	North Sea
Denmark	Møller (1990)	North Sea
Germany	Kelletat (1992)	Sylt Island
Japan	Walker and Mossa (1986)	various
Portugal	Granja and de Carvalho (1995)	north-west
UK	Bray and Hooke (1998a)	Poole/Christchurch
	Bray *et al.* (1992)	Bournemouth
	Bull *et al.* (1998)	Llandudno, Wales
	Hall *et al.* (1995)	Poole
	Wrigley (1991)	Morecambe
	Owens and Case (1908)	Dover/Holderness
USA	Douglas (1987)	Murrel's Inlet, Carolina
	Fletcher *et al.* (1997)	Hawaii
	Griggs (1990)	Monterey Bay, California
	Hall and Pilkey (1991)	New Jersey
	Komar (1998)	California
	Leidersdorf *et al.* (1994)	Santa Monica, California
	van Dolah *et al.* (1984)	Murrel's Inlet, Carolina
	Walker (1987)	Santa Barbara, California

involved are similar to those of groynes, but here the aim is to stop longshore drift completely, with accretion of sediment up-drift of secondary importance. There are also occasions when 'jetty' is used as synonymous with shore-attached breakwaters. In order to avoid confusion with detached breakwaters (see Chapter 6), discussion of shore-attached breakwaters will be considered under the heading of jetties.

The key factor with both groynes and jetties is that they are only effective on coasts where there is longshore sediment movement. The direction of this long-shore movement can be readily detected by investigating the build-up on the up-drift side of the structure; it may also be associated with erosion on the down-drift side due to sediment starvation (Figure 4.1). As this book focuses on the defence of the coastline, concentration will be on groynes as it is these structures which have the primary defence role. Many impacts of jetties occur beyond the surf zone, and so will be considered only inasmuch as they impact on the coast.

Groynes

Groynes are, together with sea walls, one of the more traditional forms of coastal engineering and, as such, have a long history of use. Owens and Case (1908) record their use at Dover, southern England from the seventeenth century, and the Royal Commission on Coast Erosion (1907) records a scheme at Spurn Head, Yorkshire in 1850. These structures are a common sight today

a) Drift aligned beach

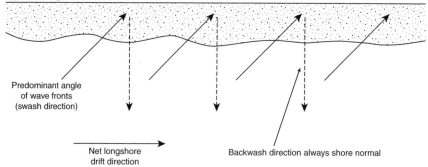

Predominant angle
of wave fronts
(swash direction)

Net longshore
drift direction

Backwash direction always shore normal

b) Swash aligned beach (groynes)

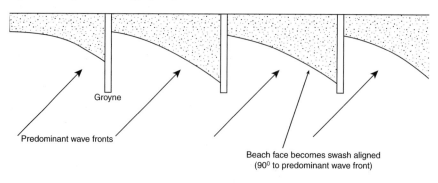

Groyne

Predominant wave fronts

Beach face becomes swash aligned
(90^0 to predominant wave front)

c) Swash aligned beach (jetty)

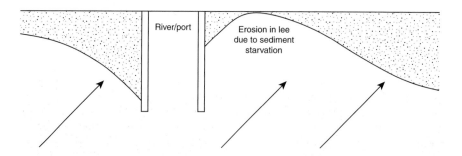

River/port

Erosion in lee
due to sediment
starvation

Figure 4.1 Longshore drift and impact of structures on sediment movement and
    beach form
    a) Uninterrupted longshore movement producing a drift-aligned beach.
    b) and c) Groynes and jetties interrupt sediment movement and
        produce a series of swash-aligned beaches, with beach faces at right
        angles to swash wave direction. Erosion, due to sediment starvation,
        may also occur down-drift.

on many beaches (Plate 4.1) which experience longshore drift, although the advent of beach feeding and its increase in popularity has meant that installation of new structures has decreased since the 1970s (Viles and Spencer 1995); in those which have been built, rock boulders rather than traditional wood or metal piling have tended to be used.

In optimising their effectiveness, groyne dimensions, i.e. the spacing, length, height, shape, as well as construction material are all important. Of these variables, two are perhaps more critical than the others. First, we saw in Chapter 2 how longshore drift predominantly occurs within the swash zone. Groynes are designed so that each groyne compartment fills with sediment, after which sediment transportation along the coast is resumed. It is important, therefore, that groynes should not be so long as to stretch out beyond this zone, as the complete blockage of the surf zone will prevent renewed longshore movement, and could well mean the loss of sediment to deep water due to the development of rip currents. Also, because of the excessive interruption of sediment, the upstream and downstream impacts will be greater, with a larger area of sediment entrapment up drift, and a greater erosion area down drift. Data from the Atlantic coast of the United States (USACE 1984) illustrates this point. Groynes which extend to $-3.0\,$m mean low water (penetrating well beyond the surf zone) will trap 100 per cent of sediment moving along shore. Groynes extending to between -1.2 and $-3.0\,$m will trap 75 per cent of sediment, while groynes extending to less than $-1.2\,$m will trap only 50 per cent. Such

Plate 4.1 Groyne field composed of a series of parallel shore normal structures, Bognor Regis, Sussex, UK.

differences will have significant impacts on down-drift erosion rates due to the variation in the amount of sediment bypassing the groynes; this could lead to management problems elsewhere along the coastline. As a general guide, the length of a groyne should be 40–60 per cent of the width of the surf zone in order to interrupt enough sediment for beach accumulation, but allow sufficient to proceed downdrift to minimise the impact of sediment starvation. An example of the impact of groynes on beach accretion is given in Box 4.1 and Figure 4.2. A further important consideration here is the type of sediment which forms the beach. Shingle tends to be transported as bedload within the swash zone (upper breaker zone) and so much of the available sediment can be intercepted close to the shore. Sand, being finer, may also be transported in suspension, and so the available sediment load is distributed over a wider area of the breaker zone. Therefore, in order to intercept sandy sediment, the structures need to be longer.

A second important issue relating to groyne structures is miscalculation of the spacing between each groyne, which will lead either to a failure to retain sediment or to erosion within the groynes, depending on whether the gap is too narrow or too wide (Figure 4.3). The correct spacing between groynes is generally taken to be approximately 4 times the length of the groyne (Komar 1998), or 2 to 3 times (USACE 1992). The difference in estimates between authorities reflects the fact that they are not exact values, because there are many modifying factors to the calculation of spacing, such as wave approach angle, sediment

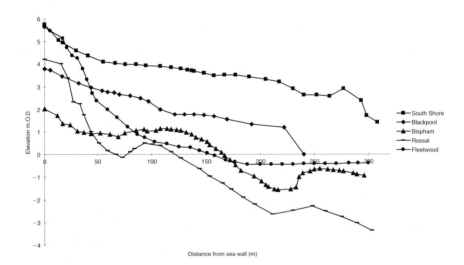

Figure 4.2 Impact of groynes on beach levels along the Fylde coast. Note steeper upper beach profiles in profiles from Rossal and Fleetwood (groyned), compared to those from Bispham, Blackpool, and South Shore (no groynes), caused by retention of sediment in groyne compartments (Data collected from the Fylde coast, north-west England, during winter of 1998) (See diagram in Box 4.1 for location information).

Box 4.1

Impact of groynes on sediment movement along the Fylde coast

As coastal sediment budgets are dependant on sediment movement, and longshore drift may be a significant component of this, the fact that groynes interrupt longshore sediment movement can have a significant effect on a coastline for some distance both up and down drift. Up drift, sediment may be impounded and so beaches will build up. Within the groynes, sediment will be trapped and the beach levels increased. Down drift, the entrapment of sediment within, and updrift of, the groynes will lead to a deficit in the sediment budget which may be reflected in either increased erosion of the hinterland, or beach lowering.

The ability of groynes to trap sediment and the impact on down-drift beach form can be demonstrated with the example of the Fylde coast, north-west England. In Box 3.1, the impact of sea walls on this coast were demonstrated. In addition, however, this coastline also demonstrates the impact of groynes. The Fylde is a linear stretch of coastline stretching for 20 km in a north-south direction, but is managed by two different local authorities. The management strategy in the northern part of the coast includes the use of groynes, while in the southern part, all groynes were deliberately removed by the local authority in order to free up sediment movement and reduce down-drift sediment starvation, which was causing a serious erosion problem threatening a dune field to the south. Because of this, it is possible to study sediment accretion patterns for adjacent parts of the coastline in the same sediment cell but under different management strategies. Figure 4.2 illustrates the impact which groynes have had on upper beach elevation. Plots for Fleetwood and Rossal represent beaches which are protected by groynes. As would be expected, the levels of the upper beach are high due to retention within the groyne field but, once offshore, levels drop rapidly outside the groynes' influence. On the sections without groynes (plots for Bispham, Blackpool and South Shore — see Map 4.1 for locations) the beach profile is much flatter and generally shows a gentle, uniform seawards slope. Hence, in this example, the presence of the groynes has resulted in artificially high upper beach levels, and a steeper profile due to sediment retention, but a significant drop to seawards.

Figure 4.2 also demonstrates another aspect of groyne impact, that of sediment starvation down drift. The three profiles where groynes are absent (Bispham, Blackpool and South Shore), are all sub parallel but at different elevations. The lowest profile (Bispham) is south of, but immediately down drift of the groynes, and could be expected to show the greatest impact from the groynes due to down drift sediment starvation. The next highest (Blackpool) lies further south still, and the highest profile (South Shore) is the furthest from the groynes (see Map 4.1 for locations). This demonstrates, therefore, how sediment starvation due to groynes can cause beach lowering down drift; in effect, the beach at Bispham acts as the main sediment source for the beach at Blackpool, which in turn provides sediment for the beach at South Shore. It is also clear that once free from groynes, beaches can rapidly

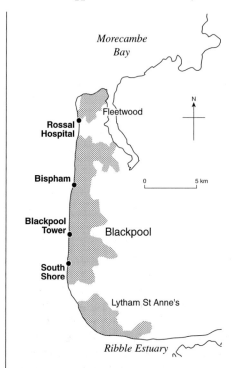

Map 4.1 Fylde coast

regain their elevation, although this is at a cost of sediment removed from other sources, such as the profile at Bispham. In this example, there are no hinterland sediment sources until South Shore (dunes) (this location can be seen in Plate 3.1) due to extensive sea wall development, hence sediment supply must come either from offshore, or from the beach itself.

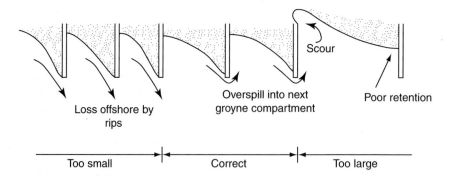

Figure 4.3 Importance of groyne spacing in preservation of beach levels. Too small will produce offshore loss due to rip currents, correct spacing (see text) will allow filling and overspill, too great a spacing will produce poor retention and scour.

concentration and grain size, and water depth. In the case of sediment, for example, the difference between sand and shingle beaches means that for sand, the ratio of length to spacing should be in the order of 1:4, but on gravel beaches, it needs to be in the order of 1:2. Similarly, the relation of wave approach to the shoreline is also important. In situations where wave fronts approach almost parallel to the coast, a large spacing can be used. When the angle is more acute, the spacing will need to be less (USACE 1992). It should be stressed, however, that when wave fronts do approach almost parallel to the shoreline, the generation of longshore currents will be minimal and so groynes may not be an effective solution to management problems in these situations. Recent research reported by the UK Ministry of Agriculture, Fisheries and Food (MAFF 1995a) has questioned the approach of using length to spacing ratios, and favours instead an approach based on the calculation of 'effective groyne length' from cross-shore distribution of longshore transport and beach crest orientation.

This reassessment of the effectiveness of the traditional approach to groyne construction was facilitated by a series of findings based on wave tank modelling at a 1:50 scale (MAFF 1995a). Results emphasised the importance of onshore/offshore transport with respect to the movement of sediment past groyne fields. Clearly, sediment which is transported in the seaward part of the surf zone has the potential to be moved inshore or offshore by shore normal currents, thus enhancing or depleting accretion within the groynes. Other key issues have been identified:

- the efficiency of groynes varies as the beach builds up — as the beach form changes, so its relationship with waves and currents will change, thus altering the dynamics of beach/wave interaction
- the possibility under high wave energy conditions of sediment, normally transported as bedload, being transported in suspension may bring into question design height calculations in exposed areas — in areas where significant sediment movement only occurs under high energy conditions, groynes may need to be designed to standards beyond significant wave height
- in situations of limited up-drift sediment replacement (as in the case of the Fylde coast discussed in Box 4.1), scour is greater down drift of timber structures than around rock — with less sediment available in the sediment budget, greater quantities have to be sourced from the existing beach, rather than by continuous longshore transport

These findings allow for greater modification of groyne field design to increase their sympathy with wave and sediment dynamics. By incorporating of such measures, it is possible to reduce the detrimental impacts of these hard defence structures.

Accepting the issue of increased sediment movement in suspension during storms, dimensions such as height are mainly of importance with respect to accretion of sediment on the up-drift side of the groyne. When this sediment reaches the top of the structure, it will spill over into the next compartment and resume its transfer down the coast. If the groynes are too high, therefore, a lot of sediment will need to build up before down-drift transfer will start, and this could lead to down-drift starvation. If the height is insufficient, sediment will over-spill at lower levels, resulting in too little sediment retained and leaving the beach protection less effective. In effect, the height and profile of the groyne will act as a template for beach shape. Given that the natural profile for a beach is to decrease in elevation seawards, the heights of groynes should reflect this. Carter (1988) suggests a height of 0.3–0.5 m above the level of desired beach height, while Bird (1996) suggests that the height should slightly exceed the limit of the swash on normal tides, and decrease in elevation seawards in accordance with the beach profile. An important consideration, however, is that the eventual height of the beach will be a function of the stable angle of repose for the sediments. Shingle can form stable beaches at angles up to 30°, but sand beaches will only be stable at angles around 1° (Pethick and Burd 1993). As shingle can be stable at steeper angles than sand, the height of groynes can be greater, and the length less, than for sandy systems.

Groynes are rarely built as individual structures, but as part of a series along the coast. Providing the calculations mentioned above are correct, these groyne fields (Plate 4.1) will trap sediment and fill up, at which point longshore drift will be resumed by the movement of sediment around the seaward end and over the top of the structures. A full groyne field will, therefore, provide a wide beach to attenuate wave energy and reduce erosion and flooding at the coast. The ultimate goal of building groynes is that they should fill with sediment, and become buried and not seen at all. However, due to the problems of getting the construction right for environmental conditions which may show considerable variation over time, i.e. waves, this situation is rare. Perhaps the closest field situation occurs along the North Sea coasts of the Netherlands and Germany, where many groynes have been buried in sediment and transport occurs along the coast uninterrupted (Davis 1996).

Jetties

In contrast to groynes, the primary role of jetties is to stop longshore drift completely for management of ports and harbours. As such, they are much larger structures and generally built singly well beyond the surf zone in order to protect and stabilise harbour mouths or inlets from sediment encroachment. When discussing groynes, we stated the importance of not blocking all longshore sediment movement by only partly blocking the surf zone. The fact that jetties stop longshore movement means that areas down drift start to erode due to sediment starvation (see below) and therefore the usage of jetties has serious implications for coastal areas.

Because they are built as single structures, calculations regarding length to spacing are not a factor here. Strength is important, however, because the structures extend further out to sea into deeper water and so receive greater wave forces. In order to estimate these factors in the design of the structure, high-water levels are used in conjunction with wind data to estimate wave base and subsequent wave heights, and also the significance of wind–wave generation and impact. Low-water levels are used in the designing of the toe protection part of the structure (USACE 1986). In this respect, therefore, the design of these structures show many similarities to sea walls.

Shore normal structures and their methodologies

We have already established the areas and reasons for which groynes and jetties are used. Given that the coastal environment is very variable, there needs to be a range of designs for different environmental conditions. We have already hinted at this with respect to variation in the length and spacing requirements for groynes between sand and shingle sediment on a beach. In the case of jetties, because they often penetrate into deeper water, variations in waves and currents mean that their structure has to be more finely tuned to the local environmental conditions. Their length and shape can therefore vary considerably.

Groynes may be built from a variety of materials, such as metal, wood, concrete or rock boulders. The material and design chosen will reflect functional performance, budget, durability, and required lifetime. Traditionally, wood has been a common material (see Plate 4.1), although concrete and steel have also been extensively used. Such structures tend to be cheap and easy to install with materials readily available. Design typically involves pile-driving vertical pieces into the beach to support the 'wall' structure which prevents the longshore sediment movement. Despite their relative ease of construction, their lifetime is relatively short and deterioration in structure is paralleled by a deterioration in functional ability. Hence, continuing effectiveness may require frequent maintenance. Another design aspect relates to their vertical design and solid nature. Although necessary to trap sediment effectively, these structures present a solid barrier to waves and thus tend to reflect wave energy unless prevented by design measures, such as rubble toe structures. Reflection can cause scouring of beach sediment and so lead to loss of accumulated material, similar to the processes previously described in association with sea walls.

In recent years, new groyne construction has increasingly tended to favour the use of rock boulders (rubble mound structures) over other materials; this is becoming common across many aspects of coastal defence, largely because of durability and low maintenance costs. Recent research reported by the UK Ministry of Agriculture, Fisheries and Food (MAFF 1995a) has indicated that there appear to be behavioural differences between rock and timber structures. Under conditions of limited sediment moving along shore, scouring is common around timber structures during storms but not around rock structures, a result of differences in wave energy reflection and absorption between the two design types. Similarly, up-drift scouring under any wave conditions is also less where rock groynes are used. Despite these scour-related problems, however, the study found no overall variation in performance between rock and timber structures of similar size. Rock can also be used in the construction of groynes out of gabions (wire baskets filled with small pebbles or cobbles). Although effective in low energy conditions, these constructions are not really suited to the open coast as they would be subject to rapid damage.

As groynes are solid structures, it is important to consider how waves impact upon them. Typically, consideration of wave energy absorption tends to be concerned with sea wall design. However, energy adsorption capabilities provide a further reason for the switch from wood to rock. Because of their irregular surfaces, wave energy dissipation is significant on rubble mound structures; this will not only reduce reflection, and thus scour, but also encourage the increased deposition of sediment due to reduced energy conditions. Another operating advantage over sheet piling designs is that they are slightly flexible and can still function effectively after some deterioration.

Apart from construction material, the other major variation in groyne design lies in whether they are open or solid. The most common form of groyne is the solid type. Given that these structures are used to prevent longshore sediment movement, this is to be expected. Over the past decade or so, however, the use of permeable or semi-permeable structures has increased as a way of increasing the entrapment and retention of sediment on beaches, but reducing the impacts on tide and wave currents. Thus, while sediment accretion on beaches is increased by the structures, their semi-permeable nature means that some material is still able to pass down drift, thereby avoiding complete sediment starvation downdrift of the groyne. While in principle this method sounds fine, counter to it is a report from Carter (1988) which states that semi-permeable structures could, if too permeable, increase turbulence and tidal scour; this would reduce the volume of trapped sediment and, thus, the effectiveness of the structure. Clearly, we need to consider what is the best design type for individual scenarios, and to remember that each beach is different and a full understanding of its functioning is essential before any defence work is implemented.

When groynes were first used, the sight of a beach building up sediment was sufficient reward for its constructors. However, we now acknowledge the fact that many other factors are important, such as the adjacent coastline outside the groyne field, as well as the coastal processes overall. Ongoing research into the behaviour of currents and waves at the coast have led to increased effectiveness in design and construction. In recent times, the use of groynes has developed so that it now overlaps with the concept of offshore breakwater structures (see Chapter 6). Offshore breakwaters are placed off the coast to protect the shore from wave attack. While groynes will produce beach accretion by successfully interrupting longshore sediment movement caused by shore parallel currents, in areas where shore normal currents are also well developed, sediment loss offshore will reduce this sediment retention within groyne compartments. As a result, it may be advantageous to deal with the need to trap beach sediments (with groynes) and to protect from shore normal currents (with offshore breakwaters) at the same time. The addition of lateral projections at the seaward end of the groynes will serve this purpose, so that the structure will not only interrupt longshore currents to facilitate sediment trapping, but also protect the accreting beach from wave attack. This will also solve some of the problems relating to rip currents seen in Figure 4.3. The resulting structure resembles a 'T' or, more commonly, 'Y' in plan, and is often referred to as a fish-tail groyne (Plate 4.2). These structures have been effective in accumulating sediment where they have been used, although in some cases the amount of accumulation has been below that predicted. Bull *et al.* (1998) detail a scheme involving three fish-tail groynes at Llandudno in north Wales. Although this scheme resulted in adequate protection from shore normal currents, resulting sediment accumulation was dominated by fine sediments (silts and clays) which has reduced the appeal of the local beach for amenity usage. This problem has resulted from the low energy conditions caused by the new structures, which are calmer than predicted at the design stage. Another scheme involving 15 fish-tail groynes at Morecambe, north-west England, is detailed by French and Livesey (2000). In this example, rapid sediment accumulation occurred after construction but, as with the previous example, this accumulation was dominated by fine grained sediments, again reflecting the low energy conditions caused by the structures. This represents a shift from the predominantly sandy sediments which used to occur here. It has been shown in this case that these coarser sediments are

Plate 4.2 Large fish-tail groyne fronting Morecambe. The combination of a rock groyne, and offshore structure (attached) has successfully encouraged sediment accretion and beach development, Morecambe Bay, north-west England.

currently accumulating offshore as a sandbank, but nearshore energy levels are too low for additional shorewards sediment transport.

The design and construction of jetties will follow the same principles as for groynes, although due to both their larger size and increased strength needs, the cost of construction is generally greater. Despite this, the materials used tend to be the same as those used in groynes, but again with a recent increase in the use of rock for building rubble-mound structures. However, due to their larger scale, more elaborate engineering techniques are available to engineers, including the use of large caissons or block walls. When discussing jetties, we enter the area of offshore structures and marine engineering. This topic is extremely complex and it is not possible to do it justice in the context of this volume. Detailed descriptions of methods, construction materials and designs are given in USACE (1986, 1992).

Importance of terminal groynes and transition zones

On several occasions it has been stated that one of the main impacts of groyne fields is the downdrift increase in erosion which typically follows their construction. This issue will be discussed more fully below, but it is necessary here to include one important design feature which can serve to reduce the severity of this problem.

Given the expense often invested in groyne fields by management authorities, the loss of sediment from their area of authority is an unwanted occurrence. In order to combat this, the *terminal groyne* has become a common inclusion in the design, particularly where beach feeding has also been used. Terminal groynes are often constructed at the end of a groyne system if losses are large, i.e. longshore sediment movement is rapid. Such structures are large, high and penetrate further into the surf zone to prevent sediment losses to the system. In this respect they are similar to jetties. Although it is perhaps understandable that an authority which has invested a lot of money in maintaining or providing a beach does not want to see their investment literally disappear along the coast overnight, the impacts of terminal groynes may be significant for neighbouring authorities down drift and could be considered as rather antisocial.

A much more neighbourly approach is to end a groyne field with a transition zone (Bruun 1952, USACE 1992). Given that down-drift sediment starvation is caused by the groynes, if the end of the groyne field is introduced gradually rather than suddenly, then impacts can be reduced and spread over a greater area. We have seen how groyne length can be used as a control in the amount of sediment retained on beaches. If this effectiveness in sediment trapping was to be gradually reduced, then gradually less sediment will be held in the groynes and more will move on down drift. In practice, this can be achieved by making groynes shorter along a converging line from the last full length groyne to the natural coastline (Figure 4.4). If groynes are made shorter, then according to the length:spacing ratios issues discussed earlier, they also become closer

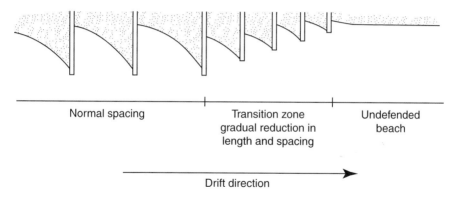

Drift direction

Figure 4.4 Transitional groyne field as a means of reducing impact of groynes on down-drift coastlines.

together. Typically, the line of convergence is around 6°, although there are detailed mathematical equations to calculate this for each location, taking into account such factors as beach elevation, width and sediment grain-size (see USACE 1992).

Impacts of groynes and jetties

A common feature of both groynes and jetties is that they are constructed in one of the most highly dynamic environments known. They are designed to influence the fundamental processes of longshore sediment transport and current activity, so it is inevitable that significant post-construction environmental modifications are going to occur. The logic and reasoning behind a decision to interrupt longshore sediment movement in order to build up beaches and produce effective shore protection is sound for the part of the coastline which is being protected. In reality, however, this only applies to the area of coast covered by groynes but, over the coast as a whole, the ideal is rarely achieved because sediment accumulation on the up-drift side of the groyne generally equates to sediment erosion from the down-drift side. Experience has shown that in many areas where groynes have been used, their outcome has rarely been entirely satisfactory, and some have been major failures. It has also become apparent that in some cases groynes should not have been used in the first place as they have adverse impacts on sediment movement. One such example is at Sandy Hook in New Jersey, as reported in Bird (1996). Here, a sea wall and protective groyne field were built as the main coastal defence structures along 11 km of shoreline. Following construction, major erosion occurred down-drift of the terminal groyne. The input of over 150 000 m³ of sand during a beach recharge programme in 1977 reduced the problem in the short term, but this sediment was also rapidly removed by longshore drift. The choice here was to recharge beaches again, extend the groyne field further, or opt for another coastal management strategy. Given that long-shore drift is so great, and beach sediment is critical to the sediment budget, it has been argued that a far better method would be to acknowledge the original mistake, remove all the groynes completely, and operate a beach recharge scheme at the up-drift end (Nordstrom *et al.* 1979, Nordstrom and Allen 1980). A similar problem concerning groynes and longshore drift on the Fylde coast in north-west England was dealt with in part by the removal of the offending structures, with a favourable response in beach levels down drift, even without any beach feeding (see Box 3.1).

With managers now beginning to question the wisdom of installing groynes, it is tempting to ask why the technique became so popular. The answer lies in the fact that groynes are seen as the best way of holding sediment on the beach. Historically, however, without the necessary understanding of the implications for processes along the down-drift coastline, such as we have today, groynes were generally regarded as a 'no-problem solution' to beach sediment loss by longshore transport. With our increasing knowledge, it has become apparent that significant impacts do exist, and this increase in knowledge has produced a greater understanding of the suitable use of groynes and the construction methods used. The main areas of groyne impacts on the coastal environment are considered below.

Interruption of shore parallel currents (longshore transport)

Given that the aim of groynes is to promote the build-up of sediment by inter-rupting longshore drift, impacts associated with this may be regarded as an occupational hazard. However, the problem caused by reducing longshore drift should be balanced against the benefit of increased beach levels. It is important to note that site characteristics and impact severity vary; in some cases, the impacts experienced are not acceptable. In some circumstances, groyne fields may intercept almost all longshore sediment movement, so that little or none will be able to pass beyond the groynes to the unprotected coast down drift. In this situation, the local sediment budget will be in serious deficit and new sedi-ment sources will be required in order to replenish supplies. As a direct result of this, erosion will be triggered down drift of the groyne field, in effect meaning that the original erosion problem has not been solved, but merely laterally transferred along the coast. The situation is similar to the flanking effect at the end of sea walls (see Chapter 3). The easiest solution to this problem is to extend defences along the newly eroding coast. However, little thought is required to realise that this will only serve to move the erosion problem still further along the coast and represents a classic example of dealing with the symptoms and not the cause, and is one which can easily lead to the 'progres-sive defence scenario' detailed in Figure 1.2.

The impact of groynes on the trapping of longshore sediment has already been illustrated (Figure 4.2). It is clear, therefore, that as well as enhancing sedi-ment accretion within the groyne field, these structures can also be responsible for considerable impacts down drift in the same coastal cell. These problems may be partly offset by allowing some sediment to pass through the groyne field, and also to reduce downdrift starvation by installing a transition zone (Figure 4.4). In the case of jetties, down-drift sediment starvation still arises but, given the greater size of the structures and the fact that they are designed to interrupt longshore drift completely, the impacts may be that much greater (Figure 4.5). As some larger jetties may be up to a kilometre in length, the volumes of sediment which can be trapped can be enormous and lead to significant deficits down drift. To illustrate this point, an inlet south of Ocean City, Maryland was stabilised by jetties shortly after 1933, following its original formation by a hurricane. Since construction, the up-drift jetty has intercepted approximately $140\,000\,\mathrm{m^3\,a^{-1}}$ of sediment, preventing it moving further along the coast (Davis 1996). As a result of the lack of sediment supply down drift of the inlet, erosion has increased in order to balance the sediment budget. This has caused the coastline to retreat by about a kilometre inland. Another commonly cited example of how jetties can interrupt longshore drift is that of Santa Barbara Harbour in southern California. This coastline experiences a strong west to east drift which was interrupted in 1930 by the construction of a jetty. Immediately upon completion, accumulation of sediment begun up drift of this structure. Within 7 years, sediment had built out to the end of the structure and had started to impinge on the harbour by forming a sand shoal at its entrance. Considering such a large amount of sediment was being intercepted by the jetty, the removal of this volume from the sediment budget down drift of the structure had serious implications. Wiegel (1964) estimates that $100\,000\,\mathrm{m^3\,a^{-1}}$ of sediment is being eroded from the down-drift coast to compensate for this budget loss, producing serious management problems. In this case, the solution has been to dredge the sediment shoal and place the material on the down-drift beach as a recharge. Other examples can be found where problems of sediment entrapment by jetties has been overcome by physical removal. Heavy plant has been used at Newhaven Harbour in Sussex, UK and at Shoreham Harbour, also in West Sussex. In the latter example, bypassing of sediment started in 1992 and has occurred annually, involving volumes between 10 and 20 thousand tonnes each year (Shoreham Port Authority, personal communications).

Increasingly, pumping of sediment is being used to transfer material mechanically around jetties to recharge the beaches down drift. In this respect, the technique has similarities to beach recharge (see Chapter 7) although here, it is a local scheme to push sediment past an obstacle, rather than a capital scheme to create beaches. A range of methods are available, ranging from floating dredgers which remove the sediment and either physically move or pump it around the obstacle; to land-based mechanical methods which dig the sediment and move it around either by road or by bucket hoist, or by automated pumping methods. The techniques in detail are very complex and need to reflect the hydrodynamics of each case in turn. It is not possible to prescribe a method for general application but, instead, it is essential to undertake detailed coastal process monitoring studies in order for the most suitable method to be

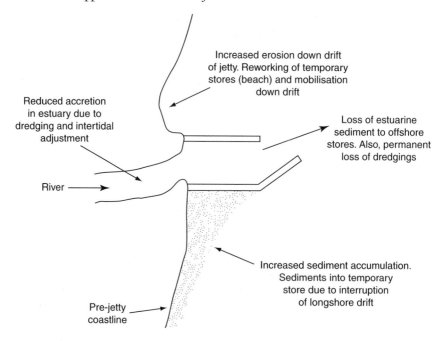

Figure 4.5 Impacts of jetties on coastal form and sediment budget. Accretion up drift of jetty to form swash-aligned beach is marked by down-drift erosion and losses offshore.

adopted, and for this to work as planned. Clearly, such discussions are beyond the scope of this book — a detailed account is provided by USACE (1991). In the above example from Santa Barbara, automated methods worked initially but failed in 1952 due to system overloading during storms.

Modification of channel processes by jetties

A further problem with jetties arises when they cause modification to the channel which they are protecting. World-wide, rivers are the most important contributors to the coastal sediment budget. As such, river mouths are import-ant for the delivery of sediment to the coast, so any management of the mouth which impacts on this may also have knock-on implications for sediment supply and movement at the adjacent coastline. In the natural situation, movement of sediment downstream supplies sediment to the coast where it can either be deposited immediately, in the form of estuarine mudflats or salt marshes; to seaward as a delta; or transported in the longshore currents as a sediment supply down drift of the inlet mouth. In any form, this represents a direct contribution to the coastal sediment budget.

The interaction of jetty and river mouth dynamics is critical, therefore, in the functioning of the sediment supply and transport processes operating there.

Jetties either side of an estuary mouth may accelerate estuarine deposition if spaced too far apart because, if the inlet is too wide, current velocities of the river water entering the sea may slacken; this will cause deposition within the inlet or estuary itself and, without management, silting up. Clearly, any sediment deposited in this location will not be available to the coastal budget. In the Bann estuary, Northern Ireland (Carter 1988) channel modification by jetties led to the formation of a sand shoal in the harbour entrance due to the slackening of currents; this needed removal by dredging. Conversely, jetties may be built too close together; in this case the river currents will increase due to the space constriction, thus increasing their capacity to transport sediment. While this will produce natural sediment scouring and maintenance of the inlet, the sediment within the water will be transported far out to sea as a plume, quite possibly to a position beyond the limits of the longshore currents. At St. Mary's Inlet, Florida, construction of two rock jetties of 5 km and 3.5 km on the north and south sides of the entrance respectively, has channelled flow through the entrance with increased velocity. Estimates suggest that 24 million m^3 of sediment has been deposited offshore as a sand shoal beyond currents which could deliver it to the beach (i.e. below wave base). The formation of this shoal has also interfered with any currents which could have allowed natural sediment bypassing to occur.

Problems caused by jetties may be exacerbated further because of the deflection of longshore currents seawards by the obstacle. Again, this means that if sediments are forced seawards, they will not be available to supply the coast immediately downstream of the inlet. In a broader view, such changes associated with alteration of the hydraulic regime of the river mouth may also trigger further physical, biological and water quality issues. A wide jetty spacing in which water currents become slack may well provide an area of water with reduced flushing rates and the subsequent accumulation of river-derived pollutants. In association with this, further impacts may be felt by faunal and floral communities in the vicinity.

Loss of sediment from groyne fields offshore due to shore normal currents

We have seen how groynes may be of use to the coastal manager when shore parallel currents produce longshore sediment transport along a stretch of coastline. Groynes are of little use, however, where the sediment movement is mainly in response to shore normal currents, i.e. onshore/offshore movement. Without shore parallel currents (longshore drift) there will be little sediment to intercept in the first place, so groynes should not be regarded as an effective management technique. Such misuse of structures occurred at Virginia Beach on the east coast of the USA (Bird 1996), where groyne construction in the late 1940s failed to accrete sediment and even a beach feeding scheme did not halt the erosion problem. It was subsequently discovered that the predominant direction of sediment movement along this section of coast was in response to shore normal rather than shore parallel currents, and so groynes had little chance of being an effective defence technique. In this case, there was little shore parallel sediment movement. In other situations, however, longshore drift may be important but with shore normal currents still resulting in significant offshore removal of sediment. In such cases, it may be necessary to modify the groyne design to cope with both transport directions, such as by the use of fish-tail groynes (Plate 4.2).

Even when shore parallel currents are well developed, groyne construction may actually enhance a beach erosion problem by generating its own shore normal currents (rips). The result will not only be the transportation of sediment offshore, but also its loss to the nearshore transport system, possibly to offshore sinks, representing a net loss to the coastal sediment budget. Under normal conditions, as sediment builds up in a groyne compartment, the beach face becomes swash aligned (aligns parallel to the predominant swell wave direction). If, during storms, the predominant wave direction is not from this direction, and especially if it comes from a 90° angle to it, a rip cell can become established against the down-drift side of the groyne (Figure 4.6). This leads to the rapid transfer of material offshore and, potentially, to deep water beyond normal wave base, thus representing a net loss to the sediment budget (Silvester 1974). The degree to which such rip cells develop depends on the angle of storm wave approach in relation to the beach face angle and clearly should be an area of detailed research into the local wave climate at the defence planning stage.

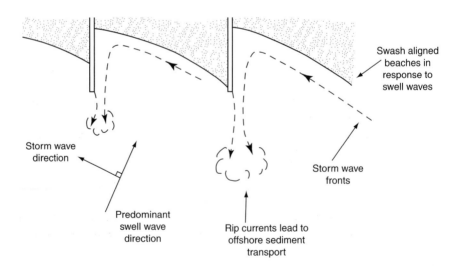

Figure 4.6 Transfer of material offshore from groyne fields as result of rip current generation. Such losses are increased by storm waves approaching the beach from different directions to swell waves.

The chance of a beach/wave misalignment occurring can also be a function of the groyne spacing (Komar 1998), with greater spacings being more vulnerable to rip-cell development due to the greater probability of storm waves hitting the beach face obliquely, and also the greater space for current acceleration (Figure 4.6). Under certain conditions, it may also be possible to generate rip cells at either end of the groyne compartment (Gaughan and Komar 1977). This can occur when incoming (incident) waves interact with outgoing (reflected) waves from the beach profile. Edge waves can become established due to such interactions, and if the wave frequency is the same, strong circulatory currents can develop which may result in strong shore parallel currents developing along a beach in both directions.

*   *   *

The types of impact described above have, for sake of clarity, to be treated individually. Many areas will, however, reveal a combination of interconnected impacts which function as the natural system attempts to regain an equilibrium following changes induced by groyne construction. Douglas (1987) illustrates this with respect to Murrells Inlet in South Carolina, USA. Historically, the inlet used to become choked with sand but jetty construction in 1980, comprising a rubble mound structure, allowed the inlet to become stabilised and safe for navigation. Longshore drift in this area is bi-directional, in that the net direction of movement will periodically reverse, and may operate in one direction over a time scale of several years. Using a photographic reconnaissance, Douglas has highlighted a series of impacts which have occurred simultaneously. The main system responses were:

- zones of beach accretion of up to 150 m for approximately 1.5 km north and 5 km south of the jetties, reflecting the reversals in drift direction;
- zones of beach erosion of up to 15 m between 5–6.5 kms south;
- onshore migration and breakdown of ebb tide shoals following construction of the jetty across the ebb tide channel and subsequent diversion of ebb currents away from the shoal, thus allowing waves to become dominant in sediment movement;
- formation of a new ebb shoal in accordance with redirected ebb currents. New shoals being fed by sand transfer from behind the southern jetty. This shoal is developing rapidly both in the offshore and longshore directions;
- between 1978 and 1982, about 28 300 m^3 of sand had accumulated down drift of the southern jetty, and about 14 000 m^3 against the northern side.

This project has shown the association of impacts following jetty construction. It is argued by Douglas that the trapped sediment represents a permanent loss to the sediment budget, and may well account for areas of beach erosion.

Benefits of groynes and jetties for the coastal environment

It is clear from the last section that groynes can have significant impacts on the environment in which they are constructed. Before all proposed groyne developments are dismissed on the basis of negative impacts, however, it is important to consider impacts as a part of an assessment which includes benefits. Using the previous example given by Douglas (1987), it is clear that significant impacts have occurred which are causing problems: some of these need to be considered in a temporal framework. The issue of ebb tide shoal position, for example, relates to the repositioning of a subtidal sand body in response to a new current position. Before the jetty was built, the original shoal was also developing and, as it did, was changing shape and form in response to sediment addition; due to this change, water would flow differently across it, thus ebb tide flow was being altered. The processes remain the same, but in a different position. Thus we need not necessarily regard this new development as a negative impact, unless it occurs in an area where it is of detriment to other processes or activities. A similar phenomenon has occurred in association with the fish tail groynes at Morecambe. Here, changes in areas of sediment accretion caused large volumes of sediment to accrete on previously rich shell fish beds, this affected local ecology and economics.

Leaving such considerations aside, the prime advantage of groynes lies in their ability to accrete sediment and, in so doing, build up beaches and provide increased wave attenuation during storms, and increased amenity value to a coastline for the benefit of tourism. Clearly, such benefits may be considered as offsetting some of the disadvantages mentioned earlier.

Wave attenuation

It was mentioned in Chapter 2 that a wide beach is efficient in the attenuation of wave energy due to the friction caused as water flows across the sediment surface. As groynes trap sand, therefore, beach levels will build up and wave energy attenuation will increase accordingly, so that the energy received at a sea wall will decrease. This phenomenon can be inferred from the two fish-tail groyne examples previously quoted in this chapter. The fact that the sediment regime has changed from sand to silt/clay indicates a drop in energy level, produced partly by the sheltering effect of the groyne, and increasingly by the continued accretion of sediment between the structures. Any structure reducing wave energy can lead to important financial savings with regard to sea wall construction, because the lower levels of wave energy received will mean that the design criteria can be scaled down. The details pertaining to wave attenuation are discussed more fully in the context of beach feeding (Chapter 7).

Increased sediment stability and amenity value

Perhaps one of the ultimate advantages of using groynes is that they promote beach accretion and hold it in place. Irrespective of the problems caused elsewhere in the coastal cell, this factor alone may be considered as having overriding importance, especially if the beach concerned fronts a tourist resort. Under such circumstances, the maintenance of a healthy beach is important in order to protect not only the hinterland from erosion, but also the socio-economic factors associated with tourist use of the beach for amenity purposes. In many circumstances, the importance attached to the socio-economic factors overrides any concerns regarding coastal processes and holistic management.

Such high beach levels are critical not just for primary tourist usage. Groynes are used for beach accretion and increased sediment stability along large stretches of the Dutch and Danish coast where dunes are also common. The maintenance of high beach levels serves to protect the dunes from excessive draw down by storms (see Chapter 8), thus preserving their integrity, as well as their conservation and amenity value. In the case of the Dutch coast, the protection of dunes is also critical for the preservation of a major aquifer.

Other benefits of groyne usage

Given a situation where some degree of human intervention is necessary to maintain coastal stability and hinterland protection, groynes may offer further benefits over other coastal protection measures. First, compared with other hard engineering methods, groynes are relatively cheap structures to construct. Given the range of materials available, and the fact that construction principles are simple, costs can be kept low. Beach feeding is an alternative and may produce the same results, but this does not eliminate the problems of continued longshore transport of material.

A second advantage of groynes over other hard methods is that they provide protection to the coast without fundamentally altering the characteristics of the surf zone. Wave height and length remain the same before and after construction, leaving this aspect of coastal dynamics unaltered. In contrast, methods such as offshore breakwaters protect the coast by reducing wave energy, thereby causing additional changes in coastal dynamics.

Groynes and sea level rise

With ongoing sea level rise, problems of a fixed coastline will invariably affect groyned coasts. In the worst case scenario, major alteration of longshore currents, due to deepening wave base and changes in sea bed morphology, could make groynes untenable if shore parallel currents are altered too greatly. Perhaps the most immediate impact relates to the ability of groynes to trap sediment. It has been stated that trapping ability is partly dependant on the amount of the longshore drift which the groyne interrupts, i.e. how far the groyne penetrates into the surf zone. This degree of penetration could increase if the wave base were to increase as seas deepen. As a result, it is possible for groynes to trap more sediment with ongoing sea level rise, and thus accrete beaches faster. Ultimately, this must benefit beach resilience because of increased sediment and increased attenuation. However, considering the process of longshore drift holistically, any increased trapping of sediment within groyne compartments will be mirrored by a corresponding decrease in sediment passing beyond the groynes.

Although the above outcome appears favourable, a series of complications need to be considered. First, with deepening wave base, greater amounts of energy will be experienced at the beach faces. This may further enhance swash alignment, but may also increase beach draw down and changes in profile. Therefore, while groynes trap more sediment, the beach profile may flatten, thus reducing the protection afforded to the hinterland or sea wall. A second issue relates to the material from which the groynes are built. We have previously said that wooden structures are more reflective than rock. This will link to any increase in wave strength and activity, with greater reflectivity equating to greater scour around wooded structures and, as a result, a greater threat of undermining and groyne failure.

Summary

It is clear from this chapter that groynes and jetties have been used extensively in both contemporary and historical situations. It is also apparent that their construction can have major impacts on the coastal zone. Given this, it is pertinent to ask why engineers build them when they have such knowledge of their impacts.

Historically, groynes have been built in many areas because they were regarded as a suitable method to trap sediment on beaches. Given the limited knowledge available at the time when many of these schemes were constructed, this process cannot really be criticised; after all, a method which could be easily seen to increase beaches at a time when coastal tourism was in a state of rapid expansion could not be considered as a bad thing. However, increasingly we are seeing a shift from this approach. Today, many groynes are being removed from the world's coastlines. One reason for this is that they are reaching the end of their useful lives, and need replacing. Given the introduction of the softer techniques (see Part III), many of these methods, especially beach feeding, are preferred to groynes. A second reason is the recognition of the importance of unhindered sediment transport along the coast and the holistic nature of good coastal management. Engineers now realise that a natural coast is cheap to maintain. As a result of this philosophy, many defence authorities are removing groynes without replacing them, thus freeing up sediment. The example quoted for Blackpool (Box 3.1) demonstrates this well; evidence at the location clearly shows that problem areas of sediment starvation due to the original groynes have now largely recovered.

Groynes provide a defence methodology which is tried and tested, and which still has an important role to play in coastal management. We are seeing a shift from the traditional wood and concrete structures to the use of rock. These tend to be more substantial in form but have advantages over their counterparts. This does, however, give rise to a newly emerging trend in coastal management. Consistently in this, as well as earlier and succeeding, chapters there has been mention of rubble mound structures, rock toes on walls, etc. There is a danger here of 'jumping on a bandwagon'. It has been shown that rock structures work; but by using such methods as frequently as we do, we may be in danger of falling into the same trap as our predecessors did when they discovered that groynes work. Many coastlines today are taking on the appearance of rock-strewn boulder fields, and we need to exercise caution not to overdo things, and investigate all possible technologies.

While various technologies exist to replace the use of groynes, jetties do not have such alternatives. Given that the role of a jetty is to protect a harbour mouth from sediment drift, and to protect the navigable channel, a structure built out to sea is the only logical way to do this. As such, impacts from these structures look as if they need to be 'put up' with. We cannot, however, overlook the fact that many erosion problems have arisen as a result of jetty construction. Increasingly, research is focusing on the development of sediment bypassing methods in order to allow the longshore movement of sediment to be maintained as much as possible. Such approaches may well be the main hope in reducing impacts from this quarter.

Summary of the benefits of groynes and jetties

- Increased beach levels provide improved hinterland protection
- Better beach development for amenity usage
- Improved protection and access for ports and harbours
- Protects coast without altering wave dynamics

Summary of the problems with groynes and jetties

- Interruption of longshore sediment movement
- Increased erosion down drift
- Alteration of hydrodynamics at estuary mouths (jetties)
- Deflection of longshore sediment pathways offshore (jetties)

Recommended usage

- To protect and build up beaches in areas of high longshore sediment movement
- To increase and protect amenity usage where other solutions are not suitable
- To protect harbour entrances

5 Cliff stabilisation

Introduction

To many people, a cliffed coastline provides dramatic sea views and coastal scenery, and represents all that is permanent about the coast. It is true that some cliffs remain relatively unaffected by coastal processes, but others change rapidly and play a key role in contributing to the sediment budget, which subsequently serves to build up beaches. The western tip of Cornwall, UK for example, is typified by tall bold cliffs of granite or metamorphosed country rock (hornfels) (Plate 5.1a) which, because of their hardness, erode extremely slowly and provide very little sediment. In contrast, soft-rock cliffs, such as those of Holderness in eastern England, or those of the Dorset coast (Plate 5.1b) erode rapidly, measured in metres per year. As a result, they provide a lot of sediment which forms an input to the sediment budget (see Figure 2.6). This is subsequently transported by currents and moved along shore or off shore. However, the distinction between hard and soft rock cliffs in terms of sediment supply is rather too simplistic for our purposes, because it is the *overall* strength of the cliff which controls erosion rates, rather than the material of which it is composed. Thus, it is not just the material from which a cliff is composed, but geological structure, amount of subaerial weathering, and degree of wave attack which control the rate of cliff erosion.

The fact that there are a variety of factors relevant to cliff stability means that cliff management is a complex task. Clearly, the degree of wave attack is important in a coastal setting, but this is not the only factor which has to be considered. Cliffs are morphological features exposed to the air as well as the sea and, as such, are also subject to subaerial processes. Waves may erode the base of a cliff and, once a cliff fall has occurred, wave energy is focused on this fallen debris which is ultimately removed by longshore processes. While this is occurring, however, the cliff remains active due to the continued degradation and weakening caused by ongoing subaerial processes. Surface water, whether from precipitation or surface drainage, will continue to seep into the rocks, attacking minerals, lubricating joints or increasing pore water pressure within the cliff itself. In colder climates, the freezing of water within the rock will cause them to open and the overall rock strength to decrease.

Plate 5.1 Influence of rock type over cliff morphology and sediment input
    a) Hard, wave-resistant cliff composed of metamorphosed country rock (hornfels) and granite (lighter coloured rock), Porthmoer Cove, Cornwall.
    b) Soft, easily erodable cliff composed of clay. The lack of sediment on the beach reflects the minimal input of coarse clastic debris suitable for beach accretion, Dorset coast.

It should be borne in mind, therefore, that just because a cliff may be protected from wave attack at its base, whether by fallen material or a defence
structure, does not mean that it ceases to be active. This is clearly an important
consideration in the management of cliffed coasts. A good example of this
occurred along the Scarborough coast of north-east England, where a large cliff
slide occurred because of subaerial processes, largely pore water pressure,
despite the presence of a sea wall (see Box 5.1). What happens to water which
enters the cliff is an important element in the consideration of subaerial
processes in cliff stability. From a geotechnical perspective, the force which pore
water pressure can exert on rocks is important when considering cliff stability.
Where water can freely drain from a cliff, pressure cannot build up because the
pathway taken by the water is, in effect, an open system (some pressure increase
may occur during heavy rain if the volume of water within the cliff exceeds the
capacity of the 'transfer system'). In cliffs which experience cracking (such as
clays), or where there are different rock types of varying permeabilities, the
pathways taken by water may be 'blind', i.e. they have an opening but no exit.
In this situation, the addition of more water, such as would occur during rainfall, would increase the pressure within the rocks. This is known as pore water
pressure, and is a common cause of cliff failure. Increases in pore water pressure
may also occur if drainage pathways become blocked, such as when a wall is
built along a cliff face. McGown *et al.* (1987) studied such a phenomenon in
cliffs composed of London Clay in Herne Bay, Kent. Measurements revealed a
steady increase in pore water pressure over a 20-year period following coastal
protection works. The Holbeck Hall slide in Scarborough, details of which are
given in Box 5.1 is another example.

While failure mechanisms are important from a management perspective, it is
also important to consider the importance of cliffs in coastal sediment supply,
because defending cliffs will result in a corresponding reduction in sediment
supply. For every length of cliff prevented from eroding, a corresponding reduction in sediment input to the sediment budget occurs. As an example, Clayton
(1989a) discusses sediment input from the cliffs of Norfolk, eastern England; he
states that input from erosion of these cliffs averaged $500\,000\,\text{m}^3\text{a}^{-1}$ of which
around $350\,000\,\text{m}^3$ was sand and gravel and important in the coastal sediment
budget. Coastal protection along this coastline reduced erosion, and the annual
rate fell by $100\,000\,\text{m}^3\text{a}^{-1}$, producing a deficit in the coastal sediment budget.
This illustrates one of the greatest conflicts currently experienced by the coastal
manager. On the one hand, there is the need to protect coastal property, but on
the other, there is the maintenance of healthy beaches and protection of property down drift. If eroding cliffs provide large volumes of sediment to beaches,
then which is to be the priority, a house on top of a cliff, or beaches and coastal
protection down drift?

Gravel cliffs at Holderness, on the North Sea coast of the UK fall into
this issue. The supply of gravel and sand is of critical importance for many
countries bordering the North Sea. Historically, the tendency has been to
allow cliffs to erode, even though this has meant the loss of many villages to

Box 5.1

The Holbeck Hall landslide, Scarborough

The loss of the Holbeck Hall Hotel due to a cliff fall during the 3 and 4 June 1993 was a sensational media event, achieving international press and television coverage. The case was interesting because it occurred in spite of a sea wall along the base of the cliff. This helps to illustrate some of the issues raised previously: in that wave attack on cliffs is not the only mechanism by which cliff failure can be induced. The original sea wall was built in 1800, and extended in the early 1900s so that a completely defended frontage was in place by 1908 (Clements 1998) (Plate 5.2a). In addition to the wall, the cliffs had been regraded to an angle of 30° over a horizontal distance of 105 m (Clements 1994), and drainage installed. As a result of the fall, the cliff recessed inland by 50 m with 15 m back scars and formed a toe penetrating for over 100 m across the beach (Plate 5.2b). In total, around 1 million tonnes of rock and sediment were displaced.

The main issue which arises here is that, despite cliff protection measures, failure was still caused by rotational movements. These movements undermined the wall and pushed it seawards, and so must be the result of land-derived forces. Clements (1998) details two main types of rotational failure along this coastline. First, multiple, deep-seated slides put pressure on the sea wall from the landward direction, in effect, pushing it over (bear in mind here that sea walls are generally designed to withstand wave attack, and thus their maximum strength is from the seaward direction); second, single deep-seated failures, where the failure plane penetrates below the sea wall, produce a seawards rafting of the wall structure as part of the slide.

In the case of the 1993 failure, it is not clear which of the above occurred, but one eye witness did recount seeing the sea wall tipping back as it moved seawards, indicating that it was rafting on the slide before becoming buried by material from above. The main cause of the slide was a combination of pore water pressure and rock type. The cliffs are composed of boulder clays with some sand horizons, overlying interbedded mudstones and sandstones. The area had experienced a series of dry summers, so the clays had become dried and deep-seated cracks had formed. During the wetter winters, water could penetrate the cracks and increase pore water pressure, and also form perched water tables. A series of heavy rainfall events in late May, coupled with the use of soak-aways for rain-water discharge from cliff-top properties (including the hotel), increased the volume of water in the cliff, thus further increasing pore water pressure beyond the critical failure threshold. Many of the existing cliff drainage installations failed due to the breaking of pipes and dislocation of drainage routes in the early movements (a similar process occurred in the Spittals, adjacent to Black Venn, Dorset, also leading to the loss of property). This would have released large volumes of water from the perched water tables, and from within the pipes, straight into the cliff face.

At the present time, the toe of the slide has been protected by armour-stone (Plate 5.2b), with the higher levels of the slip having been further

Plate 5.2 The Holbeck Hall landslide. Despite the presence of a sea wall, this major landslide still occurred due to a deep-seated failure plane which rafted the wall seawards
  a) The vertical sea wall at Scarborough which fronted the failed cliff.
  b) The newly protected toe of the Holbeck landslide.

graded to 24° in the upper levels, and as low as 6° towards the base. An extensive drainage system has also been installed. While the armourstone will protect the slip from marine erosion, the regrading should act as a 'diluter of gravity', and the drainage serve to reduce pore pressures. Thus, the slip should become stable. However, such methodologies were used before the slip occurred; perhaps it would be safer to acknowledge the vulnerability of the area and zone it in the planning regulations as favoured for no new development, and also for no replacement development as remaining properties reach the end of their useful lives (see Figure 5.10 and discussion on zoned/staged retreat in Chapter 12).

the sea (see Parker 1980 and Prestage 1991 for examples). Increasingly, pressure is being put on planners to protect homes and property by defending cliffs, even though the loss of sediment input would be significant. In compensation for cliff loss, however, the sediment derived from cliff falls and reworked by the sea has meant that large areas of low-lying land have received protection from healthy beaches. In Chapters 7 and 9, it will be seen how such a healthy beach can attenuate wave energy and protect the hinterland. Along many coasts, eroding cliffs are one of the main suppliers of this material. What we have, therefore, is a series of sediment inputs from cliff erosion which are reworked by the sea to provide sediment to build up beaches and protect other areas of the coast. As far as coastal erosion management is concerned, this is the fundamental relationship which needs to be considered. In Box 5.2, the case of Fairlight Cove, Sussex is discussed in respect of a large offshore breakwater built to prevent cliff erosion. This has successfully prevented erosion, but has resulted in the loss of $9\,750\,m^3\,a^{-1}$ to the sediment budget, which is being compensated for by increased erosion further down the coast, typical of the end effects noticed around hard defence structures.

Box 5.2

Cliff protection at Fairlight Cove, Sussex, UK using a non-segmented offshore breakwater (sill)

Fairlight is a small village on the south coast of England located on top of eroding cliffs which are composed of poorly consolidated sands and gravels. In the 1980s, the rate of cliff erosion increased due to a reduced longshore supply of shingle which formerly supplied the fronting beach and afforded some degree of protection to the base of the cliff. The increased erosion rapidly cut the cliff face back inland, causing the loss of houses, gardens and infrastructure. The reason for the reduction in sediment supply is thought to be, primarily, harbour construction works in Hastings, about 5 km up drift, behind which a large shingle bank has accumulated; and, secondly, general improvements to other up-drift coastal defences. Land-use surveys and cliff erosion prediction were used to investigate the severity of the problem, and to plot erosion contours at 7-year intervals (Penning-Rowsell *et al.* 1992). Different protection methodologies were then assessed using costbenefit techniques and, having considered the range of alternatives, including doing nothing, it was decided to construct an offshore breakwater (sill) stretching for 500 m along the cliff base. This would contain 120 000 tonnes of Scandinavian granite, and cost £2.5 million. On completion in the late 1980s, the scheme secured the immediate future of 47 houses (Stevens 1995). Plate 5.3 provides an aerial view of the scheme.

Plate 5.3 Offshore breakwater (non-segmented/sill) protecting cliff-top development, Fairlight Cove, Sussex (Reproduced with permission of Rother District Council).

The sill appears to be successfully reducing cliff erosion and providing significant protection of the cliff. Rates of retreat have fallen from an average of $1.14\,\mathrm{m\,a^{-1}}$ between 1981–7 (pre-construction) to $0.36\,\mathrm{m\,a^{-1}}$ between 1992–7 (post-construction) (Halcrow 1997). The base of the cliff has now naturally stabilised, although substantial processes are still ongoing in the upper cliff area. The reduction in erosion since completion is due to a combination of reduced wave activity, retention of cliff-fall sediment behind the sill forming a perched beach (see Chapter 7), and the build-up of shingle via longshore drift. Despite variation of magnitude with time, the structure has had impacts on both cliff and beach processes up and down drift. On completion, the rate of long-shore movement across the protected cliff frontage decreased with the result that cliff erosion at the eastern (down drift) end did increase between 1993 and 1997. This can be clearly seen in Plate 5.3, where a marked area of erosion can be seen at the eastern (right-end) end. As a result of increased shingle supply longshore due to amendments in Hasting's harbour wall, material is now bypassing the sill and moving to the east, at pre-construction rates (Halcrow 1997). During storm events, however, there appears to be transfer of material over the sill to the lee area, from which it is not possible for it to return to the longshore sediment budget. This, therefore, represents a net loss of sediment to the coastal system, but will benefit the increased stability of the defended cliff, and will stop once the area has filled and become a perched beach.

In terms of sediment budgets, the defending of the Fairlight cliffs represents a net loss of sediment to the coastal budget. Using simple volumetric calculations the loss of input to the sediment budget can be estimated:

Length of scheme = 500 m
Average height of cliff = 25 m
Rates of erosion = $1.14\,\mathrm{m\,a^{-1}}$ before construction, $0.36\,\mathrm{m\,a^{-1}}$ after construction

Therefore:

Pre-construction volume of sediment = $14\,250\,\mathrm{m^3\,a^{-1}}$
Post-construction volume of sediment = $4\,500\,\mathrm{m^3\,a^{-1}}$
Volume lost to sediment budget = $14\,250\text{--}4\,500 = 9\,750\,\mathrm{m^3\,a^{-1}}$

Thus, the coastal sediment budget is now in deficit by $9\,750\,\mathrm{m^3\,a^{-1}}$ as a result of this cliff defence project. This neatly illustrates the issues raised concerning coastal defence construction and budget impacts. Furthermore, increased deposition of sediment in the lee of the structure means that this sediment is no longer being deposited in the areas in which it would have been pre-construction. As a result, the sediment gains behind the structure are offset by losses elsewhere. Such losses may be sufficient to produce further localised sediment deficits and the onset of erosion on neighbouring beaches, such as is evident by the flanking effect at the eastern end of the sill.

While sediment supply is important, another problem relates to the associated loss of property which occurs as cliffs retreat. Where development is high, the land will be of such high value that the cost-effective option will be to defend the cliffs. Given this, it is necessary to understand ways of achieving this. Cliffs differ in comparison with other coastal morphologies in that the erosion problem can be manifest in a variety of ways depending on the composition of the cliff and, as has been seen, is not always a product of marine processes. Clay cliffs may slump or even flow, and limestone cliffs may fail by block fall. Because of the fundamental difference in forcing factors and failure style, the approach to protection has to vary accordingly. Because of this, it is necessary to understand the various modes of failure before we can start to look at methods of protection.

Role of cliffs in coastal sediment supply

The lithology and structure of cliffs exert not only basic controls over how cliffs fail, but also the sediment supplied to the beach. There are three primary controls over cliff sediment supply (Figure 5.1):

Coastal processes	**Geological processes**

Waves - height/period
        - approach angle
Cell circulation/rips
Tidal properties
Storm exposure/surge
Sea level
Volume of sediments
Beach slope (grain-size)
	Composition - hardness
	              - talus production
	              - sediments
	Bedding and jointing
	Folding/structure
	Height/slope
	Ground water/pore pressure
	Sediment composition

Human processes	**Biological processes**
Development (drainage) Development (construction) Land-use Defences	Vegetation Fauna (burrowing)

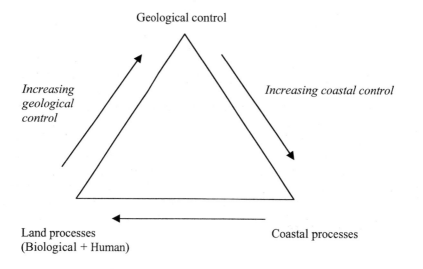

Figure 5.1 Primary controls over cliff erosion.

1   coastal processes;
2   geological structure of the cliffs, including cliff composition;
3   land-based activity, including human and biological processes.

Our main concerns here are the first two, although the third will be incorporated into subsequent discussion where necessary, as it does link with defence methods and issues arising from these.

Geology

Cliffs can be composed of any rock type and are, by definition, erosional features. However, just because a coastline has cliffs today does not mean that they have all formed under contemporary wave and climate conditions. All active cliffs produce sediment to some extent, although the volume is dependent on the rock type and the rate at which erosion processes occur. Pethick and Burd (1993) provide an indication of typical retreat rates for different lithologies (Table 5.1). These rates are only 'typical' in that the amount that any lithology fails is also dependent on the forces acting on the cliff, and the degree to which the rock is jointed or cemented. The last two criteria, although important, are geological factors largely beyond the scope of this work. Interested readers can consult any general geology or sedimentology text book. For our purposes, a sandstone, for example, with a strong cement will be more resistant than one which crumbles in the hand. Similarly, a limestone with many joints and cracks will fall much more easily than one with only a few.

Of greater importance from a defence point of view is that different lithologies behave and fail in different ways. Clay cliffs, such as those which typify the Dorset coast of the UK, may slump and produce a lobe of sediment stretching from the cliff across the beach, which the sea then reworks. Because on high-energy, wave-dominated coasts, where most active cliffs are located, clay and fine silts are rarely deposited on beaches, eroding clay cliffs supply little sediment for beach build-up. They may, however, impact on sediments reaching the beach by longshore drift if the slumped lobes cover a beach and persist for any length of time, because in such cases the lobes will act as large groynes or jetties and cut off longshore movement. This is happening along the Dorset coast of the UK, where the large Black Ven landslide complex is continually moving seawards, maintaining a lobe of sediment protruding out to sea (Chandler and Brunsden 1995). Using digital elevation models (DEMs), it has been shown

Table 5.1 Typical cliff recession rates for different lithologies (Data compiled from Pethick and Burd 1993; Sunamura 1983).

Lithology	Recession rate ($m\ a^{-1}$)
Granite	<0.001
Limestone	0.001–0.01
Shale/flysch	0.01
Chalk	0.1–1.0
Sandstone	0.1–1.0
Glacial till	1.0–10.0+
Recent pyroclastics	10.0–100+

that at one locality within the complex, sediment has been continually drawn from the cliff top to the frontal lobe, which has maintained its morphology despite wave attack. Again, this process can be related to the earlier discussion on the importance of subaerial weathering processes. At Black Ven, the land-slides are being driven not by marine processes directly, but by a continuous seaward flow of fine sediment, mobilised by a combination of subaerial processes. In contrast, sandstone or till cliffs contain coarser material that is of more use in the building of beaches; limestone cliffs, igneous and metamorphic cliffs, on the other hand, where failure is more joint than grain size controlled, produce large boulders which, because of their size, are not easily removed from the fronting beach. Examples are common, and are typified by some of the beaches along the coast of north Cornwall, such as the one shown in Plate 5.3.

Another important consideration in cliff failure mechanisms lies in the fact that not all cliffs are composed of single lithologies. Such cliffs are referred to as composite and, from a management point of view, are sometimes difficult to handle. The exact shape and form of the cliff depends not only on such factors as strength of the waves, strength of the rock, and structure of the rock, but also on the relative hardness of different rock types. In a cliff composed of a hard rock and a softer rock, (sandstone and shale are common combinations), erosion of the softer rock by wave action could leave the harder unsupported, which will also subsequently fail. Plate 5.4 illustrates this phenomenon with a

Plate 5.4 Geological controls over cliff recession. Vertically bedded cliff with differential erosion exploiting the weaker clay layers. Note the beach contains fallen blocks from the cliff, as well as sand, Widemouth Bay, Cornwall.

cliff composed of vertically bedded sandstones and shales near Bude, north Cornwall, UK. When dealing with such cliffs it is often useful to adopt the principle that a cliff is only as strong as its weakest part. Figure 5.2 shows three scenarios: uniform rock type, less resistant at top, and less resistance at base (Figure 5.2a). In Figure 5.2b, erosion exploits the weaker strata and the resulting profile varies accordingly. From a management point of view, different defences may be required here because of the different ways in which the cliff fails. In Figure 5.2bi, the cliff is retreating uniformly and failure will depend on the rock type and strength. Mechanisms of failure could be joint or lithology controlled, and include rock fall, wedge failure, or slumping. In Figure 5.2bii, harder rock at the base of the cliff is harder to erode than the softer material above, which may succumb to either wave erosion (depending on its height) or subaerial weathering. The resulting failure may include slumping of the top layer over the bottom, or falling of blocks. This cliff morphology is typically referred to as 'slope over wall'. Protecting these cliffs is difficult because erosion is concentrated in the top part of the cliff and involves subaerial processes. Finally, in Figure 5.2biii, undercutting of the cliff base by the sea removes supporting material and allows the top, more resistant material to fall to the beach. In this situation, the problem is more easily remedied because the base of the cliff can be protected by installing a hard barrier to prevent the waves from hitting the cliff base.

Structure

All of the scenarios discussed in Figure 5.2 deal with horizontal strata. Figure 5.3a and b show dipping rocks. In (a), the direction of dip is inland; with the

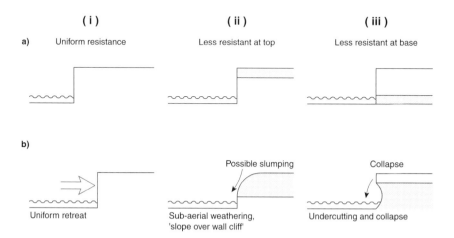

Figure 5.2 Relationship between different rock types, and their interrelationship in cliff stability and morphology (After French 1997).

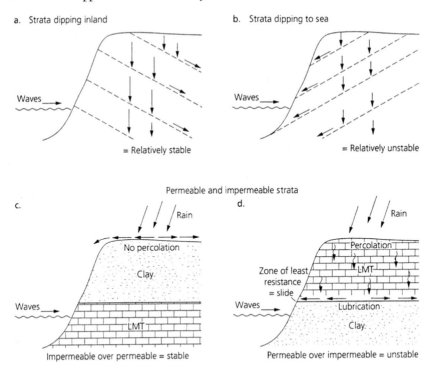

Figure 5.3 Role of structure and rock permeability in cliff stability
a + b)   landwards and seawards dipping strata
c + d)   permeable and impermeable strata.
N.B. LMT = Limestone.

bedding planes representing the greatest plane of weakness, sliding is unlikely as all the forces of movement are directed into the cliff. In such situations, the cliff can achieve relative stability with the prime factor in erosion being the rock strength rather than the structure. In contrast, in (b), the direction of dip is seaward, so that there is nothing to stop the rocks sliding in the seaward direction, causing rock slides and wedge failures. The rates of failure are increased in areas where drainage is poor and pore water pressure correspondingly high. Hence, in the cases of dipping strata, seaward dipping rocks pose greater management threats than do landward dipping rocks. The treatment of such areas is complex because of the multiple failure planes (theoretically each bedding plane). Solutions include pinning or support of the most vulnerable areas (see later).

Figure 5.3c and d represents another scenario that is partly linked to problems discussed earlier. The relationship between rocks of different erosion resistance has already been shown, but another issue is the relationship between permeable and impermeable rocks, as this controls which rocks water can and cannot pass through. Where water can pass through one layer and not the one below it, then it can either accumulate within the cliff, building up pore water pressure, or flow out along the bedding plane between the two layers, in effect acting as a lubricant. In either case, the likelihood of failure is increased. In Figure 5.3c, impermeable rocks over permeable means that water cannot penetrate the cliff from above, and so no downward movement occurs. If the impermeable rock is jointed, or as in the case of clay, desiccated, then any water which does soak in can naturally soak away through the permeable rock beneath, unless pore water pressure is sufficient to induce failure (e.g. the example of Scarborough detailed in Box 5.1). In the opposite extreme, Figure 5.3d demonstrates the movement along the boundary between the two rock types, leading to a situation where failure is likely, particularly if the strata is dipping seawards. Such failure mechanisms are common in alternating limestone and shale sequences, such as occur frequently along the Dorset coast of southern England.

Cliff failure mechanisms and approaches to coastal defence

In a simplistic way, we have seen how variations in rock type, permeability, and structure can control how cliffs behave, and consequently how they may be defended. Although the study of cliff failure mechanisms is a complex research issue in its own right, we need to step back slightly in order to maintain our focus on defence issues and subsequent impacts. Sunamura (1991) provides extensive details regarding cliff erosion and develops many of the points raised above.

The first major consideration regarding the nature of cliff falls is that they are episodic. It is common to read in the literature figures for cliff erosion quoted as 'x m a^{-1}' suggesting that cliffs retreat by the same amount each year. This interpretation is a mistake, because a cliff fall may be followed by no further erosion for a number of years. We have already stated that wave attack at the base of a cliff triggers many cliff falls. Following a cliff fall, the base of the cliff is covered by a pile of fallen rock, which acts as a protection against further erosion. Before any new fall can occur, waves have to remove this material (Figure 5.4). Once this is done, waves can then restart the undercutting of the cliff, producing further falls. It should be remembered, however, that during this process of talus removal, subaerial processes continue which, given suitable rock type, can initiate further cliff failures (e.g. Black Ven, Dorset).

As already stated, it is the nature of the cliff fall that is of concern here because it is this that defences need to combat. There are various forms which cliff falls can take (Figure 5.5a–f), depending on lithology and structure. Rock which is massive (i.e. lacks internal planes of weakness) (Figure 5.5a) will be undercut if wave energy is sufficient, and then fall under gravity due to the unsupported weight of the overburden. In contrast, similar mechanisms at work in well jointed rock may also lead to toppling failure, where the vertical joint opens into a fissure, peels away from the cliff and collapses (Figure 5.5b).

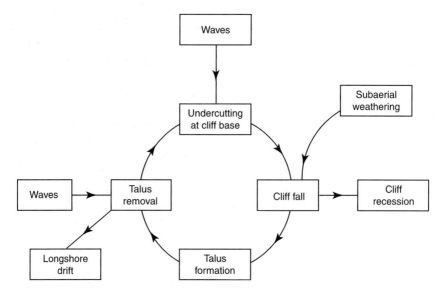

Figure 5.4 The cycle of cliff erosion.

Subaerial processes, such as freeze–thaw and material falling into the opening fissure may also assist the opening of the wedge, leading to eventual collapse. This type of failure is common in limestone areas, and is well exemplified on the Isle of Portland in Dorset, UK (Figure 5.5b). Also joint controlled, wedge failure may occur where equally weak joint planes intersect on the cliff face and a block falls away. The primary control in all of these types of failure is under-cutting; prevention of this by toe protection may inhibit rock falls and toppling failures. Wedge failure is more complex, being due to a combination of gravity and lubricated joint planes. In this case, pinning may be more effective to hold ('bolt') the rock together. Planes of weakness from cliff tops may be opened up by desiccation, tectonics or plant growth. Once this occurs, water may enter and subaerial weathering, such as freeze–thaw and increased pore water pressure starts to occur and failure planes occur across structural boundaries (Figure 5.5c). Other planes of weakness occur naturally within rock. Bedding planes which dip seaward, as we have already discussed, offer a series of potential failure planes (Figure 5.5d).

   In contrast to the examples discussed above, sediments in soft rock cliffs may change physical state when water is present (i.e. liquefaction); the resulting cliff failure may occur not as a coherent mass, but due to flowing or slumping onto the beach. The distinction between the two is that a slide (Figure 5.5e) remains as a relatively coherent mass and may rotate or fall along a failure plane, whereas a flow (Figure 5.5f) moves incoherently and becomes a jumbled, 'flowing' mass with no internal structure. These failures are common in clay cliffs and are

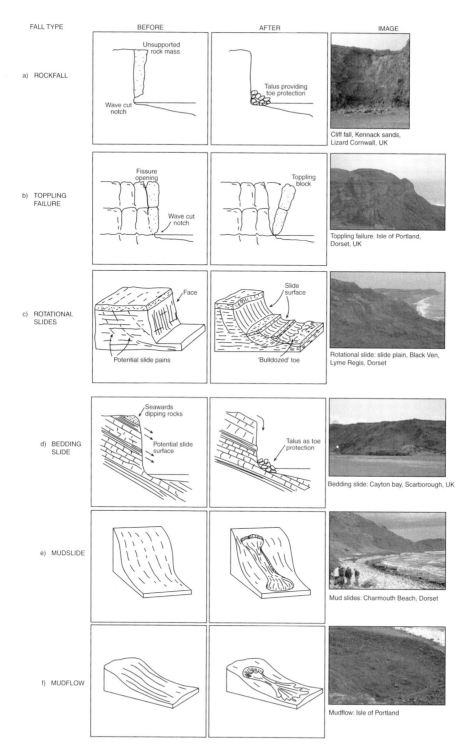

FALL TYPE	BEFORE	AFTER	IMAGE
a) ROCKFALL	Unsupported rock mass / Wave cut notch	Talus providing toe protection	Cliff fall, Kennack sands, Lizard Cornwall, UK
b) TOPPLING FAILURE	Fissure opening / Wave cut notch	Toppling block	Toppling failure. Isle of Portland, Dorset, UK
c) ROTATIONAL SLIDES	Face / Potential slide pains	Slide surface / 'Bulldozed' toe	Rotational slide: slide plain, Black Ven, Lyme Regis, Dorset
d) BEDDING SLIDE	Seawards dipping rocks / Potential slide surface	Talus as toe protection	Bedding slide: Cayton bay, Scarborough, UK
e) MUDSLIDE			Mud slides: Charmouth Beach, Dorset
f) MUDFLOW			Mudflow: Isle of Portland

Figure 5.5 Relationship between cliff morphology, geology and failure mechanism. Note how different rock types will fail in different ways, thus necessitating different management methods.

particularly prevalent along the Dorset coast of England. Black Ven is a classic example of this type of failure, while the neighbouring Spittles Complex just east of Lyme Regis typically shows slumping.

Predicting retreat

Before any assessment of cliff defence can be made, it is important to under-stand how the cliff is 'behaving': in other words, how it is failing (see above) and also its rate of erosion. Knowledge of the erosion rate makes it possible to predict future cliff line positions, and therefore assess the longer-term risk of allowing erosion to continue. Lee (1998) quotes three approaches to the problem:

- those based on historical rates and current behaviour;
- those based on simple empirical relationships (Bruun rule) (see Chapter 2);
- those which model behaviour using a process/response approach.

Irrespective of the approach taken, the basic requirement is to have historical data. If historical rates of cliff retreat are not known, it is impossible to predict future behaviour, or even to assess the severity of an erosion problem. Herein lies one of the key problems, as for many areas of cliff erosion measurements are few and far between. Many rates of cliff recession are quoted using map evid-ence or, for more recent periods, air photographs. These are only snapshots in time, their availability depending on when photograph surveys are undertaken, or when maps are revised and updated, which for the coastline may be rarely.

Where sufficient data are available, by taking historical erosion rates, say for example $x\,\mathrm{m\,a^{-1}}$, it is tempting to say that the future rates will be the same, so that in 5 years time the cliff line will be $5x\,\mathrm{m}$ further inland, or in 10 years time it will be $10x\,\mathrm{m}$ inland. This falls into the simple trap of assuming that erosion happens by the same amount every year (i.e. has a linear relationship with time). However, a cliff fall will protect the cliff and prevent any further erosion until it is removed by waves. Only a detailed set of erosion figures can give further insight into this problem. Far more importantly, however, simple linear extrapo-lation into the future assumes two things. First, that the cliff will continue to fail in the same way and by the same amounts. This assumes that variables such as water content, jointing and cementation of grains, remain constant throughout the cliff. Second, it assumes that the baseline conditions, such as wave activity and sea level, remain fixed. In reality, it is not safe to assume either!

Whilst it is not practical to measure the first assumption (bore hole or geot-echnical surveys can provide answers here but are expensive and generally cannot be justified); estimates of sea level changes are known, and for many coasts, sea level is not static; thus, we immediately have a further factor to plug into our cal-culations of future estimates. Per Bruun, as seen in Chapter 2, was one of the first to relate sea level rise to erosion trends (Bruun 1962), and although there has been much misuse of the 'Bruun rule' equation, the approach is still valid as a way of understanding how coastal systems respond to variations in sea level. Bray and Hooke (1997) have developed these ideas into a predictive model for soft cliff retreat, producing equations that develop the simple linear extrapolation techniques to include sea level rise. Such an approach provides more accurate estimate of future cliff position, although there is a range of 'unknown' factors that need to be considered when discussing predicted coastal positions. In its simplest form, the model predicts future retreat rates (R_2) on the basis of historical retreat (R_1) experienced under a given rate of sea level rise (S_1), and extrapolating into the future using predictions of future sea level rise (S_2). R_2 is then calculated using the following formula:

$$R_2 = (R_1/S_1) \cdot S_2$$

By calculating the predicted rate of cliff recession at a given time in the future (T_2), then integration between the present (T_1) and T_2 will provide an estimate of total cliff recession between the two dates. Normally, T_2 would equate to the predicted rate of sea level rise, such as 'x mm by 2100'. With these two erosion rates, it would be possible to integrate to estimate erosion between any intervening dates, thus plotting predicted cliff positions at regular intervals, such as every 10 years. These ideas are developed as a cliff management application later in the chapter. By using such models, greater predictive ability is increasingly available to the coastal engineer, although it should be stated that the science of prediction is really still in its infancy.

Despite the obvious limitations, coastal erosion models provide a way of predicting what is likely to happen to a cliff, given constant rock and structural conditions, and providing the sea level rises by the amount predicted. With such tools, it is possible to estimate the threat to cliff top property, and the areas most in need of protecting. What cannot be predicted, however, are one-off events — sudden storms, or impacts caused by future defence planning further up drift.

Methods of cliff protection

People who live on cliffed coastlines regard eroding cliffs as needing attention from coastal engineers, and ideally, erosion as needing to be stopped. This outlook is understandable: if people see the edge of a cliff getting closer to their property during successive years, then they feel threatened, both personally and financially. Cliff erosion is rarely seen as anything other than a problem which should be made to go away but, in reality, cliff erosion is as natural a part of coastal processes as waves hitting a beach. We have consistently taken the approach that defences should be used sparingly, and free movement of sediment should be allowed to continue wherever possible. However, given that there are situations where cliff protection is necessary, it is important to review techniques and their applicability to different cliff types. At this point, the issues surrounding the actual decision-making process relating to whether the cliff should be defended or not will be put to one side, but returned to later.

So far we have seen the different ways in which cliffs fail, and the corresponding range of problems which the coastal engineer has to deal with in order to protect coastal development from cliff falls. There are a range of approaches to the problem but, as has already been suggested, the best approach will depend on the lithology and structure of the cliff itself (Table 5.2). The UK Department of the Environment (DoE 1996) outline the main objectives which any cliff protection scheme should aim to achieve. These are:

- reduction of pore water pressure by drainage;
- reduction in destabilising forces by removing potential landslide material, particularly from the upper part of the slide (cliff grading);
- increased stability by the addition of weight to the toe, or increased shear strength along potential failure planes (toe protection, pinning, drainage);
- support for unstable areas (toe protection, walls);
- prevention of basal erosion (walls).

Clearly, not all of these objectives are equally relevant in all cases, although they are often used in combination either to make the cliff stronger, or to prevent the forces of erosion acting on it. Thus cliff protection measures can be grouped into those which aim to increase the cliff's strength and, therefore, resistance to failure (cliff strengthening), and those which aim to reduce the impact of waves on the cliff (wave impact reduction). In addition, two further methods may be employed, first the promotion of erosion in some areas but not others (Hard points), and second, an increasingly common approach to coastal erosion management — do nothing.

Cliff strengthening

In basic terms, cliffs fail when the gravitational forces exceed the coherent strength of the material composing the cliff. Clearly, it is not possible to reduce the pull of gravity (although it can be 'diluted' by cliff grading), but it is possible to increase the coherent strength of cliff faces. There are four ways in which this can be achieved.

Pinning

Figures 5.3b and 5.5d highlight the problem of cliff failure along bedding planes (or other planes of weakness) with a seawards dip. The weakest parts of this cliff are these planes and this is where failure is most likely to occur. To increase the strength of such cliffs, therefore, cohesion in these areas has to be increased in order to prevent slipping. In some examples, the slippage may be caused by lubrication from percolating water, in which case, this will need to be removed (see discussion on drainage below). In other cases, bolting the rock layers together can increase strength. This technique, known as 'pinning' is

Table 5.2 Applicability of different cliff stabilisation methods to cliff failure mechanisms.

Method	Cliff Strengthening				Wave impact reduction			Retreat
Failure	Pinning	Grading	Drainage	Vegetation	Sea walls	Toe protection	Beach feeding	
Rock fall	✓	✓	✗	✗	✓	✓	✓	✓
Toppling failure	✓	✗	✓	✗	✗	✗	✗	✓
Wedge failure	✓	✓	✓	✓	✗	✗	✗	✓
Slide	✓	✓	✓	✗	✓	✓	✓	✓
Rotational slump	✓	✓	✓	✗	✓	✓	✗	✓
Flow	✗	✗	✓	✓	✗	✗	✗	✓

commonly used in engineering, particularly in road cuttings, but increasingly it is seen as a way of preventing coastal land slips.

In coastal slip situations, the technique can be employed to bolt bedding planes together (Figure 5.6a), or to increase the strength of shear planes, such as rotational slips. While this technique does not prevent wave attack, and therefore undercutting and so on, it will reduce the threat of mass movement along planes of weakness induced by subaerial processes, and thus reduce the average cliff retreat rate. In Bouley Bay, New Jersey, USA, rock pins, in conjunction with netting (Figure 5.6c) and subsequent re-vegetation, were used to reinforce over steep slopes (Warner and Barley 1997). This problem centred around the fact that the unconsolidated cover on the cliff slope was steeper than its natural angle of repose, and internal cohesion and shear strength was all that was preventing the material from slumping. Following pinning, there has been no slippage in the area.

A further example of pinning, although this time in well jointed rock (chalk), can be seen in the stabilisation works carried out at East Cliff, Dover, UK. The White Cliffs of Dover are a famous landmark, but inspection revealed extensive degradation around the Dover Castle area, with English Heritage expressing concern for this landmark (Daws and Elson 1990). The cliffs are composed of Middle and Upper chalk (Cretaceous), and include numerous layers of marl and tabular flints. The main concern is the well-developed sets of joint planes, one of which dips southwards (seawards) at around 70°. Failure is largely due to subaerial (freeze–thaw) action, as the cliff face is now cut off from the sea following development of Dover Harbour which now largely fronts it. Rock bolts were extensively employed to reinforce the joint planes, with secondary methods including regrading and grouting.

Grading

Cliffs fall under the influence of gravity. The steeper the slope, the greater the risk of a collapse occurring, and so by reducing the angle of cliff slope (dilution of slope), collapse frequency will be reduced. The angle required (natural angle of repose) will be a function of the rock type, water content and structure of the cliff. In effect, the technique attempts to mimic the formation of this natural angle of repose, i.e. the slope angle at which, given constant environmental factors, the cliff would achieve its dynamic stability. Tainter (1982) indicates that a clay slope with high water content may need a gradient of less than 40 per cent to become stable (22°), while the same material when dry may be stable at a 45° slope. Along the shores of Lake Superior, red clay cliffs do not become stable until reduced to an angle of 33.3 per cent (18°) (all angles measured from the horizontal). Along the Dorset and Devon coasts of the UK, attempts at regrading have been met with varying success. Dennis *et al.* (1975) discuss the regrading of clay cliffs to 33°, coupled with other stabilising techniques, such as planting and drainage. May (1977) discusses how regrading has been effectively used at Bournemouth, UK, and indicates how artificially main-

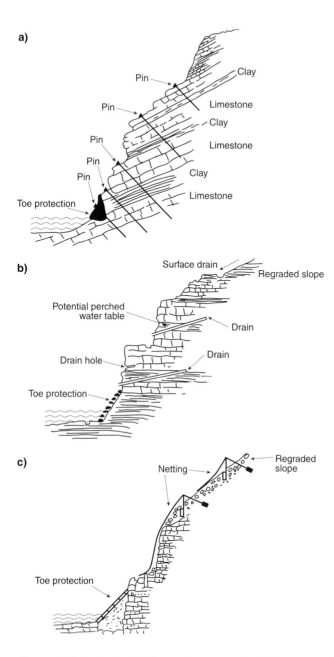

Figure 5.6 Methods of dealing with structural cliff failures
   a) Rock bolting (pinning) to 'clamp' rock layers together and increase resistive strength of potential failure planes
   b) Cliff draining to reduce water content, and thus lubrication of failure planes and pore water pressure
   c) Netting to reduce impact of falling blocks
   (Modified from Fookes and Sweeney 1976).

taining cliffs as close to their natural angle of repose as possible is the preferred policy of the local authority. As with pinning, this method is not used in isolation, but in association with drainage and seeding with vegetation. One case where regrading was carried out as a single technique is at Llantwit Major, South Wales. Here, the threat of cliff falls due to undercutting by waves was threatening the safe use of the beach by holidaymakers. Consultants suggested a scheme designed to regrade the cliffs to an angle of 23° by blasting. Unfortunately, due to poor design, the desired outcome was never realised, and the threat of falls remains. In this situation, the idea to regrade the cliff was methodologically sound, but ill-conceived, with a lack of understanding of how the cliff system was operating in response to natural processes (Williams and Davies 1980).

Exactly how these stable angles are achieved can vary. Cut and fill methods are most common, with material removed from the top of the cliff and placed at the bottom, thus reducing the overall angle of the cliff (Figure 5.7a). This serves both to reduce the gradient and to push the cliff base seawards; in effect, it adds a toe protection to the base of the cliff and forces the cliff base into the surf zone. Such a technique was used at Whitby, north-east England (Clark and Guest 1991), where the cliffs are typical slope over wall structures, with bedded Jurassic limestones overlain by poorly consolidated glacial tills and fluvio-glacial sands and gravels. Problems included continuous cliff erosion that led to the loss of the coast road, lifeboat station and various access points, and was starting to threaten houses. Cut and fill was used to reduce the angle of the cliff top, the material being placed at the foot of the cliff where it was stabilised by toe protection. In addition, some parts of the cliff were drained because of the acknowledged contribution of groundwater and pore water pressure to slope failure (see below).

Two variations can be developed from the cut and fill approach, that of fill only (Figure 5.7b), where material is brought into the area and used to re-grade the cliff without any material being removed. Conversely, cut only could be used to reduce the cliff face (Figure 5.7c), where large volumes of material are removed. This latter method is the only one where the shoreline remains static, the previous two push it seaward. The material brought in should be free draining, otherwise slumping may occur, and given that it will be subject to wave attack, will also need toe protection. Drainage is also an important prerequisite for all of these schemes.

In some situations, reshaping methods need not be as drastic as those outlined above, and simple terracing may be suitable if slopes are shallow enough. The use of gabions to facilitate drainage on each terrace will minimise the total groundwater flow through the cliff, converting some to surface flow from where it can be drained via channels. Gabions not only have some advantages in cliff protection, such as their ability to protect the base of a cliff like a sea wall, they also have the added advantage of allowing water to pass through them, thus reducing the potential of increased pore water pressure. Counter to this, however, gabions are prone to damage by water borne debris, as well as by foot traffic.

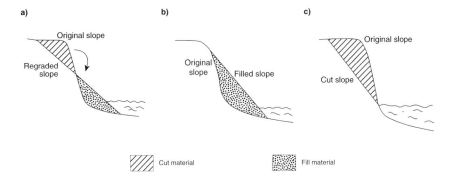

Figure 5.7 Cliff grading to increase cliff stability
    a) Cut and fill to reduce slope gradient, by transferring material from the top of the slope to the bottom.
    b) Fill only, in which material is imported to reduce slope angle
    c) Cut only, where material is exported from the site.

Drainage

Water flowing through a cliff, whether in pore spaces or in cracks and joints, can accelerate cliff failure, either by lubrication of potential failure planes, or through increased pore water pressure. Cliffs which contain alternating permeable/impermeable strata may develop perched water tables above the impermeable layers. This may instigate movement along these surfaces, producing lubrication, or increase pore water pressure. Similarly, in clay cliffs, reduction of pore water pressure can reduce the formation of tension cracks and failure planes that may develop into slump planes and slide surfaces. Hence, by draining a cliff prone to such failure mechanisms (Figure 5.6b), water can be removed from the cliff sediments faster, thus reducing pore water pressure or the amount of water flowing along planes of weakness and, in turn lubrication of failure surfaces and tension cracks.

There are a range of possibilities for installing drainage although all have the same underlying principal — removal of water from the cliff in as short a time as possible. Techniques include the use of agricultural drains to pipe water out of the cliff, and the provision of lines of easy flow, such as gravel-filled trenches. Other possibilities include the modification of cliff top drainage to prevent the input of storm water from soakaways draining domestic or industrial developments. In Whitby (Clark and Guest 1991), the extensive regrading scheme discussed above was supplemented with the installation of cliff drainage. The failure of the upper till layers was largely caused by water seepage and pore pressure. The redirection of surface runoff, reduction in groundwater within the cliff, and the prevention of increased pore water pressure around till/fluvio-glacial till boundaries, have increased stability within the profile. Similar methods of cliff drainage were employed in Chesapeake Bay, USA (Leatherman 1986), where a cliff face was drained to increase its stability and prevent collapse onto shorefront buildings. In Clacton-on-Sea, Essex (Harris and Ralph 1980), a new drainage system was designed to intercept water and prevent increased pore water pressure in the cliff itself. Here, a system of vertical and inclined drains was used to channel water rapidly through and out of the cliff face.

Vegetation

An alternative approach is to increase the strength of the surface layers to reduce the risk of failure. It has long been established that vegetation roots are effective in binding sediments and preventing their erosion. Similarly, they are also effective in preventing sediments from slumping on slopes (quarrying companies often use vegetation to stabilise waste tips, and civil engineers are developing seed 'sprays' to help stabilise roadside cuttings). Tainter (1982) suggests that the planting of vegetation on slopes subject to slumps due to high water content can also help in removing water through root uptake; water uptake is particularly high in the case of woody vegetation. Furthermore, surface runoff can be reduced or slowed, reducing gulleying and sediment transfer down to the beach.

This method, although useful for some forms of cliff failure, will only protect the top layers of sediment/soil, and will not stop deep seated failures. For this, planting needs to be used in conjunction with other methods, such as drainage and regrading. As a result, vegetation planting on cliff slopes is a supplementary practice, used to protect rehabilitated slopes. May (1977) describes how planting of a range of species was used at Bournemouth following regrading and terracing of the sandstone cliffs. As a result, the cliff face is now covered with a range of vegetation, dominated by grasses and gorse. Similarly, revegetation of cliff faces was used in Chesapeake Bay (Leatherman 1986), and at Bouley Bay, New Jersey (Warner and Barley 1997).

Wave impact reduction

So far we have concentrated on methods which prevent cliffs from slipping by removing water and reducing the gravitational effects on slopes by regrading. In effect, these methods alter the physical properties of the cliff sediments, such as by increasing shear strength. All of them, however, are pointless if, after construction, the sea still pounds the foot of the cliff, eroding sediment and making the whole face liable to failure due to undercutting and gravitational instability. In order to prevent this, we have to resort to traditional hard defence methods of stopping the sea from interacting with the hinterland. The most obvious is the construction of sea walls, but other methods, such as toe protection and beach feeding, may also be adopted.

Sea walls

The aim of using sea walls to protect eroding cliffs is to prevent waves from undercutting the cliff base and triggering cliff falls due to gravitational instabilities. The many forms and properties of sea walls are described in detail in Chapter 3, and need not be repeated here. One additional issue, however, concerning the building of walls against cliff faces is that of pore water pressure. We have already identified this as a cause of cliff failure, but similarly, the build-up of water pressure behind a sea wall can lead to wall failure and cliff collapse. The problem occurred in Barton, UK where extensive drainage and revetment schemes did not prevent the slumping of the cliff forward which 'bull-dozed' the defences onto the beach. Another famous example occurred in Scarborough, UK in 1993, when a landslide destroyed the Holbeck Hall hotel (Box 5.1).

The last two examples demonstrate how sea walls alone may not be sufficient to solve cliff instability problems. However, given the right design and combination of techniques, increased stability and decreased erosion is achievable. The ideal situation, as far as local residents are concerned, maybe to remove the threat by stopping cliff recession completely with walls. However, for reasons discussed below, the cessation of cliff erosion is not favoured by managers because of the importance of cliff-derived sediment to the sediment budget. To illustrate this point, at Hengistbury Head within Poole Bay, Dorset, the construction of sea walls along the cliff base has led to a serious deficit in the sediment budget and the erosion of beaches within the bay (Bray and Hooke 1998b).

The concern with cutting off sediment supply has facilitated the development of a range of structures designed to slow down erosion, rather than halt it. Clayton (1989a) discusses a scheme to protect the eroding cliffs of north Norfolk using a variety of hard engineering techniques, including walls, revetments and groynes. While not actually stopping cliff retreat, the defences have caused a significant reduction and thus improved protection to cliff top properties. The use of open or offshore structures (Brampton 1998) is increasingly common for such an approach, and occurs under a range of scenarios, such as at Fairlight Cove, Sussex (see Box 5.2).

Toe protection

To a certain extent, the building of sea walls and toe protection serve the same purpose, and some authors will regard them as the same thing. However, in specific terms, sea walls are built along cliffs to prevent undercutting and erosion, such as rock falls; toe protection is often installed to prevent sliding and slumping by adding weight to the toe of the slump, thus preventing movement. However, there is no major distinction and the two are presented separately purely as a means of distinguishing the two types of problem/solution.

Where rotational slides or slumps have occurred, it is often sufficient to place protection at the toe of the slide. Given that a slide may be moving seawards by rotating along a slip plain (Figures 5.5e and f), placing blocks at the foot of the slide may be sufficient to stop this movement, and thus prevent the sliding block from moving further. Returning to the Holbeck Hall slide (Box 5.1), the resulting lobe of sediment which formed seawards is now protected by toe protection, comprising various grades of armour stone (Clements 1994) (Plate 5.2b). This is designed to serve two purposes: first, to prevent further seaward movement, and second, to prevent marine erosion of the toe. Other examples include the protection of slide toes at Folkestone Warren, Kent, where slips caused problems with the coastal railway line (Viner-Brady 1955); and at Llandulas, Clwyd, Wales, where protection was constructed at the foot of landslips (Wilson and Smith 1983). There is a point, however, when the flow is too liquid for toe protection to be effective (Figure 5.5f). If the slide reaches a high liquid limit, and becomes a mudflow, then no blocks at the base of the cliff would stop this relatively liquid material. In such cases, the aim must be to reduce the water content of the cliff, rather than try to prevent toe movement. One example of mudflows can be found on the Dorset coast, at Black Venn. Here, some landslides are almost continuous, in that they continue to flow seawards, maintaining their position on the beach because the sea erodes the toe at the same rate that the flow moves seawards (analogous to a 'stationary' glacier).

Toe protection may also be used in conjunction with the other techniques mentioned here to protect the base of the cliff (Figure 5.6c). Placing toe protection at the cliff base can further increase cliff stability following pinning, regrading or dewatering. It is possible to regard this as a sea wall or armour-stone, although it is typically not as substantial, and often constructed of graded stone (Plate 5.2b) or, alternatively, wooden fencing or gabions.

Beach feeding

The reduction of wave activity at a cliff base, as will be shown in Chapter 7, may also be achieved by increasing wave attenuation across the fronting beach. Earlier, the cyclicity in cliff erosion was mentioned (Figure 5.4): a cliff would be undercut until it failed, producing a pile of sediment at the cliff base which the waves would erode. During this period, no cliff recession would take place because all of the waves' erosional energy is focused on moving the fallen material. Hence, by artificially building up a beach in front of the cliff, this stage of build-up of fallen material could be artificially maintained; the sea would continuously erode the material placed on the beach and, given adequate replenishment, never actually erode the cliff. In theory, this sounds an ideal solution to many cliff defence problems. However, the detailed issues which arise out of this are relevant to the general theme of beach feeding, which occurs in many coastal situations, not just cliffs. As such, further discussion will be left until Chapter 7.

Hard points

One way of overcoming the problem of protecting isolated properties is to use a series of 'hard points' along the coast, i.e. defend small parts and allow erosion to continue in others (Brampton 1998). The outcome to this will be the creation of stable bays, ultimately providing an embayed coastline with headlands representing the areas of defence, and embayments the areas of erosion.

The idea behind this approach is not a new one, and has often been adopted in the context of defending individual properties. Pethick and Burd (1993) discuss the method, and argue that allowing embayments to form will increase the length of the coastline. Assuming that in doing so the wave energy does not change, the total amount of wave energy will be distributed along a greater length of coastline, and therefore the amount of wave energy per unit length of coast will decrease, thereby decreasing the potential for cliff erosion.

In nature, coastlines with alternating hard and soft rock will experience different erosion rates, as controlled by the strength of each rock type. The result will be the formation of natural embayments, such as those along the Dorset coast shown in Plate 5.5. On straight coasts, erosion tends to be uniform; in order to create areas which are resistant to erosion to act as headlands, these need to become more resistant to wave energy. These are the hard points, and are typically created by placing boulders at the cliff base, and allowing erosion to continue between them, thus forming the embayments (Figure 5.8). The shape of the bay will depend on the wave regime of the coast, its angle of approach, and the spacing between hard points. Barber (1984) presents a

Plate 5.5 Naturally embayed coastline, the ultimate end scenario for an artificial embayment creation scheme. Note also the creation of a tombolo behind a naturally occurring rock breakwater (see Chapter 6).

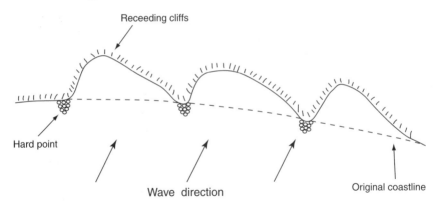

Figure 5.8 Creation of artificial embayments using hard points (Modified from Pethick and Burd 1993).

detailed technical argument in favour of this approach to coastal management, relating spacing of hard points to bay shape and size. It is argued that an approach, where natural processes are coupled with theory, i.e. forming natural coastal features in unnatural situations, may be a viable way forward in some scenarios. However, one major difference does occur, in that natural bays are coupled with a characteristic offshore bathymetry, over which waves become refracted and focused onto the coastline. With artificially created embayments, no such offshore bathymetry occurs and, as a result, the characteristic wave refraction does not take place. 'Stability' of the cliff, therefore, is hard to achieve due to the unpredictable wave foci. Perhaps the greatest issue, however, is unpredictability. Once initiated, embayments will form, but it is difficult to predict how far inland these will reach, or exactly what their final shape will be. This may be particularly important where the defence of properties is concerned; one only has to look at natural stages of embayment evolution to see what can happen if erosion continues unchecked. As bays develop, so they enlarge, until they reach a point where neighbouring bays encroach on each other. At this point, erosion may continue and cut in behind the headlands, which may become cut off to form 'stacks'.

*   *   *

It has been shown how the problem of cliff erosion may be tackled by a variety of methods. It has also been indicated, and should become clear from reading the chapter, that different cliffs need to be handled in different ways. For example, toe protection is no use on a cliff prone to toppling failure. Table 5.2 demonstrates the applicability of the methods shown to the types of cliff described (Figure 5.5), while Table 5.3 cites a range of examples from around the world.

Table 5.3 Examples of different cliff management programmes from a range of locations (For full references, see References at the back of the book).

Country	Authors	Location
Japan	Walker and Mossa (1986)	Byobugaura Beach
Portugal	Alveirinho Dias and Neal (1992)	Vale de Lobo
	Correia *et al.* (1996)	Quarteira
UK	Arthurton (1998)	various
	Clark and Guest (1991)	Whitby
	Clayton (1989a)	Norfolk
	Clements (1994)	Holbeck Hall, Scarborough
	Daws and Elson (1990)	East Cliff, Dover
	Dennis *et al.* (1975)	Charmouth, Devon
	Harris and Ralph (1980)	Clacton-on-Sea, Essex
	McGowan (1998)	Holderness
	McGown *et al.* (1988)	various
	Mason and Hansom (1988)	Holderness
	May (1977)	Bournemouth
	Pitts (1983)	Dee Estuary
	Riby (1998)	Holbeck Hall, Scarborough
	Viner-Brady (1955)	Folkestone Warren, Kent
	Watts (1998)	The Naze, Essex
	Wilson and Smith (1983)	Llandulas, Clwyd
USA	Jones *et al.* (1993)	Jump-Off Joe
	Kuhn and Shepard (1984)	California
	Leatherman (1986)	Chesapeake Bay
	Sayre and Komar (1988)	Thompson Island
	Warner and Barley (1997)	Bouley Bay, New Jersey, Cl

Impacts associated with cliff protection

In everything discussed in the last section, the prime motive was to reduce or halt cliff erosion with the ultimate aim of protecting cliff-top developments. This may appear to be a clear case of engineers stopping natural processes in order to allow people to populate and use the coast. However, it is also important to look at the other side of the argument. In many countries, a large proportion of the coastal sediment budget, i.e. that material available to build beaches (see Chapter 7 for the significance of this), comes from cliff erosion. Hence, large-scale protection of cliffs will reduce sediment input and cause beaches to experience a great loss of available sediment. Although we are now in a position to suggest ways in which we can stop cliff erosion, we must therefore ask ourselves a series of questions. Do we want to stop it? What will be the knock-on implications for the rest of the coast if we do? Is it better to allow a few houses to fall into the sea in order to protect a village or town down drift? These are not easy questions to answer, particularly the last one, and their very suggestion has fuelled many heated debates at public meetings. In order to be able to give a balanced view, we need to put the case for not protecting, alongside the case for defending which has already been presented.

Sediment budget

As seen in Chapter 2, the coastal sediment budget is fundamental to the health of coastal sedimentary landforms. Negative budgets will by characterised by erosion, and the removal of sediment inputs is one way in which budgets go into deficit. Cliff inputs (along with rivers, although the relative importance of each will vary from location to location) are often fundamental in sediment budgets. The cliffs along the Holderness coast, for example, are very active and are retreating by between $1.2-1.7\,\mathrm{m\,a^{-1}}$ (Clayton 1989b), but locally reaching $8\,\mathrm{m\,a^{-1}}$ (Pethick 1992). They are, in fact, thought to be the most rapidly eroding cliffs in Europe. Similarly, in Newport, Oregon, large-scale erosion has occurred since the 1860s and has resulted in rapid cliff retreat and loss of people's homes (Sayre and Komar 1988). As a result of this rapid erosion in both examples, the contribution of material to the coastal sediment budget is great, providing much material for beach build-up. Other examples exist: at Thompson Island, Massachusetts, USA (Jones *et al.* 1993), cliffs have a similar geological composition to those at Holderness and average cliff recession here ranges from $0.4-0.6\,\mathrm{m\,a^{-1}}$. This, coupled with cliff height and length data, suggest an annual input to the local sediment budget from these eroding cliffs of around $6\,500\,\mathrm{m^3\,a^{-1}}$ (calculation based on the total input between 1939–77). Such a volume of sediment is significant in the context of longshore sediment movement. Were these cliffs to be defended, then the local sediment budget would be deficient by this amount, and would attempt to compensate by eroding material from elsewhere, such as beaches. At Thompson Island, the beaches are already eroding, contributing a further $5\,500\,\mathrm{m^3\,a^{-1}}$, so they are unlikely to be able to compensate for the loss of cliff input, indicating that other sediment stores further down drift may become erosional.

The dilemma for the coastal manager, faced with such a problem, is that the erosion of these cliffs provide such an important supply of sediment that to stop the erosion through defences would result in a serious deficit to the sediment budget, the subsequent likelihood of initiating erosion down drift and, there-fore, new threats to many more properties. Countering the earlier point about heated public debate: when it is decided not to defend a cliff, what would be the outcome of a meeting where a town's residents are told that their beaches would disappear and the frequency of flooding and erosion increase because the main source of beach sediment (an eroding cliff) has been stopped by cliff defence so as to protect a few farms and hamlets?

The process of down-drift initiation of erosion has already been seen (Figure 1.2), and from this it is easy to see how the problem can escalate. Work by Correia *et al.* (1996) has demonstrated this phenomenon as occurring along the Algarve coast of Portugal. Here, the rates of cliff recession have recently been estimated at $3.0\,\mathrm{m\,a^{-1}}$, having increased from $0.7\,\mathrm{m\,a^{-1}}$ following sea defence construction up drift. These new erosion problems, however, need not necessarily occur at the point of defence construction, as the newly built defences merely cut off the sediment source. It is the areas where this sediment would have been deposited which will feel the effects of its loss, as there will no longer be the necessary supply of material to maintain beaches, dunes, or marshes.

Eroding cliffs that do not contribute to the coastal sediment budget

From what has been said so far, the reader could be mistaken for thinking that all eroding cliffs contribute to the build-up of beaches. This is not actually the case, *per se*, although all eroded cliff sediment is important somewhere. At several points we have mentioned clay cliffs, both in respect of their modes of failure and the implications of landslides on longshore drift. In Chapter 2 we discussed wave and tide-dominant coastlines. Cliffs are found on wave-dominated coastlines because of the high levels of energy needed for erosion to occur. However, reference to any coastal morphology textbook will show that muds and clays do not occur on such coastlines, but are transported by currents into slack areas, such as offshore, or into estuaries or sheltered coastal embayments. Clearly clay cliffs do not contribute to beach accretion. Plate 5.1b and Figure 5.5e and f show eroding clay cliffs, but the beaches fronting these contain cobbles and boulders, demonstrating that cliff erosion makes little or no contribution to beach accretion (clay cliffs often contain sand, silt or gravel layers which may add to beach sediments). It may be tempting to think, therefore, that protecting these cliffs will not have an impact on beach build-up, and so there are no problems with doing it. While this is true, any respectable coastal manager will want to know where the fine material is going. Along the Holderness coast, for example, the material eroded from the cliff includes sands and silts, but also clays. While the sands do remain on the beach, the finer grained material can be traced moving south into the Humber estuary, as well as moving across the North Sea to continental Europe. A study by de Ruig and Louisse (1991) indicates that the Dutch coast receives around 0.4 million $m^3 a^{-1}$ of sediment from Holderness, via anticlockwise circulation around the North Sea. Hence, these fine sediments are often critical in the sediment budget of estuaries, and are responsible for the supply of sediment to marshes and mudflats.

Beach loss

It should be clear by now that many beaches owe their survival to inputs of sediment from cliffs. Cliff defence will affect beaches in two ways: first, as already discussed, by cutting off the supply of sediment and second, by the same processes that beaches are often lowered following the construction of sea walls — wave reflection, because walls are commonly used to protect cliffs. As these issues have already been investigated in Chapter 3, they will not be discussed again here.

Another issue is where beach lowering in front of a cliff is not related to reduced cliff sediment input, but to a reduction in other sources related to reduced longshore drift caused by defence schemes up drift (see Figure 1.2). Because of the protection that beaches offer cliffs, a lowering can result in increased cliff erosion from increased wave activity. At Alum Bay, Isle of Wight, the owners of a local theme park became concerned about increased cliff erosion following the loss of beach material during the early and mid-1990s. The beach, comprising coarse flint pebbles, had lowered by 1–2 m (McInnes 1998).

Cliff management by non-intervention

By keeping the coastline natural and unaffected by human intervention, natural processes are allowed to occur as they would without modification and so make the management process easier. We have discussed how erosion can be regarded as a sign of coastal instability, and that the very process of erosion represents the attempts of an unstable coastline to regain a dynamic equilibrium. It can be argued, therefore, that the best way to achieve such a state would be to allow coastal processes to find their own stability, and that a policy of non-intervention will be the most effective way of allowing this to occur. The 'do nothing' option thus represents a method of cliff management which achieves its objectives by allowing nature to take its course. Considering all the evidence supplied in the earlier parts of this chapter, it can be seen that there are arguments both for and against the defending of cliffs. On the one hand, cliffs are important suppliers of sediment to the coastal sediment budget; it is important to maintain this input in order to gain adequate coastal protection. On the other hand, because people have developed coastlines, some cliff erosion is now threatening this development. While it is true that when development occurs, people rarely build up close to a cliff edge, a village built several centuries ago several kilometres from the sea may now find itself on the brink of a cliff edge.

Rationale behind non-intervention policies

Because of the way in which the planning process works for coastal defence schemes, the 'do nothing' option is automatically considered in all cases because it provides a base line against which the benefits and problems of other schemes (such as sea walls, groynes, beach feeding etc.) can be compared, i.e. what are the advantages of building scheme 'a' compared to not building it. Where there is development or high-value land use, the benefits of a scheme will clearly outweigh the benefits of not doing it and the 'do nothing' option will not be considered further. In cases where the advantages of defending a coastline are minimal, then doing nothing becomes a viable option. This need not mean, however, that if there is a property located on a cliff top, that cliff will be protected. There are many examples of individual properties being left to fall into the sea (Plate 5.6), some of which are discussed below.

Plate 5.6 Abandoned house on well jointed chalk cliffs of northern France.

One example is on the east coast of the UK. The Holderness coastline is composed of unconsolidated glacial tills that experience average erosion of around $1.8\,\mathrm{m\,a^{-1}}$ (Brickle *et al.* 1998). Historically, this stretch of coastline has seen the loss of around thirty towns and villages over the past 200 years (Prestage 1991). Over recent years, several major towns and industrial locations along this coast *have* been defended, but much of the cliff top area is low-value farmland and has not been afforded such protection. The policy of the defending authority here is to 'do nothing', meaning that retreat and loss of farmland will continue. In addition, however, several isolated farmsteads are also to be allowed to fall into the sea, as their protection is not considered cost effective. The reasons for adopting this policy here are twofold. First, to construct sea defences along the entire 45 km coast would be cost-prohibitive, although off-shore structures (see Chapter 6) have been suggested as an alternative. Second, the significance of these cliffs for the regional sediment budget is considerable, with Holderness being the single largest input of sediment to the North Sea, helping to build beaches not only in the UK, but in many other North Sea nations as well, particularly the Netherlands.

Holderness is not the only coastline where 'do nothing' is the adopted line of coastal defence. In Suffolk, UK, coastal planners have taken this policy and developed a planning strategy that aims to eliminate future defence problems. A line (the 'Red Line') drawn on the map predicts the future position of the coastline in 2068. At Dunwich, this line is 150–190 m inland from the present cliff, and marks the predicted new cliff line in 2068, assuming uncontrolled erosion occurs. No new permanent buildings are allowed to be built seaward of this line (Garford 1999), and it is acknowledged that property on the seaward side of the line will be lost to the sea. While reducing the problem of inappropriate development, it has also caused major property blight for any house currently seaward of the line.

In the UK there has been a tendency to identify areas of low land value and allow natural erosion to take its course. Rarely will the 'do nothing' option become the 'preferred' option in areas of large-scale cliff top development. One location did occur in Fairlight Cove, Sussex (see Box 5.2) where the original intention was to leave the coast alone, but local pressure reversed this decision. Despite this, however, there was still the tendency to leave individual landowners to the mercy of the sea. The issue of land loss and compensation will be discussed later; one has to look towards the United States to discover a more resident-friendly approach to using 'do nothing' as a preferred management option. Griggs (1995) investigates a series of examples from California. In Monterey Bay, rapid cliff (bluff) erosion was threatening a large cliff-top property, with defence costing around $7 million, plus subsequent maintenance costs. Instead of defending, it was decided to opt for a 'do nothing' approach, but instead of just abandoning the development to the sea, a relocation scheme was established at a cost of $3 million ($2.3 million for reconstruction of a replacement building, and $0.5 million for demolition and salvage). This represents a net initial saving of $4 million without recurrent maintenance costs, yet still provides the amenities of the cliff-top development. Similarly, at Lighthouse Point, coastal erosion was threatening the stability of the lighthouse. Costs to protect the cliff were far in excess of those to reconstruct the lighthouse and re-route infrastructure, and so the cliff was left to erode and the lighthouse rebuilt inland.

Similar problems with a lighthouse (Belle Tout) occurred on the chalk cliffs of Beachy Head in Sussex, UK. In what turned into a national media event, the building was not knocked down and rebuilt, but physically moved inland, allowing continued erosion of the cliffs below. By digging under the structure, the building was jacked up and hydraulically pushed about 65 m inland on rails. Given current rates of cliff recession here, this should provide a further 50–60 years of protection. It should be pointed out, however, that this exercise was funded not by coastal defence authorities, but by a lighthouse preservation society.

The decision — to defend or do nothing

Having stated that 'doing nothing' is a method of coastal erosion management on a par with any other form of defence, how is it possible to determine whether it is the best method or not? Subjectively, it can be argued that such an approach should be allowed to occur wherever possible, because it maintains the naturalness of the coast, as well as ensuring continuation in sediment supply to the sediment budget. The decision as to whether to do nothing or do something lies in the planning of coastal defences. French (1997) summarises the procedure for deciding on which method to adopt. Basically, the decision will be governed by cost, environmental considerations, and amenity uses.

- What is the value of the land to be protected (house, hamlet, village, town, industrial)?
- What is the nature of the problem (severity/rate of erosion/wave climate)?
- What will be the cost?
- What are the main components of the sediment budget?
- Can the loss of cliff sediment be compensated for in other ways?
- What are the implications of defending (erosion elsewhere/edge effects)?

The first question asked by coastal managers is 'What is the land worth?' In other words, a cost/benefit analysis will determine whether the area should be defended or not. The basic equation will compare the value of the land, and any structures on it which will be lost in a given time period, to the cost of the various defence techniques. If the cheapest form of defence (excluding doing nothing) is greater than the value of the land, then the coast is not likely to be defended, and 'do nothing' will become the preferred option, i.e. the land is not *worth* defending. Clearly the case will not be open and shut and other considerations need to be incorporated into the equation. One of the failings of this approach, however, is that the way in which the hinterland is costed may not be a true estimate of its actual worth. What price do you put on social structure, emotional ties to a property or value of life?

Management problems associated with 'doing nothing'

At first glance, it may appear that leaving the coast alone to erode at its own rate does not incur any management problems because there is no human interference. If this were to be really the case, the argument could well be justified. However, in the case of coastal management, the decision to adopt a policy of 'do nothing' reflects an active decision on behalf of the defence authorities, therefore this method, as any other, needs to be monitored and reassessed in the light of subsequent events.

One of the key issues is coastal managers' lack of direct control over the outcome. In adopting the decision to allow cliffs to erode, the manager can only use predictive techniques to assess future erosion rates and landward extent of erosion (such as those discussed by Bray and Hooke 1997). If these proceed according to prediction, then no direct management intervention should be needed. However, if rates were to increase, or a major storm strip a fronting beach of sediment, thus making the environmental criteria used in the predictive models invalid, then the method may need to be re-evaluated and changed. In this sense, the management issues relate to monitoring the coast to make sure its behaviour is proceeding as predicted. In reality, few coasts behave according to any 'average' pattern and so any prediction-based response needs to be carefully observed.

One of the most emotive issues relating to this technique is that by adopting a policy of non-intervention, the coastal manager is sanctioning the loss of land to the sea. The main issue here is that if that land is owned by a private individual or company, then they will suffer a direct loss if that land is allowed to erode. Similarly, there is the issue of collective loss, such as a community building (school, church, theatre), or infrastructure, which have indirect impacts on local inhabitants. This brings in the issue of compensation. In the United States, coastal authorities tend to approach the problem by calculating how much defences would cost and how much movement of property and infrastructure would cost, and thus calculate net savings. This represents a sensible model by which to tackle the issue because by not defending coasts, coastal authorities save a lot of money. Unfortunately, not all national strategies are like this. In the UK there is no such compensation scheme or funds for relocation. The case from Beachy Head where the lighthouse was moved inland was funded not from government funds but from a lighthouse preservation charity. Properties on the Holderness coast do not attract compensation, and there have even been occasions where owners have been forced to demolish their properties at their own expense. For example, one landowner who has recently lost her farm and livelihood has been asked to pay £3,500 for demolition of her £250,000 farmhouse and buildings, without any form of compensation. The UK government fears huge demands if they were to start paying compensation and argue that, as landowners, the occupants of buildings should cover the cost of demolition and relocation.

In addition to the issue of compensating for calculable assets, the problem of how to compensate a person or family for the social and personal loss. Every property has a monetary value that can be assessed but, in addition, a personal value to its owner, particularly if their family has lived in it for many generations. Similarly, each collection of properties has a social and community value. Rarely is much attention paid to these issues and, as a result, it is quite common for people, no matter whether they receive financial compensation for loss of property or not, to suffer loss as a result of such defence policies. Hence, while there are many scientific arguments for adopting a non-interference strategy, the social and community issues can be significant, and should perhaps play a more important role in assessing such a scheme's suitability for adoption.

The issue is clearly one that is not sufficiently understood and needs to be more adequately resolved than it currently is. The example of Fairlight village in Sussex is an example of how a favourable outcome can arise (Box 5.2, Plate 5.4). Local managers initially adopted a decision of 'do nothing' for this coastline: local authorities would carry the costs of relocating infrastructure, but individual houses were to be abandoned when amenities were cut off by erosion, or the house became threatened by erosion. Private surveys by the Fairlight Coastal Preservation Society persuaded a change of policy, leading to the construction of the offshore sill now in place along the coast (Plate 5.4). However, this was an example of group pressure forcing a change of heart by planners. In other situations, owners of individual properties do not have as much 'clout' in these situations.

Environmental benefits of 'doing nothing'

If we are to adhere to our original 'statement of intent' with regard to policy on coastal defences, then we need to argue that the more natural a coastline is following the adoption of a management scheme, the better will be the outcome of the scheme for both ease of management and aesthetics of the coastline. Clearly, the idea of doing nothing and allowing a cliff to erode fulfils these requirements nicely.

Using this technique gives us, by definition, a 'natural' coast and in a manner that is relatively cheap. In the UK, this is certainly the case because there is no compensation paid to landowners. In the USA, the cost benefits are slightly less but still tangible. Being a technique of non-interference, there will be no maintenance costs, as there is no initial capital investment in either structures or sediment.

In addition, such a policy also fulfils the sediment budget requirements as inputs are maintained. In many cases, it is the importance of the sediment budget which makes this method the favoured option. In the UK, Holderness is a good example of this, but others exist along the North Norfolk coast, and also the Suffolk coast, where the 'Red Line' approach has been used to predict future coastline positions. Clearly, such a technique fulfils all environmental requirements, making it the ultimate in environmental friendly coastal defence. This should not be surprising, however, because it allows natural processes to operate as they would normally do. Ideally, all coasts should be managed this way; in reality, this will never happen due to constraints imposed by development and land use.

When to stop — the identification of set-back lines

It is clear that the decision to 'do nothing' as far as cliff protection is concerned should not be a permanent one but, along with any defence strategy, should be revised and, if necessary, reconsidered if erosion rates are above those predicted. In order to decide on the suitability of non-intervention, the coastal manager has to predict the future position of the shoreline and judge just where it will be in a given period of years. Only by doing this can he or she determine the threats to land use and development. Figure 5.9 represents a hypothetical situation where predicted shoreline positions have been plotted for 25, 50, 75, and 100 year erosion predictions. The method by which such a map is drawn

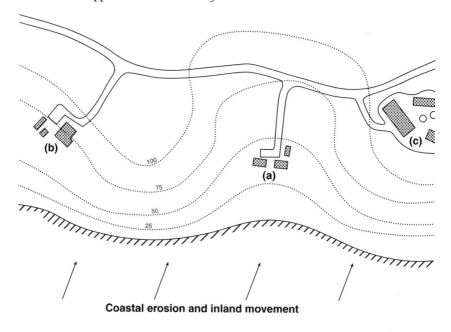

Coastal erosion and inland movement

Figure 5.9 Predicted retreat lines as used in the management of cliffs under the 'do nothing' scenario. Techniques for construction are detailed in the text.

may well follow that of Bray and Hooke (1997), using calculations outlined earlier. Given a policy of non-intervention, it can be assumed that the development labelled (a) will be threatened sometime after 25 years, and would have disappeared in 50 years time. Similarly, development (b) would have disappeared in 75 years, as would part of the road system. Development (c) would still remain after 100 years, although it would be necessary to revise its road link sometime before this.

This type of model can allow the determination of a longer-term plan by which to manage this cliffed coastline most effectively. One scenario may be to adopt the 'do nothing' strategy for at least 25 years. After this, a decision will need to be taken with regard to development (a). One option would be to allow this to fall into the sea and reappraise the situation as development (b) becomes threatened. At this time, it will also be necessary to consider increased costs of relocating the road in order to maintain access to (c). Clearly at this point, the 'costs' of doing nothing actually increase as money will need to be spent on relocating infrastructure. By adopting such a method, the 'do nothing' approach can be used with reasonable confidence, providing that regular monitoring proceeds to verify the predictions made. This is because the whole approach is based on one major assumption, and that is (in our example) that the four predictive lines plotted are accurate. This accuracy will depend purely on the reliability of the models and data used, but constant monitoring of the situation will allow this to be verified (and will also allow the models themselves to be improved in the light of 'real time' observations).

Once the predictive work and decision making in terms of defence strategy has been done, the need for some revisions may be prevented by not exacerbating the situation by allowing any further development to occur on this stretch of coastline. Integrating the coastal management plan with the planning policies of the region can achieve this. Obviously, the coastal manager will not want a developer to build a new housing estate close to the coast where erosion will threaten it in 25 years time. Such a policy is well exemplified on the Suffolk coast of the UK, where Waveney District Council, as part of their coastal management plan, have adopted the predictive model approach to coastal protection. In this case, the predictive line is for 2068 (75 years from policy adoption) and any planning proposal for seaward of this line is refused (Suffolk District Council, personal communication). This represents a coastal position around 150–190 m inland of its present position in some localities. In addition, the area is likely to see the loss of property to the sea as a result of non-intervention. A further issue here has been the dramatic impact on property prices, and the general inability to sell property in some places. Such logic is not always employed, however, and many examples exist of inappropriate development along cliffed coasts. An often quoted example of how planners failed to recognise the problems of eroding cliffs is that of Jump-Off Joe in Newport, Oregon (Sayre and Komar 1988, Viles and Spencer 1995). The cliffs on the Newport coastlines are mudstones and siltstones with bedding planes dipping seawards at angles of 10–30°. As a result of this, and their internal structure and tectonic setting, they are particularly prone to slumping and slips, triggered partly by subaerial processes and partly by wave action. Major slump blocks can be found in the area and are indicative of the erosion history of this coastline. In 1982, a condominium was actually built on one block that had slumped in the winter of 1942–3. Toe protection was installed to stop further movement. The planning proposal was accepted following a geologist's report that stated minor rates of recession and no evidence of slumping. In 1982, building started, and included cliff drainage to remove excess water from the cliff sediments. In 1983 wave erosion had undermined the toe protection and the developers went bankrupt. Following this, ground movement due to reactivation of slip planes caused the drainage pipes to rupture, and the increased water and pore pressure facilitated a new slump which caused the foundations to fail. In 1986, remaining buildings were demolished. Despite this, the area is still zoned as land suitable for future housing development.

This is an extreme example, and even a basic knowledge of cliff morphology and process would raise doubt as to the sense of such a scheme. Furthermore, the idea of building on top of a slump block, whether active or not, is nonsensical given the issues of increased weight and loading which must occur. It represents much about what is wrong with planning: in face of need for development and to cater for residential desires, building was carried out in an area clearly unsuited for it. Whether or not the landslip had moved recently was not the issue. This was a site of known slips, containing rocks which was unsuited for development given their structural configuration and known marine erosion history. Planners should have thrown the proposal out at an early stage. For the record, the geologist in question had his certification revoked on grounds of gross professional incompetence and negligence.

The concept of predicting erosion lines in association with cliff management is increasing in popularity as a means of supporting the non-intervention approach to coastal management. If planners acknowledge that a piece of coastline only has a life expectancy of 50 years before it will be eroded, then they should not authorise any activity to occur on that land which will last longer than this time period. In that way, the coast will be free to erode the land when the time comes and maintain the coast as a natural system without any need for defence construction. This approach has been adopted by the US Federal Emergency Management Agency (FEMA), and is also used by Fletcher (1992) when discussing the impacts of sea level rise on the US coast. By pre-dicting erosion rates, taking into account sea level rise (see Bray and Hooke (1997) for rationale and methodologies) it is possible to zone the coastline into areas of 'life expectancies' and to use this as a basis for development plan-ning (Figure 5.10). FEMA termed these erosion zones 'E-zones' and used

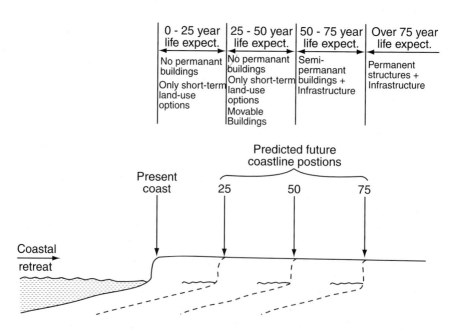

Figure 5.10 Erosion zone (E-zone) classification on eroding coastlines and zones of permitted development (Modified from Fletcher 1992).

them in coastal management strategies, such as the National Food Insurance Programme (NFIP).

Cliff defences and sea level rise

The impact which sea level rise will have on cliffs will vary according to differences in geological structure and rock type. Increasing depth of water, greater wave base, and resulting increases in wave energy will increase the undercutting of cliff faces leading to more frequent gravity failures. What is unsure from the scientific literature is whether the increased rate of cliff failure would be even across the range of failure mechanisms. Clearly, increased undercutting leads to toppling failure and gravity collapse, but whether similar increased failure rates would also apply to slumping and sliding is not certain. Grainger and Kalaugher (1987) claim that the presence of mudslide toes on a beach where sea levels are rising will encourage increases in ground water elevations, facilitating greater surging movements and the reactivation of currently dormant landslide areas. Also, clay cliffs will need to migrate further inland to reach a stable angle of repose, and so they may show a greater response to sea level rise (Bray and Hooke 1997). This would indicate that some large-scale increases in movement may occur on clay coastlines under sea level rise. If this should be the case, the increased input to the coastal sediment budget would be minimal, as the fine grained sediments produced will be too fine to be of much use on high energy coastlines. Despite this, it is also apparent that, while other coastal morphologies, such as dunes and beaches, show a quite rapid response to sea level rise, the response time of cliffs is much longer (Nicholls *et al.* 1995).

Although cliffs may not be a defence structure in their own right, their erosion, and any increases due to sea level rise, does have implications for coastal defence management in two ways. First, erosion of cliffs represents one of the key inputs of sediment to the sediment budget. Increased erosion will lead to increased sediment within the coastal system; some authors (see Carter 1988) argue that this increased sediment, if of the right grain-size and properties, could increase sediment input to beaches sufficiently to offset losses due to increased wave energy. As a result, we could actually see increased beach accretion down drift of cliffs. The second defence issue follows directly from the first. While increased erosion will increase the sediment available to the sediment budget, it will also increase the rate at which the cliff moves landwards, and so lead to greater calls for protection from clifftop properties as they become threatened. Pye and French (1993b) look at this problem from the perspective of the English coastline. Given predicted rates of sea level rise, and erosion rates of cliffs, it is estimated that in order to protect high-value clifftop land which will become threatened by increased cliff erosion by 2040, it will be necessary to construct over 25 km of new sea walls along cliffs currently undefended, in order to prevent erosion from causing property to fall into the sea. This will impact considerably on the sediment budget, have cost implications for defence authorities, and may also instigate the progressive erosion scenario discussed earlier.

A further aspect of sea level rise and cliff response relates to those cliffs which are currently stable. At their natural angle of repose, cliffs will maintain stability all the time the external forcing factors remain constant. Clearly, sea level rise represents a change in such forcing factors which may lead to the reactivation of cliffs. The impacts of this will be the same as those above. Initially, sediment supply would increase, but threats to clifftop development would rapidly lead to calls for increased defence. This would be further complicated because, human nature being what it is, a cliff which has become stable is seen as prime 'sea-view' development land, and so the clifftop development is likely to be considerably greater than on active cliffs.

Summary

This chapter has introduced the problems associated with a major coastal land-form type. We have seen how cliffs vary in their lithology and structure, and how this means that they fail in many different ways. This, in turn, means that cliffs provide the coastal manager with one of the greatest headaches as far as coastal defence is concerned, because the range of methods available for defence is wide and varied, and also extends to non-interference methods involving leaving the cliff to erode naturally.

In many ways, it is important not to defend cliffs if at all possible because of their importance in sediment supply. In reality, however, the old story rears its head again — people have built on top of cliffs so protection is necessary. Increasingly, managers are adopting the approach of letting houses fall into the sea because of the importance of maintaining sediment provision, but from this a new set of issues in coastal management arise — how to equate the obvious needs of the coastal system with those of human emotions. Is it right to let houses fall into the sea when their owners have lived there for many years? This is a question whose answer may contain as much emotive input as it does scient-ific; but the fact that allowing these few isolated houses to fall into the sea to leave a cliff undefended will protect a whole village from erosion down drift may make the decision easier. This does not help the occupants of those houses, however. Also, to rub salt into the wound, it is not unheard of for those same occupants to receive no compensation, and to be given a bill for the demolition of their own homes. In such cases, defence authorities should, perhaps, recon-sider their management policies.

Increasingly, serious consideration is being given to non-interference as a policy of coastal management. Also, it is no longer true to say that any coast on which people live or work will automatically warrant protection. Isolated houses or small villages may well be allowed to fall into the sea if the overall cost-benefit analysis works against them. This marks a large-scale shift towards a coastal process-based (proactive) response to erosion, rather than a protective (reactive), development-based response. We can no longer protect any coast regardless of use, due both to financial reasons, and also the increased awareness of the impacts of protective measures.

The ethics of coastal defence is a major issue in its own right, but the realistic coastal manager should be aware of all the issues which he or she will encounter. We are in a situation where most cliffs can be protected if necessary. It may be that we reduce erosion rather than stop it. This approach certainly gains favour with geologists, whose research and teaching often rely on being able to observe fresh exposures in cliff faces. It is interesting that much of the push for maintaining active cliffs has come from this sector, especially in geologically important areas, such as the Dorset/Devon and Isle of Wight coasts of the UK. Indeed, a text containing proceedings of a conference on coastal defence and earth science conservation, held at Portsmouth University in 1996, has been often quoted in this chapter (Hooke 1998), and contains contributions by many of the active practitioners in the discipline. A further consideration with regard to taking a 'do nothing' approach is that many members of the public often regard its adoption as negative. People commonly adopt militaristic ideologies by claiming that to allow the sea to continue to erode, we are 'giving in' and 'retreating in the face of the enemy' (the sea). There have been many occasions when members of the public quote these ideas in interviews, or at public meetings to discuss planned defences. But when looked at logically, what are the options? It is not possible to continue to squander money on coasts to stop all erosion because there is abundant evidence to prove that this will affect other areas, which will themselves need protection (see Figure 1.2). It is unfortunate that in some cases this means that people's property will be lost to the sea, but what are the alternatives? We could protect the Holderness coast in the UK and save a series of isolated farmsteads and small villages, but should this be at the expense of towns in the Humber, the East Anglian coast and parts of Essex, as well as parts of the Dutch coast. In this respect, the coastal manager is in a 'no win' situation; whatever the decision, there will be losers. The key is to compensate people adequately for their loss. Finally, with regard to the militaristic principles quoted earlier and claims of negativity, any military leader will tell you that in order to win the war, there are times when retreat is necessary in order to advance elsewhere!

Summary of benefits of cliff protection strategies

- Increased security for clifftop development
- Increased security for beach users
- Increased economic security for coastal towns

Summary of problems of cliff protection strategies

- Reduced sediment supply to coastal sediment budget
- Reduced exposure for scientific study (geology/palaeontology/etc.)

Recommended usage

- In areas where alternative sediment supplies are available to compensate for losses from cliffs
- In areas where clifftop development is too valuable to allow cliff failure
- In areas where continuous mud flows may significantly impact on long-shore drift

Summary of benefits of the non-intervention approach

- Increased naturalness to coastlines
- Increased conservation potential of habitats
- Improved protection against sea level rise
- Cheap method
- Maintenance of sediment supplies to budget

Summary of problems with the non-intervention approach

- Loss of property and livelihoods on land eroded
- Compensation issues
- Social and cultural losses
- Predictability/unpredictability of coastal behaviour

Recommended usage

- In areas of low land values where other defence structures are cost prohibitive
- In areas where inputs to sediment budget are critical

6 Offshore structures
Breakwaters and sills

Introduction

Many of the methods discussed previously have focused on techniques which overcome incident wave energy but are attached to the shoreline. A possible alternative to protecting the shoreline occurs in the form of offshore structures which actually reduce the wave energy which hits the coast by providing a barrier offshore. Typically, these structures tend to be constructed close to the shore in shallow water, causing waves to break and expend much of their energy away from the beach area. Because of this, the area between the shoreline and the structure tends to be one of reduced energy conditions in which sediment can build up, thus protecting the coast and possibly also extending the beach.

Although this technique is a relatively new approach to coastal protection, the principals of reducing wave energy at the coast are not new (compare the use of jetties at harbour mouths, for example). The principal of extending the idea to the open coast is thus a logical step in wave abatement approaches to coastal defence. There is also a natural analogue to this approach in the form of coral reefs and offshore sediment banks. Reefs are a good natural analogue to submerged offshore structures and provide effective wave baffles. In effect, they play the same role as the structures to be considered in this chapter. Their significance, however, is often only realised after damage has occurred either naturally (e.g. by storms) or anthropogenically (e.g. by mining), and increased wave energy causes coastal problems. Similar to coral reefs, but more mobile, another form of natural offshore breakwater is the sand or shingle bank; these also provide significant wave attenuation and subsequent coastal protection. The effectiveness of these structures can be seen from the example of Hallsands, Devon, UK (Job 1993). Between 1897 and 1902 dredging for aggregate in Start Bay, Devon, in order to obtain gravel for the construction of Plymouth Harbour, removed a large offshore shingle bank. It had been predicted that natural recharge would occur and the area regenerate. However, no such replacement of material occurred and a combination of spring tides, 13 m waves and north-easterly gales caused severe damage to the village (Hails 1975). As a result of increased wave exposure caused by the loss of the shingle bank, 97 per cent of beach volume was lost with elevations dropping by up to 6 metres (Worth 1909); the village experienced loss of properties and was subsequently abandoned to the sea.

Both reefs and sediment banks reveal the extent to which nature can produce structures which protect coastlines. In the above examples, the degree of protection was not fully appreciated until the natural structures were removed. Engineers, however, soon realised that if removing natural barriers leads to increased wave attack, then by mimicking them with artificial structures offshore, it is possible to reduce it.

The design of artificial offshore structures varies considerably (see below). One common factor is that they need to be tough and resist the full force of the waves. Carter (1988) states that these structures are the offshore equivalent of sea walls. This suggests that they will receive the same wave forces; in fact, because offshore structures are sited further out to sea, they fall within an area of deeper water, and so energy attenuation by friction at the sediment/water interface is less and wave energy greater. Carter (1988) also reports how wave energy can be so great that waves 400 m long and 11 m high have been known to move interlocking concrete tetrapod units of 40 tonnes which had been used to construct an offshore breakwater. This further highlights a problem with these structures: despite much preconstruction modelling and testing, damage is common (see Table 6.1) and repair an ongoing process. This led Magoon *et al.* (1984) to refer to offshore breakwaters as the modern enigmas of coastal engineering.

The history of the use of offshore breakwaters is quite simple. Their original use was largely as protection and shelter for harbours and ports, and it was not until the 1930s that the first offshore structures were built on the open coast specifically for shoreline protection purposes (Komar 1998, Magoon 1976). However they did not become popular as a tool for coastal engineering until the 1980s. The history and usage of breakwaters can, in fact, be traced back to the ancient Egyptians (Tanimoto and Goda 1992) although the greatest advances have been made since the late eighteenth and early nineteenth century. Tanimoto and Goda state that the modern rubble mound design first appeared in the Cherbourg (France) breakwater of 1784, although considerable problems surrounded this structure and large-scale damage occurred even before construction was complete. Partly for this reason, rubble mound structures were superseded by vertical structures, such as that built in Dover (UK) in 1847; although even these were not immune from damage, with examples in Catania (Italy) in 1933, and at Alger and Leixoes (Portugal) in 1934. Design evolution continued with the use of caisson structures, and also the return to, and subsequent modification of, the rubble mound design with larger blocks and protective surfaces. Rubble mound structures are currently the most frequently used design, with considerable research continuing into their effects on the environment, and the effects of waves on the structure. Recent structures in the UK (Sidmouth, Elmer and Happisburgh, see below) have all had unforeseen environmental problems; the main reason for this is suggested (MAFF 1996) to

Table 6.1 Examples and causes of offshore breakwater damage and failure.
(Compiled from a range of sources quoted in the text).

Year	Location	Cause
1905	Byzerte, Tunisia	Inappropriate design, wave overtopping
1926	Valencia, Spain	Storm wave regime underestimated, sediment scouring
1928	Antofagasta, Chile	Storm wave regime underestimated, construction
1930	Catane, Italy	Storm wave regime underestimated, inappropriate design, construction
1934	Algiers, Algeria	Storm wave regime underestimated, sediment scouring
1955	Genoa, Italy	Storm wave regime underestimated, construction
1965	Arviksand, Norway	Storm wave regime underestimated
1971	Antalya, Turkey	Storm wave regime underestimated, construction
1976	Niigata, Japan	Poor foundation
1976	Bilbao, Spain	Storm wave regime
1978	Sines, Portugal	Storm wave regime underestimated, construction
1980	Arzew el Djedid, Algeria	Poor design and construction materials
1981	Tripoli, Libya	Storm wave regime underestimated
1981	Akranes, Iceland	Storm wave regime underestimated

be that the optimal design has not yet been achieved for UK coastal waters, which have a higher tidal range than other locations where these structures have been successfully used. Ongoing research involving wave tank and modelling studies is in progress at various institutions in order to address key issues of wave energy transmission across and through structures, scour, and structure-induced current production.

In its simplest form, the offshore breakwater may be a series of rubble mounds placed off the coast. In its most complex, it may be a carefully engineered structure containing a variety of wave return and energy absorbing structures. In addition to rubble mounds, breakwaters built in lower energy environments may take other forms, such as floating structures, submerged/emergent, or segmented/continuous. These varying types are discussed in detail below, with a range of examples given in Table 6.2.

Offshore structures and their usage

We have already determined that the reason for constructing a defence offshore is to prevent the waves from hitting the coastline with full force. This result will be a low energy area between the structure in which sediment can accumulate and therefore beach width increase. Figure 6.1 details the main groups of structures, and Figure 6.2 details the various methods of construction. The

Table 6.2 Examples of offshore breakwater impacts from different locations (For full references, see References at the back of the book).

Country	Authors	Location
Denmark	Bruun (1985)	Various
Iceland	Bruun (1985)	Akranes
Israel	Nir (1986)	Various
Japan	Tanimoto and Goda (1992)	Various
	Toyoshima (1972)	Various
Libya	Gunbak (1985)	Tripoli Harbour
Portugal	ASCE (1982)	Port Sines
	Bruun (1985)	Port Sines
Puerto Rico	Fast and Pagan (1974)	South west Puerto Rico
Romania	Spătaru (1990)	Black Sea coast
Singapore	Wong (1981)	Singapore
Spain	Berenguer and Enriquez (1988)	Various
	Martin *et al* (1999)	Various
Turkey	Gunbak and Ergin (1985)	Antalya Harbour
UK	Fox (1997)	Sidmouth, Devon
	MAFF (1996)	Various
	Moody (1997)	Elmer, Sussex
	Pethick and Reed (1987)	Dengie, Essex
USA	Anglin *et al.* (1987)	Chicago, Illinois
	Bruun (1985)	Various
	Dean and Pope (1987)	Redington, Florida
	Dean *et al.* (1997)	Palm Beach, Florida
	Douglas and Weggel (1987)	Slaughter Beach, Delaware
	Good (1993)	Cameron, Louisiana
	Inman and Frautschy (1966)	Santa Monica, CA
	Magoon (1976)	Winthrop, Massachusetts
	Nagashima *et al.* (1987)	Holly Beach, Louisiana
	Noble (1978)	California
	Pope and Rowen (1983)	Lake Erie, Ohio
	Sonu and Warwar (1987)	Santa Monica, CA
	Walker *et al.* (1981)	Lorain, Ohio
	Wiegel (1964)	Santa Barbara, California
	Wiegel (1993)	Maumee Bay, Ohio
Various	Komar (1983)	Various

important consideration when deciding which material to use in building off-shore structures, is that it needs to be strong enough to survive in the wave climate concerned. Various methods used have included tyres, concrete bags and caissons, although the rubble mound structure is by far the most commonly used (Figure 6.2).

Nagashima *et al.* (1987) looked at the difference in performance between structures of different design and material. Tests at Holly Beach, Louisiana used five different designs of structure:

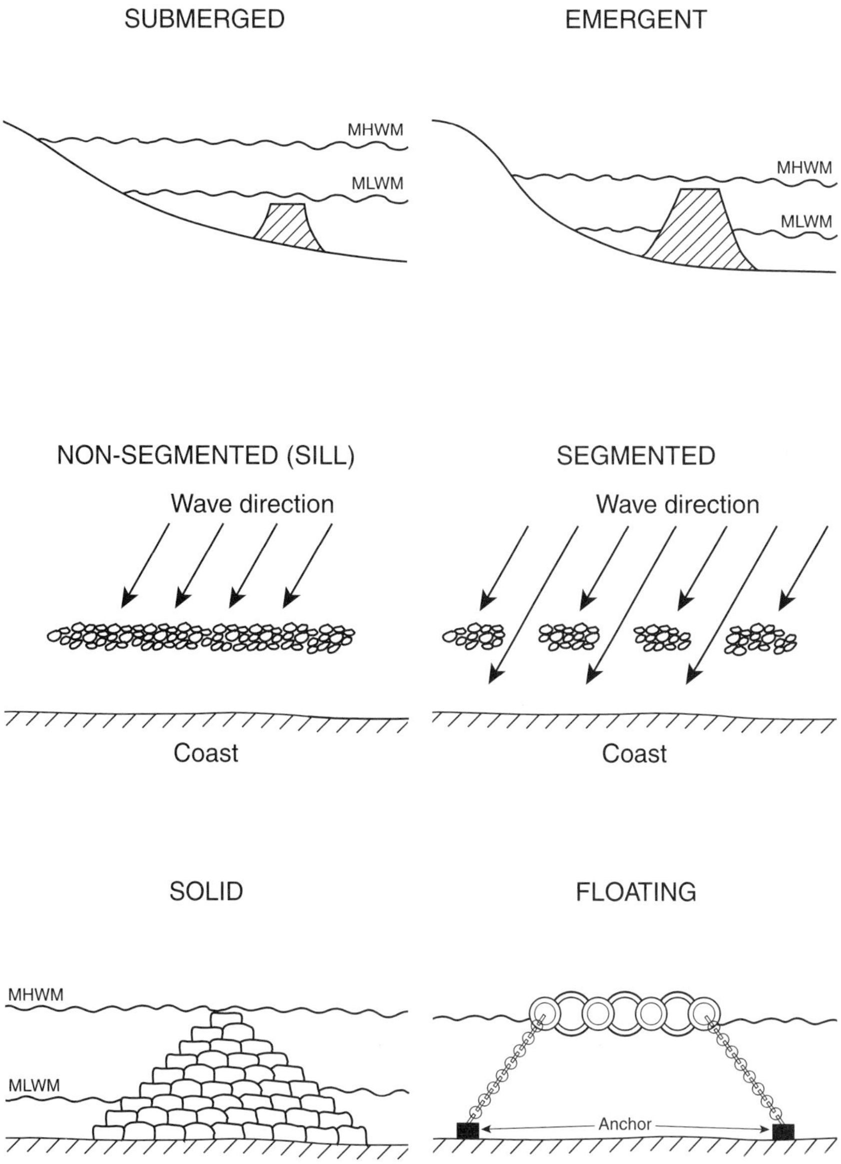
SUBMERGED
EMERGENT
MHWM
MLWM
MHWM
MLWM
NON-SEGMENTED (SILL)
SEGMENTED
Wave direction
Wave direction
Coast
Coast
SOLID
FLOATING
MHWM
MLWM
Anchor

Figure 6.1 Main types of offshore breakwater structure (Modified from various sources).

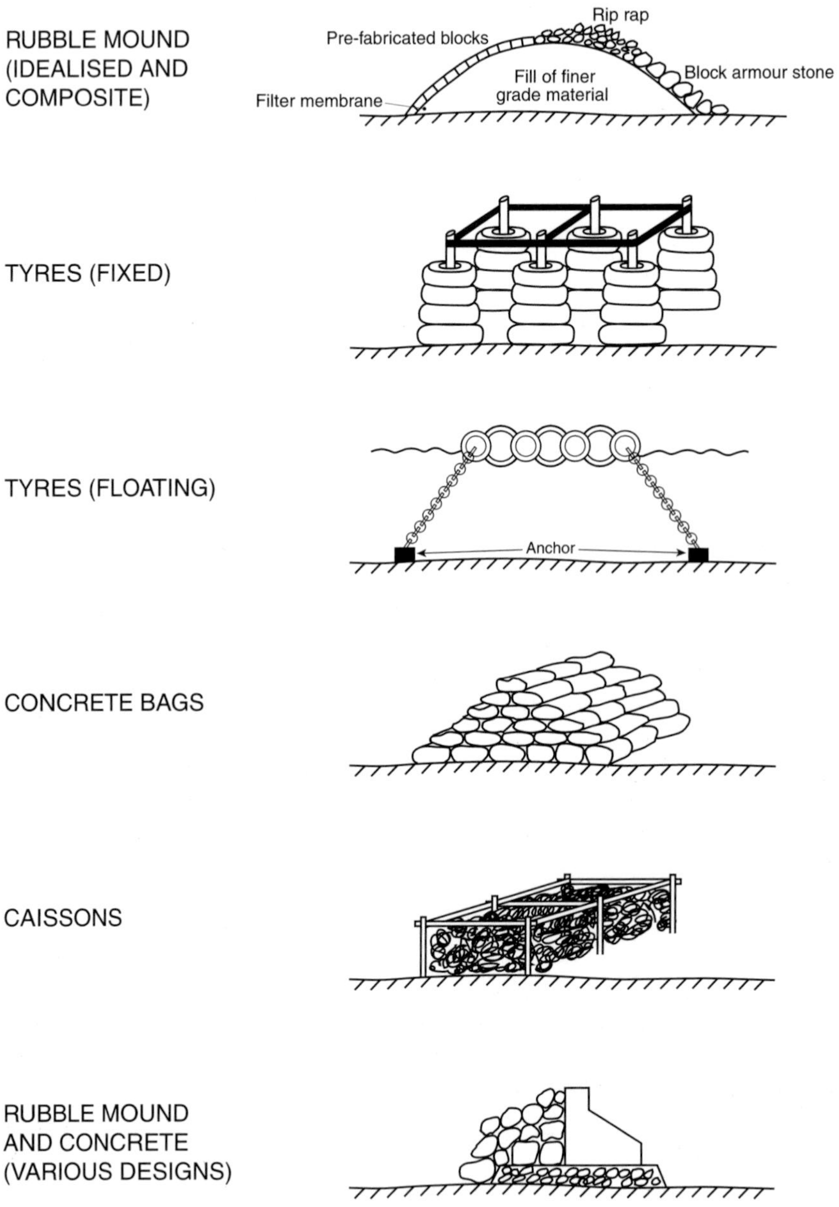

RUBBLE MOUND
(IDEALISED AND
COMPOSITE)

Rip rap

Pre-fabricated blocks

Block armour stone

Filter membrane

Fill of finer
grade material

TYRES (FIXED)

TYRES (FLOATING)

Anchor

CONCRETE BAGS

CAISSONS

RUBBLE MOUND
AND CONCRETE
(VARIOUS DESIGNS)

Figure 6.2 Methods of construction of offshore breakwaters (Modified from various structures).

1) rip-rap structure with rock;
2) rip-rap structure with two rows of timber and used tyres;
3) three rows of timber and used tyres (in line);
4) three rows of timber and used tyres (staggered);
5) single row of timber and used tyres.

The greatest amount of sediment built up behind structure (1). The greatest wave attenuation was observed in structure (3), while the least attenuation occurred in (2). The most damage was seen in the rip-rap structures, although this was put down to inadequate grade material used for the prevailing wave climate. This would indicate trends in sediment accumulation, but it should be remembered that breakwater (1) was the first in the series and thus accumulated the most sediment through longshore drift. Wave attenuation information, however, is more indicative of performance.

Submerged versus emergent

As the name suggests, a submerged structure is one whose top lies below the level of mean low water and is permanently covered. Some authors use this term as synonymous with artificial reefs. Increasingly this is being accepted as the norm, although technically speaking the artificial reef is a deeper water structure largely for ecological benefit and has no real potential for wave energy reduction. Readers should be aware, however, that the two terms may be interchanged in some sources. Clearly, the effectiveness of such structures for wave protection is not as great as emergent structures because waves can pass over the top of them. On high energy coasts, however, these structures will lie above wave base and so will play some role in protecting the coast from high energy waves, thus giving some degree of coastal protection. Experimental and field work carried out by the US Army Corps of Engineers (Dally and Pope 1986) has shown that submerged structures can effectively reduce wave height and are particularly effective in reducing large waves. The degree to which this is achieved is a function of the width of the crest, and the depth of water. Clearly, relating this back to wave behaviour, the depth of water will govern the depth of effective wave base, while the width of the crest will govern the amount of energy lost through baffling and friction. A scheme constructed at Palm Beach, Florida (Dean *et al.* 1997) produced a wave height reduction of 15 per cent for high energy waves. These levels of reduction were greater than predictions had suggested and may indicate the importance of additional factors, such as interaction between individual waves within the vicinity of the structure, and wave breaking over the structure. In addition to wave height reduction, submerged structures also have an impact on offshore sediment movement. This may be beneficial, in that transport offshore by rip currents is prevented, although interference with nearshore currents and the interception of beach building waves may have greater impacts on beaches. In the example from Florida, significant changes in current activity also occurred; this led to the development of what Dean *et al.* describe as 'breakwater-induced longshore currents' being responsible for negative impacts on sediment accretion. Submerged structures have the additional advantage in that being submerged, they remain out of sight, and so have particular appeal on tourist coasts where aesthetics are important.

One of the main role of submerged structures is to prevent the offshore loss of sediment due to strong backwash and rip currents (USACE 1992). For this, they are typically placed several hundred metres seaward of the coastline with a crest parallel to the shore. This was the intention in an example from Palm beach (Dean *et al.* 1997), but observations contradicted the prediction, in that the longshore currents mentioned above produced an increase in erosion when compared to preconstruction rates. Sediment build-up was observed at the shoreward foot of the structure, but the development of strong longshore currents between beach and offshore breakwater removed this and transported it to the south, effectively producing a scour channel parallel to the shoreward face. This situation is analogous to a scheme in Morecambe Bay, north-west England (French and Livesey 2000). Here, the original design of the scheme included offshore breakwaters, although modelling studies actually predicted the likelihood of strong longshore currents, largely because this stretch of coast had well developed current movements anyway and construction was likely to reinforce this. In order to overcome this problem, the offshore breakwaters (in this case emergent rather than submerged) were modified to include attachment to the coast — producing a series of fish-tail groynes (see Chapter 4). The scheme at Palm Beach highlights two important factors with respect to the design and construction of submerged structures. The depth of water above the breakwater is critical; if it is too great, too much water will be able to flow over the structure and this can lead to the formation of the strong longshore currents observed here. As in the case of Palm Beach, these can be sufficient to offset any sediment accretion which may occur. Partly connected to this is the second factor: the strength of the longshore currents varies inversely with distance from the shore. The outcome of this scheme was to remove the breakwater and apply beach recharge techniques instead (see Chapter 7). Much of the material used in breakwater construction was subsequently re-employed in the construction of a series of rock groynes. This factor was also demonstrated by Owens and Case (1908) for an example at Weymouth, UK. Here, an offshore structure comprising a long, shore parallel rock breakwater was constructed towards the end of the nineteenth century. Failure to take account of the longshore currents in the area caused accelerated sediment loss from the shoreline. The outcome here was, as in the case of Palm Beach, the removal of the structure.

Emergent breakwaters are constructed to a height closer to that of high water and so tend to be visible as the tide falls. Under most tidal conditions, they will provide a solid barrier to incoming waves and so provide significant protection to a coastline. Because waves hit them directly, they dissipate and deflect wave energy before waves reach the shore. This promotes lower energy conditions landward of the structure and may promote sediment accretion. A row of breakwaters between the shoreline and wave fronts will produce a leeward side in which wave activity is low, thus promoting the accretion of a wide beach. In certain conditions, this process may be developed to facilitate land claim. Wong (1981) describes the East Coast Reclamation Scheme in Singapore where accretion behind a series of offshore breakwaters, supplemented by land fill, provided 1 525 ha of new land. Because of their extra height, the problems experienced with water flow over the top leading to current formation, as experienced at Palm Beach, are less of a threat. However, emergent structures can still reinforce strong longshore currents and cause beach scour if such strong currents exist prior to construction (French and Livesey 2000).

Non-segmented versus segmented

When considering emergent breakwaters, construction may follow two possible methodologies. Non-segmented structures are continuous forms which provide a barrier along a stretch of coast (Figures 6.1 and 6.3; Plate 6.1) and behind which a low energy lee area may accrete sediment. Such schemes are sometimes referred to as sills and can stretch for considerable distances, such as at Fairlight Cove in Sussex (UK) where a non-segmented offshore breakwater (sill) stretches for 500 m (see Box 5.2 and Plate 5.4) and has afforded good protection to an eroding cliff.

The accumulation of sediment behind non-segmented structures will depend on the strength of longshore currents (USACE 1992). In long sills sediment commonly accretes from either end (Figure 6.3), as it is washed behind the structure by wave refraction, leaving a lagoon area in the centre. Over time,

Plate 6.1 Single offshore breakwater with tombolo formation, The Cobb, Lyme Regis.

NON-SEGMENTED

Wave direction

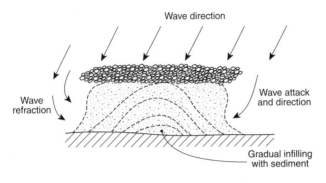

Wave refraction

Wave attack and direction

Gradual infilling with sediment

SEGMENTED

Wave direction and recreation through gaps

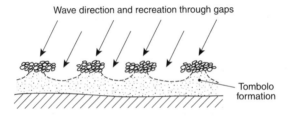

Tombolo formation

SEGMENTED

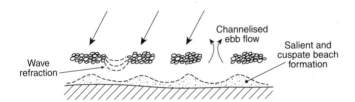

Channelised ebb flow

Salient and cuspate beach formation

Wave refraction

Figure 6.3 Non-segmented and segmented offshore breakwaters, and their associated coastal morphologies.

however, overtopping by waves and continued longshore input will lead to the formation of a 'perched beach', i.e. one which is elevated above its normal level. To seaward of the sill, there will be a rapid drop-off to natural beach level (Figures 6.4 and 6.5). Such features may well occur behind submerged or emergent structures and can significantly add to the amenity appeal of a tourist area as they provide a wide, shallow, low-angle intertidal area. The rationale behind such an approach is to reduce wave impact on the structure by forcing waves to

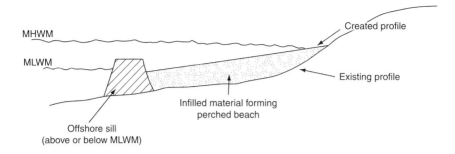

Figure 6.4 Sediment build-up to landward of offshore structures, and the concept of a perched beach. This structure leads to shallow intertidal profiles, but without the cost associated with extensive beach formation.

shallow. In the Teign estuary, Devon, UK, a perched beach was constructed as a means of reducing erosion of the cliffs. Here, an offshore wall was constructed, behind which sediment was pumped in order to afford greater cliff protection.

Without adequate space (or sediment) to build a complete beach profile, one which grades 'naturally' for part of this length is the next best option. The method does, of course, also have potential for amenity use, and this is developed further in the following chapter. The one major hazard of these features is that, unless properly forewarned, users of the created beaches do risk stepping off the edge of the perched beach into much deeper water. In addition to increased amenity usage, a perched beach will create a wide, intertidal area which will serve to provide a level of wave attenuation and energy reduction at the coast beyond that provided by offshore structures alone.

Single breakwaters which are not too long may allow the formation of a tombolo (Plate 6.1), akin to those typically formed behind segmented structures (see below). This methodology is being considered for coastal protection in the Netherlands (de Ruig 1998), although here sediment accumulation will be accelerated by artificial nourishment techniques.

More commonly, offshore breakwaters are segmented, in that a series of structures, typically between 25–100 m long, may be constructed along the coast (Figure 6.1 and Plate 6.2). Given that one aim of these structures is to protect the coast from wave activity, the gap between them is a critical design factor as this will allow wave refraction to occur from incoming waves, and also channelise flow on ebb tides. The wider the gap, the more wave energy can penetrate behind the structure; this will allow some degree of longshore current activity, and thus sediment movement, to be maintained. Under research conditions, the ideal ratio of gap width to structure length has been found to vary from 0.25 to 0.66 (USACE 1992). This variation is due to the necessity to include other factors into the design process, such as access for boats, retention of beach fill and wave climate. Whatever the gap size, waves will always pass through it and will, therefore, produce wave refraction. King *et al.* (2000)

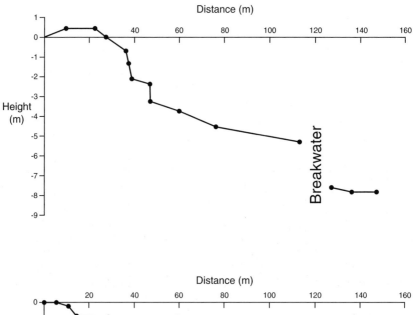

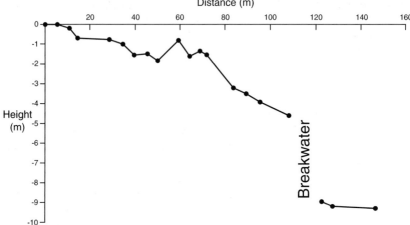

Figure 6.5 Beach topography and impact of offshore structures in creating elevated beach levels. The rapid drop seaward of the structures marks impact of wave reflection, and accumulation in the sheltered lee (Based on data from Elmer, Sussex. See Moody 1997).

report on wave refraction in gaps in a segmented structure. Given that the report by USACE (1992) gives ratios of gap width to structure length as controlling longshore current activity, this later work shows that the ratio of gap width to distance from the shore will also control the degree of erosion (scour) which occurs around the gaps. Where the ratio of gap width to shoreward distance is less than 0.8, no gap erosion will occur; where it falls between 0.8 and 1.3, some erosion may occur; but if greater than 1.3, erosion will certainly occur (Seiji *et al.* 1987).

Plate 6.2 Segmented offshore breakwater with salient formation behind, Miseno, Italy.

Gap spacing is, therefore, important for two reasons: the degree of continued longshore transport, and the degree of associated scouring. In order for each scheme to be effective, it is important for each location to be fully assessed with respect to waves and currents in order to optimise the above ratios for the benefit of the scheme and its desired objectives.

The degree to which currents remain and scour occurs will also control the degree of sediment accumulation landward of the structures. Initially, sediment accumulation behind segmented structures may form a cuspate beach pattern (salients) (Plate 6.2), or be sufficient to eventually produce a tombolo, where the beach is joined to the breakwater by sediment deposition (Figure 6.3 and Plate 6.1). This degree of sediment accretion may also be controlled by design characteristics. The degree of salient or tombolo formation will have an impact on the amount of sediment which can move behind the structure under long-shore drift. A fully developed tombolo will stop all longshore transport, as might a jetty or groyne. Because of this, in most schemes, salients are considered adequate beach forms, and using similar design controls to those above, the degree of formation can also be controlled. The ratio of breakwater length to offshore distance will govern the degree of salient development. For values over 1.5, development will be great and salients will develop into tombolos (Chastern *et al.* 1993). Values between 0.5 and 0.67 will produce salients (Dally and Pope 1986).

Hence, in summary, there are three design ratios which can be used to control the impacts of offshore breakwaters:

- Gap width: distance from shore   – degree of gap erosion
- Structure length: distance from shore – degree of salient formation
- Gap width: structure length   – degree of longshore current activity

A further factor which includes wave movement through gaps (gap width), as well as over a structure during storms, is the transmission coefficient (signified by K_T), i.e. the degree to which wave activity remains to shoreward of the structure. Work by Gardner and Runcie (1995) in connection with a scheme at Sea Palling, eastern Norfolk, UK utilised this value in the design of the structures. K_T has a value of zero when there is no wave transmission (structure 100 per cent efficient), and a value of 1 where no wave energy is stopped (totally inefficient). Modelling for the Sea Palling scheme revealed that (bearing in mind the ratios previously discussed) low values of K_T (low wave transmission) would favour salient and tombolo formation, whereas higher values where wave activity is greater would produce greater current activity and greater sediment losses to the local system.

The scheme at Sea Palling (sometimes referred to as the Happisburgh to Winterton Scheme) is a phased project of sixteen offshore structures, coupled with other defence measures, due for completion in 2012. The need arose in response to beach lowering and damage to the sea wall over a 20-year period (NRA 1991). Offshore structures were used because of the vulnerability of this part of the Norfolk coastline, particularly with respect to dunes and some lagoon areas to landward. Initial results of the scheme have shown good accretion of beaches in line with predictions. A further scheme has also recently been completed at Elmer, in Sussex (King *et al.* 2000). Here, eight structures have been built to protect low lying parts of the coastline from wave inundation and flooding (Box 6.1).

While the greatest use of offshore structures is on high energy sand or shingle coastlines, work along the mud-dominated Dengie coastline of Essex, UK has also demonstrated the effectiveness of segmented structures in these environments. In this case, the breakwaters were not constructed of rocks but of disused barges (Plate 6.3) (previously used to ship cargo up and down the Thames into London, and known locally as 'lighters'). Each lighter was 25 m long, 6.4 m wide and 2.4 m high; they were sunk in a line 350 m offshore at a depth (MHWS) of 1.5 m, with gaps of 15 m between each. This produced a 60 per cent reduction in wave energy received at the coastline and within 18 months had resulted in a sediment accretion rate of $11.5 \, \text{cm} \, \text{a}^{-1}$ (Pethick and Reed 1987).

Box 6.1

Segmented off-shore breakwater scheme, Elmer, UK

Elmer is a small village located to the east of Bognor Regis, on the south coast of England. Historically, the area has suffered from rapid coastal erosion which was partially halted by the construction of a sea wall and groyne system. The area experiences substantial wave focusing and this, along with other environmental factors, produces a regime of increased wave height and potential for flooding (Green 1992). This potential was realised in 1989–90 when severe storm activity caused large-scale damage to the existing defences (sea walls) and flooding of the hinterland. In order to remedy these problems, a system of shore parallel offshore breakwaters was constructed, and the area between these and the coast nourished with sediment (Holland and Coughlan 1994).

Construction started in 1991 and was phased over the next few years. Norwegian granite blocks of around 8 tonnes were piled to a height of 3–4 m (Jensen and Mallinson 1993). A 600 mm layer of bedstone was placed on exposed bedrock upon which the granite blocks were placed. The eight breakwaters comprising the scheme varied in length, being either 80 m or 140 m, depending on situation (Holland and Coughlan 1994), and stretched, overall, along 2 km of the coastline. A terminal rock groyne to the east of the system (downdrift end) helps regulate beach levels.

A post-construction environmental monitoring project was established to investigate the environmental impacts of the scheme, particularly sediment response and wave regime. The typical salient features associated with segmented offshore structures formed quickly and became more pronounced with time. Aluminium tracer experiments revealed that the direction of sediment addition to the salients varied according to prevailing wind and, therefore wave, conditions. A series of south-easterlies would add material to the eastern side of the features, while westerlies would add material to the west (King 1996, Cooper *et al.* 1996). Overall, with predominant wave directions from the south-west, net transport directions recorded were from west to east, with recorded rates of up to 2 m day^{-1} under the most typical wave conditions. Fluorescent pebble tracers revealed that sediment in the immediate lee of the breakwaters remained immobile during storm conditions, highlighting the degree of protection afforded by the structures, and their ability to maintain beaches. In addition, however, these experiments also revealed that under calm conditions, movement from the west into the scheme was zero, although movements out of the scheme at the eastern end did occur (King 1996, Cooper *et al.* 1996).

Sediment build-up behind the system has been rapid, initially aided by beach nourishment. The terminal rock groyne is proving successful in retaining the sediment along the defended frontage but the material being retained is no longer moving to beaches to the east of the scheme. This has meant that significant down-drift starvation of sediments is occurring, with the result that the local coastal defence authority have had to initiate a beach feeding scheme in order to compensate for sediment loss. This highlights one of the main issues with schemes which retain a lot of sediment and hence deprive other areas of their previous sediment supply.

Another feature often observed with offshore structures is the build-up of a beach in the lee, and a corresponding drop into deep water on the seaward side. These perched beaches are often associated with non-segmented structures, although at Elmer, small perched beaches are forming behind each of the segments. Work by Moody (1997) demonstrated the change in elevation either side of the breakwater structures. Figure 6.5 shows the effect for two of the structures, and reveals a drop of between 2 and 4.5 m. Clearly, some of this difference is due to the beach nourishment which has gone on in the area, but much is also due to the retention and accretion of sediment behind the structures.

Plate 6.3 Use of Thames lighters as segmented offshore breakwater to aid accretion of mudflats, Dengie Peninsular, Essex (Reproduced with permission of Cambridge University Air Photograph Library).

Solid versus floating

The most common form of breakwater is that which is constructed on the sea bed and reaches to below or above the sea surface (see discussion of emergent versus submerged). This type of structure is not only effective at reducing wave energy at the coast, but also as a barrier to offshore sediment transport. Against them, however, is the fact that they can only be constructed in relatively shallow water and even here the construction cost is still high.

In deeper water, the costs of construction would be so great as to become prohibitive and so in such situations, such as steeply sloping shelf environments, a floating structure may offer the best solution. These, as the name suggests, do not reach to the bottom of the sea, but are restricted to the top metre or so. The fact that they float means that they can adjust to high water, and are fixed to the bottom by chains or rope. However, being less substantial structures, their effectiveness against waves is less than their solid counterparts, and they are largely restricted to lower energy coasts and where waves are of short wavelength. Floating structures are typically built of tyres, chained or tied together in a way governed by the local energy conditions, wave amplitude, and the flexibility required. Piles are often considered the best way of securing the structures to the sea bed, as anchors or weights do not tend to be as resilient (Figures 6.1 and 6.2). The USA tends to be the focus for much research in this subject area. The technology's attraction is that its raw materials (tyres) are readily abundant and cheap. Estimates indicate that around two billion old tyres become available each year, and so the low cost makes the material attractive (LSUCWR 1979). However, counter to this, the structures have little resistance to high wave activity and can break apart under such conditions, leading to the littering of local beaches.

The effectiveness of floating structures in wave attenuation has been demonstrated in field situations. Gifford (1978) describes the utilisation of a floating tyre structure on the north shore of Santa Rosa Island, Florida as a means of controlling coastal erosion. Wave attenuation was sufficient here to not only reduce erosion, but to reverse the trend, such that sediment deposition facilitated the colonisation of the area by plants. Similar effects were demonstrated by Webb and Dodd (1976, 1978) in Galveston, Texas, although here, the use of tyre structures was intended to produce colonisation as part of a marsh creation scheme to stabilise an eroding coast (see Chapter 9). Spread of both natural and artificially planted *Spartina* occurred with successful wave energy reduction in what is regarded as a high energy stretch of coast. Kowalski (1974) demonstrated that a floating tyre structure reduced wave energy by 70 per cent for effective wave heights of 1 m. Stability problems in the structure did arise, however, and longer swell waves experienced little energy reduction as the tyre structure tended to 'ride' the waves. This work, therefore, tends to support the idea that such floating structures are really only effective in areas with short wavelengths, and a general absence of longer wavelength swell waves. Or, put another way, they tend to be more effective in sheltered environments, typically in estuaries or embayments where locally generated waves are the main problem.

Research from the Netherlands (van der Linden 1985) contradicts these positive results. In the early 1980s, a series of tests were carried out with floating structures by the Delta Department of the Netherlands' Ministry of Water Management and Public Works. It was found that wave attenuation was limited, particularly during storms, and that the structures themselves were prone to damage. As such, little further research was undertaken. These contrasting views most likely reflect the difference in site exposure in which the tests were carried out. Floating structures are most effective in low energy environments. The Dutch experience centred on an open coastline where storm waves occur and wavelengths are greater. Other studies, such as that previously cited by Kowalski (1974), as well as other research, has consistently shown that floating structures do not function well under high wave energy conditions, and are better restricted to sheltered areas with relatively short wavelengths.

Impacts of offshore structures on the coast

Although offshore breakwaters are being increasingly seen as a good way of dealing with coastal erosion, there are still a series of uncertainties which can produce problems, both for the structures themselves, and also for the adjacent coastline. In many countries, the overall performance of offshore structures has given great cause for concern for a variety of reasons, including enhanced erosion, generation of currents, reduced inputs to coastal sediment budgets, and changes in floral and faunal communities. These issues will be discussed in turn below but, before this, one of the most serious scenarios arises when the structure fails due to wave impact. There have been a series of examples of damage and failure regarding breakwaters, largely during periods of storm activity (Table 6.1). Although storm waves are the ultimate cause of failure, the underlying reasons are generally more fundamental and lie in design or construction. Clearly, large waves during storm events produce high energy levels and excessive wave stress, and can cause movement of boulders leading to collapse or breaching. In particular, much damage is caused by increased pressure in surface depressions caused during wave strike. This can produce problems of 'venting', in which compressed air can dislodge facing blocks and expose the internal structure to subsequent wave attack (this is also a common process in the wave erosion of cliff faces). In many cases, however, poor design, poor construction or bad foundations can be underlying causes of failure. In Table 6.1, it can be seen that one of the most common reason for failure is the underestimation of the storm-wave regime of the location. Prior to the 1940s, this generally meant that wave strength was not adequately predicted, resulting in damage to the structure. With a greater understanding of wave prediction and modelling, such problems were reduced. At Sines, Portugal, failure occurred in 1978 during a storm. The damage occurred because of movement of the prefabricated blocks used for construction; although serious, this damage was restricted to certain places along the structure. The designed maximum wave heights were not exceeded, but the cause of failure arose because of wave refraction and wave ray convergence at particular locations. Wave record analysis showed that the storm generated waves with lengths of 40 m and heights of 11 m, which proved quite capable of lifting and moving the 42-tonne blocks. Clearly, this example, and those of earlier events demonstrate the lack of detailed wave data, and the ability to model and predict accurately refraction patterns around structures. This type of failure is not restricted to the coast of Portugal. Many failures mentioned in Table 6.1 were caused by storms, but this typically involved the inability of the construction material, not the design of the structure *per se*, to withstand storm wave attack.

It is clear that offshore breakwaters do experience problems arising from a poor understanding of wave properties which can lead to their failure. Schemes which can resist storm activity, however, may also initiate their own environmental impacts which need to be considered.

Sediment build-up in lee

The area landward of the breakwater may be described as the lee or shadow zone. Sediment transport along a coastline is a function of wave and current activity, as well as sediment supply. Offshore structures have the ability to alter all of these parameters and, therefore, sediment transport. Figure 6.3 shows the patterns of sediment accretion in the lee of offshore structures. Waves entering this lee area do so either by flowing over the offshore structure, or by being refracted around or through (in segmented structures) it. Either way, original normal incident waves will tend to break at angles to the coast and, in association with longshore currents, transport material towards the centre line of the structure (Hsu and Silvester 1990). In this way, sediment builds up behind the structure to form first a salient and, ultimately, a tombolo (Figure 6.3, Plates 6.1 and 6.2). The implications for this may be far reaching but not necessarily to the coast's detriment (i.e. if it leads to the formation of a perched beach). The area of sediment accumulation represents deposition of material in a location where it had not originally done so, as we have previously discussed regarding many structures. Increased rates of sediment accretion can also have major implications for sediment infauna which, if unable to relocate within the sediment column, may experience deep burial and death. Similarly, if the original surface was rocky, such as a marine plantation surface, then the whole community would suffer if this was to become sand dominated. Both of these scenarios could have further knock-on implications for wildfowl which rely on the infauna as a food source.

Perhaps of more fundamental importance is that the construction of offshore structures prevents waves breaking on the beach; this may, as a result, hinder the generation of longshore currents, and may even cause them to stop. This, in turn, will impinge on the wave's ability to carry sediment and mobilise beach sediment for longshore transport. The building out of sediment from the coast also represents a modification to longshore current patterns. A major sediment deposit which develops behind such structures will prevent currents from taking their original path, which may produce several repercussions. First, the accreting salient may act in a similar way as a groyne, actually forming a physical barrier to longshore sediment movement. While this will serve to accelerate the build-up of sediment on the up-drift side of the salient, it could trigger increased erosion at the down-drift end. This process was observed following the development of the sill structure in Fairlight Cove, Sussex (Box 5.2 and Plate 5.4). Second, a deflection of currents around the seaward side of the accreting salient (i.e. between it and the breakwater) may constrict the current into a narrower space, and so increase its flow velocity, thus increasing potential for scour and erosion. Although this may subsequently inhibit the growth of the salient, it may also scour a deep channel along the inner edge of the structure, increasing the potential for undermining and failure.

Reduction in erosion rates and impacts on sediment budget

Given that the reduction in the rate of coastal erosion is generally the intention, it is perhaps somewhat ambiguous to present it as an environmental problem. However, each reduction in erosion rate along a stretch of coastline means a reduction in input to the sediment budget. Furthermore, it should be remembered that all the time rates of accretion are being enhanced at the coast and new areas of sediment deposition being encouraged, such as the salients and tombolos behind offshore structures, there are areas which are foregoing their normal sediment supply. We have already discussed the ideas of the coastal sediment budget in the context of other defence methods, as well as in the case of the sill at Fairlight Cove, Sussex, where the reduction in cliff erosion has produced a deficit of $9\,750\,\mathrm{m}^3\,\mathrm{a}^{-1}$ in the sediment budget (see Box 5.2).

Decreased wave activity and impacts on supratidal vegetation

Much coastal vegetation benefits from periodic inundation by salt water (salt marsh) or from spray (dunes, cliff grasslands, maritime heaths). In the case of offshore structures, the major impacts on flora comes from reduction in salt spray caused by decreased wave activity. This form of impact is predominantly of ecological importance because any loss in salt-tolerant species would be compensated for by the spread of non-salt tolerant species. In this respect, vegetation cover in the medium- to long-term would not suffer, although species diversity would. What is critical here, however, is that many maritime grassland and maritime heath communities contain spatially restricted species which are often a national rarity. Any reduction in salt spray could mean that other species invade and take over the salt-tolerant communities, causing the loss of rare habitats.

Isolation of foreshore from active coastal environment

Although the area behind the structure and that to seaward are still connected, in hydrological terms they may cease to function as a single 'unit' due to changes in shore normal and shore parallel current activity. As a result, it is possible for the area behind the barrier to experience reduced circulation, lags in flushing time and build-up of pollutants.

Wave and current activity may produce oxygenation of sea water. Reduction in such processes may reduce oxygen levels which may affect community structure. This problem could be exacerbated by pollution build-up, especially if the pollutants are organic and biological oxygen demand increases in order to break down the material. In areas where land drainage outfalls occur along the coast, it is possible for longshore currents to transfer polluted waters into the relatively still areas behind breakwaters. Clearly, in such situations, planning methods should foresee such eventualities and suitable precautions taken.

Reduced circulation of water may occur if currents are deflected away from the area (i.e. be stopped by sediment build-up) or if concentration into a scour channel occurs. This will reduce water mixing and possibly allow pollutant concentrations to develop. Reduced circulation also means decreased flushing times. Again, under certain drainage conditions, this will lead to pollutant build-up and increased concentration.

One likely outcome of a low energy lee area is the development of more fragile benthic communities. Vegetation or animals intolerant of high energy wave conditions could favour such areas and start to colonise. This could serve to increase the biological diversity of the area, but equally could produce ecological competition and produce a major community shift.

Aesthetic intrusion

Although of non-technical origin, one issue relating to emergent offshore structures is that they tend to be visually intrusive. Large piles of rock boulders, whether segmented or continuous, do have a significant visual impact (Plates 6.1 to 6.3), particularly if not sourced locally. When colonised by algae, people may regard them as a pile of green, slimy rocks rather than a coastal defence structure with valuable habitat potential. However, although visual acceptance does not detract from defence effectiveness, coastlines with particular scenic attraction or of vulnerable tourist usage may need to take account of the visually intrusive nature of the structures.

Other impacts

As with any defence scheme, it is imperative to use materials which are inert in the environment in which they are to be placed. Clearly, the use of rock will not cause any such impacts, but structures composed of anthropogenic materials could do. In various discussions in this chapter, the use of tyres has been mentioned in the construction of floating structures. There is an increasing concern that tyres may undergo chemical leaching when in the marine environment. Studies in a Scottish Loch (Allsop 1997) showed no serious risk of contamination, but this work was carried out in low temperatures and in fresh water. Under higher temperatures, loss of hydro-carbons from the tyres may be more noticeable, although the use of shredded tyres mixed with concrete in southeast Florida (Banks 1997) showed no difference in faunal/floral colonisation when compared to concrete blocks. There is no clear outcome from this issue. However, the topic has deterred some authorities from using car tyres in defence work, although more have been made wary by problems of inadequate mooring and the tyres breaking free and causing problems for water users, such as happened in Lyttleton Harbour, New Zealand (Gregory 1997).

Beneficial aspects of offshore structures

New habitats

One immediate impact of an offshore breakwater is that it provides a range of environments for fauna and flora to colonise. As with a coral reef (which may be regarded as a 'natural' offshore breakwater), a rubble mound structure will provide a useful reef habitat. Colonisation may be rapid, as in the case of Murrell's Inlet, South Carolina, where over 250 species of flora and fauna colonised within the initial 4-year post-construction period (van Dolah *et al.* 1984). It is also possible that some offshore structures are actually built with a new artificial reef habitat in mind. Hurme (1979) explains this with reference to breakwaters in South Carolina, which have become such successful habitats that they support a multimillion dollar recreation and fishing industry.

The idea of using offshore structures for ecological benefit has been extensively developed in Japan. Reef structures have been designed and placed at carefully selected locations for the benefits of commercial fishery development. Research in the UK has supported the Japanese work and has found that submerged structures can provide important habitat for marine fauna (MAFF 1995b). For example, an experimental scheme in Poole Harbour, Dorset, has indicated the suitability of using pulverised fuel ash blocks for lobster colonisation. The fundamental controlling factor here is the width of the crevices between individual blocks, and the tendency for lobsters to prefer openings of a certain size. Although this work is restricted to lobsters, it does demonstrate how this form of coastal engineering can be developed sympathetically with ecological concerns, and can be used to enhance the fishing industry. The offshore breakwater scheme at Elmer, Sussex has also been investigated to assess its ecological colonisation (Jensen and Mallinson 1993) (see Box 6.1).

Given environmental factors such as pollution, food supply and availability of species for colonisation, offshore structures are considered suitable to support a typical rocky shore community. Suitable habitats which provide protection against desiccation, sun, wind, and waves are all provided to some degree by rubble mound structures. Other work in this area has focused on the best material for promoting plant growth and animal colonisation. Fitzhardinge and Bailey-Brock (1989) report the development of benthic communities on trial breakwaters of different materials in Kaneohe Bay, Hawaii. Materials used were concrete, tyres and metal. The best community development occurred on the concrete, with the second best being metal. Only poor colonisation was found on the car tyres. As a result of this work, the authors recommend concrete as the best material for constructing breakwater structures, although natural substances which resemble this, i.e. rock, would also be good. Contradicting this, however, is work carried out by Fast and Pagan (1974). Their study involved a comparison of natural reefs and tyre reefs in Puerto Rico, and found that the greatest biomass occurred on the tyre structures with migration from the natural reefs to the tyre reefs, but not the reverse. Indeed, the biomass of the tyre structure was eight times that of the natural. Despite this greater population, species diversity was lower on the tyres, suggesting that while being preferred by some species which proliferate, there was a lower biodiversity overall, and that the natural structures had a greater range of habitats and greater community structure, as reflected by its higher biodiversity.

Amenity provision

Given that one of the main aims of offshore defence schemes is to provide shelter and enhance sediment accretion, there is potential to develop these issues for amenity benefit. Clearly, the development of a wide, sandy beach leading to shallow water free from intense wave activity has many tourist benefits, and can be achieved with a correctly installed breakwater system. Such a development has the potential to help a tourist area develop and prosper, providing the aesthetic problems are not prohibitive. Similarly, even without the sediment accretion aspects, the provision of a sheltered lagoon-type area may also be attractive for water sports. Clearly there are other aspects involved here, not least the increased hinterland development which such activity is likely to involve. It has to be remembered that any pressure for increased coastal development must be offset against the vulnerability of the coastline to erosion. After all, it should be remembered that the reason for needing the structure in the first case was to reduce erosion which was threatening development. Furthermore, it should also be remembered that an erosive coast is not one on which development should either be encouraged, or allowed.

Offshore breakwaters and sea level rise

It has been stated that offshore breakwaters are constructed as a means of intercepting wave energy before it reaches the beach. Under sea level rise, this will continue to be the case, although with deepening wave base, the amount of wave energy experienced at such structures will increase. In Table 6.1, common causes of failure are summarised, and the frequency with which wave energy is a cause is high. Such failures may well increase. In addition to deepening wave base, the general increase in water depth will also increase depth submergence in submerged structures, allowing larger waves to cross into the sheltered area behind. This has the potential of increasing sediment delivery into this area, but also to make the seas landward of the structure rougher, and potentially to experience current modification in line with the increased shoreward water movements. It is difficult to anticipate the precise impact of this, although the ratios of spacing, height and distance from the coast will be important here, as any increased water depth and current changes are likely to impact on salient and tombolo formation.

With emergent structures, the chance of wave overtopping will increase as water depths increase with, in the ultimate scenario, an emergent structure becoming submerged. However, in the shorter term, these emergent structures are going to experience greater exposure to wave energy, as will a sea wall. As such, the problems of wave reflection and scour are as pertinent here as with sea walls (see Chapter 3). Of greater importance, however, is that being in deeper water, the orders of magnitude of scour and energy are going to be greater

Summary

In this chapter, we have seen how it is possible to protect a coastline by building structures offshore. This approach is different from those we have already seen because it is tackling the cause of the problem (waves) before they reach the beach and not the symptoms (sediment loss). The advantages are that we maintain a coastline which is not shrouded in concrete, and so appears more natural. Also, if the main problem is offshore sediment loss, then it may even be possible to tackle these problems with submerged structures which cannot be seen under normal wave and tide conditions. Furthermore, by building such structures, it is also possible to promote ecological diversity through the provision of new habitats. Given the right conditions, this could lead to increased biodiversity or economic gain for an area.

This all appears idyllic, but the difficulties associated with the interference with longshore sediment transport and the sediment budget in general can make this approach impractical in many situations. Despite this, there is increasing evidence that the technique is being adopted more widely. However, a word of warning needs to be sounded. The very nature of the structures removes wave activity, thus making sections of coastline low energy and relatively 'safe' for sailing and other leisure pursuits. We need to remember that offshore structures should be used to reduce wave action for two reasons: first, to protect the mouth of a port or harbour, and second, to protect vulnerable parts of the coastline from erosion. What should not happen is for tourism planners to adopt the methodology in order to reduce wave activity along their resort. If this were to happen, then the proliferation of resorts along some parts of the coastline will result in an unacceptable development of offshore structures which will have a major impact on coastal sediment movement and the potential fossilisation of large parts of the coastline because there is not enough energy to initiate sediment movement or erosion. The impact on coastal sediment budgets would be huge, and down-drift areas would suffer dramatically. One disturbing trend which is emerging is the use of offshore structures to induce wave breaking in order to promote water sports. These 'surf reefs' are planned for a series of resorts, with what appears to be only short-term consideration. The increased trapping of sediments landward of these structures, as we have seen, can only lead to increased sediment budget problems elsewhere on the coastline.

Summary of benefits of offshore breakwaters

- Reduction in wave activity received at the coast
- Increased sedimentation and beach formation
- Reduction in coastal erosion
- Reduction of flood risk due to wave overtopping at the coast
- Reduction in sediment loss through rip cell activity
- Use of tyre structures may be a cheap form of protection, but is restricted to certain areas
- Formation of new 'reef' ecosystems and increased biodiversity

Summary of problems with offshore breakwaters

- Possible deflection and modification of longshore currents
- Retention of sediment with corresponding increased erosion elsewhere along coast
- Vulnerable to damage in storms
- Aesthetically intrusive
- Expensive to construct and maintain
- Possible scour problems through gaps in segmented breakwaters

Recommended usage

- Coastal areas experiencing erosion as a result of wave activity and excessive sediment loss by shore normal currents. Offshore breakwaters will not solve sediment source problems
- Shallow waters
- Where sediment build-up would enhance coastal resilience
- Sheltered areas with low wave energy and short wave lengths could benefit from floating structures

Part III

Soft approaches to coastal defence

Part III looks at softer approaches to coastal defence. So far we have seen how engineers can protect the coast by building solid structures to prevent erosion or to trap sediment. The main problem with this type of structure is that it often results in the land being cut off from the sea, thus preventing any land/sea interactions. In Part II we saw how this type of structure can lead to coastal problems beyond those it was intended to solve. These secondary problems become real issues for coastal managers and often require additional coastal protection measures. Hard defences are also very much a reactive response to a coastal problem, in that they are built following the identification of a problem. Hence, they represent the approach of putting a barrier between the cause and the problem, in order to effect a solution. By doing this, they isolate large parts of the coastline from the sea, thereby affecting sediment budgets and processes.

While hard defences fight against wave energy, soft engineering methods aim to dissipate it, in the same way that wave energy is dissipated in natural systems. Increasing awareness has, in recent years, led engineers to consider other methods of protecting the coastline and preventing erosion. This has brought about a major shift in approach; instead of reacting to coastal problems, engineers now adopt a more holistic and proactive approach to coastal management. A typical reactive response to cliff erosion may be to build a sea wall, but the more proactive approach would be to assess the significance of the problem, to determine whether defences are needed immediately and, if so, to produce a scheme which protects the cliff yet maintains processes and so reduces any detrimental impacts down drift as much as possible.

The introduction to Part II contained the following statement about hard defences:

> In essence, hard defence structures are used to prevent erosion of the hinterland. Two points are important:
> - These structures are designed to protect the hinterland, not the fronting beaches.
> - These structures do not provide sediment to maintain a protective beach, they just redistribute it. In fact, because many hard structures have

completely the opposite effect, in that they reduce sediment input, additional sediment sources need to be found if the sediment budget is to be maintained in balance.

The two points here identify the two main areas of concern which soft engineering aims to address. Along a coastline, wave erosion tends to be a serious problem when beaches are narrow, but less of a problem (if at all) where beaches are wide. Therefore, if we can make a wide, healthy beach, we can effectively reduce wave activity, and thus the wave energy reaching the coastline. Also, any temporal study of coastlines shows how beach profiles change with weather and wave conditions. This change needs a constant interchange of sediment between the beach and sediment store. If we can develop a management strategy which facilitates this interchange and maintains the sediment budget, then the beach can really look after itself. Not only is this more natural and sustainable, it is also comparatively cheap.

What we are saying, therefore, is that instead of using the old techniques which fight natural processes, why not use techniques which work with them? We can do this quite 'simply' by creating beaches or marshes, by adding sediment where it is needed, and by trying to keep the coastline as natural as possible. So, instead of automatically building sea walls, for example, we should adopt the more holistic, proactive approach. We can actually become quite radical here. For example, in Chapters 9 and 10 we deal with salt marsh coasts and ways to overcome marsh erosion. As with beaches, salt marshes are excellent dissipaters of wave energy so their presence is very valuable. If wave energy is a problem in an estuary, do we construct a sea wall to keep the waves out, or do we reduce wave activity by creating salt marsh, either by making new marsh on the intertidal flats or, more, controversially, by allowing land to flood? Similarly, if beach lowering is a cause of increased wave attack on a beach resort, should we build sea walls to stop the waves (Chapter 3), build shore normal structures to encourage sediment accretion (Chapter 4), or do we do neither of these, but put more sediment onto the beach and use this as a way of reducing wave activity (Chapter 7)? The other possibility of 'doing nothing' is perhaps too extreme in this example for even the most radical coastal planner to consider!

If we can start to think in this way, we start to work more sympathetically with the coast. Waves are very powerful entities and can cause a lot of change on coastlines. But this is quite natural. Forcing the coast into an artificial state will only make this change more radical because we are increasing its instability. Allowing the coast to be more natural can only make life easier because we are not constantly fighting natural processes. The coast is like a spoilt child, it will get its own way in the end!

In the following chapters, we will look at soft approaches to protecting beaches and estuaries. The reader should not assume, however, that all of the techniques discussed can be simply swapped for those discussed in Part II. As with any defence approach, each has its time and place, and quite often different techniques need to be used together.

7   Beach feeding

Introduction

Beaches, mudflats and salt marshes and are the most efficient method of wave energy dissipation and, hence, protection of the hinterland. Where erosion has moved beach material away from a threatened area, replacement sediment can be brought in artificially to 'feed' the beach. The aim of beach feeding (also referred to in the literature as beach nourishment, recharge, replenishment, restoration, reconstruction, or fill), as the name suggests, is to increase artificially the quantity of material on a beach which is experiencing sediment loss by erosion in order to allow it to provide adequate storm protection. Ironically, feeding occurs largely where sediment losses have occurred due to the construction of hard defences up drift which have had a detrimental impact on the sediment budget, leading to the prevention of sediment replacement by longshore movement of material. Most frequently, the technique is used to restore a degraded beach, although on some occasions, the process of pumping sediment into the nearshore zone has been used to create beaches in areas where none previously existed. Clearly, this is a slight anachronism because there are probably very good hydrological reasons for beaches to be absent. However, despite this, artificial beaches have been created in various locations, including Ibiza in the Balearic Islands, the south of France, and Praia da Rocha in Portugal (Psuty and Moreira 1990), and also many examples around the shores of Tokyo Bay (Koike 1990). It will probably not escape the reader that these areas are prime tourist destinations, where a sandy beach is vital to the resorts' survival. This link between beach feeding and tourism is increasingly common, although recharge is also now occurring in non-tourist areas, primarily as a means of coastal protection.

Beach feeding has become one of the most popular approaches to coastal engineering over the past decade, and many see it as the ultimate saviour both for increased coastal defence, and for tourism. Despite this, however, the technique is not fully understood simply because we do not fully understand how beaches 'work'. Because of this, many environmental and economic issues remain to be resolved. These are developed later in the chapter. Some associated methodologies are also used on beaches, namely beach reprofiling where sediment is artificially moved around the beach to produce a more stable profile; and beach dewatering, where backwash is concentrated as through-flow rather than surface flow to reduce beach sediment loss. These approaches will be discussed at the end of the chapter.

The basic technique for feeding beaches is simple and involves three stages: first, obtaining material from a suitable site; second, transporting it; and third, transferring it to the beach. Dredging is the most common way of obtaining sediment, using pipelines to transfer it to the beach if the site is sufficiently close (see van Oorschot and van Raalte 1991). Other transfer methods include transfer from a dredger offshore, dumping offshore for waves to bring onto the beach, or moving it directly onto the beach by truck. However, although the basic approach may be simple, it is not just a question of dumping large quantities of sediment onto the beach and taking no further action. Beaches have a shape and form which is controlled by local wave, wind and current activity. Such processes also control sediment grain size and the variation of the latter across a beach. It is important, therefore, to maintain as natural a beach profile as possible during construction, otherwise large scale sediment remobilisation may occur, with a high potential for offshore losses of the sediment originally placed on the beach.

All beach recharge schemes will lose sediment in the first few years, often leading to the misconception by local people that schemes fail. This is not the case, however, as much of the sediment loss is due to natural adjustment of a somewhat artificial post-recharge beach profile to a more natural one, and losses represent beach draw down and transport offshore. In order to accommodate this loss, many recharge schemes use a greater sediment volume than the beach 'needs'. In addition to the morphological problems, the sediment itself also needs to be suitable, as does grain size and, in some cases, mineralogy. These issues will be developed below.

One aspect of recharging of beaches is that the process involves large volumes of material. Table 7.1 lists a range of recharged beaches and the measures of sediment volumes used. Typically, volumes are in the hundreds of thousands of cubic metres range or above. Sources of material (see below) are generally offshore but, less occasionally, land or other sources are used. Increasingly, harbour dredgings are also being used where they are sufficiently clean and of the correct grain size range. Because of the need for such large volumes, one important consideration when recharging a beach is that the source area needs to be close to the area of recharge because the cost of transporting the sediment is relatively high. The issue of material source is one which is starting to become the subject of debate.

It will become obvious during this chapter that there is a huge quantity of information on beach feeding. Table 7.2 contains a selection of case studies; many others have not been included. The reason for this plethora of studies is that beach feeding has really 'taken off' as a means of coastal protection, and many local authorities see it as a means of increasing the tourism potential of their resort, or as a means of reducing capital expenditure on coastal defence as existing structures reach the end of their useful lives. As a result a lot of

Table 7.1 Beach recharge schemes from different world locations, with an indication of recharge volumes (Compiled from a range of sources).

Scheme	Recharge period	Volume (million m^3)
Australia		
Barron Delta, Cairns	1992	0.83
Gold Coast, Queensland	1980	2.4
Golden Beach, Caloundra	1992	0.07
Port Philip Bay	1977	0.16
Rapid Bay	1940–82	1.5
Sandringham Beach	1990	0.35
Surfers' Paradise	1974–5	1.4
Belgium		
Bredene–Klemskerke	1978	0.5
Knokke–Heist	1977–9	8.4
Zeebrugge	1980s	8.5
Brazil		
Copacabana	1969–70	3.5
Cuba		
Hicacos Peninsular	1990	0.81
France		
Cannes	1961–2	0.1
Germany		
Isle of Langeoog	1982–3	0.2
Sylt	1972	0.9
Ireland		
Rosslare Bay, Co. Wexford	1982	0.5
Netherlands		
Goeree	1971	0.6
New Zealand		
Wellington Harbour	1944–5	0.12
Nigeria		
Lagos	1975–87	1 a^{-1}
Portugal		
Praia da Rocha*	1969–70	0.88
	1983	0.15
UK		
Bournemouth	1974–5	0.7
Lincolnshire coast		7.5
Pett	1983–	0.02 a^{-1}
Seaford, Sussex	1987	1.5
Walland	1983–	0.03 a^{-1}
USA		
Atlantic City, NJ	1963	0.44
Delray Beach, Florida	1973	1.5
Lake Erie	1960–1	0.53
	1965	0.13
Lake Michigan	1974	0.18
	1981	0.06
Miami Beach	1980	87
Ocean City, NJ	1980s	4.6
Redondo Beach, California	1967–8	1.1
Rockaway Beach, New York	1975	4.3
	1977	48

*Denotes artificial beach creation, rather than recharge of degraded beach.
Note. See Dixon and Pilkey (1991) for many examples from the Gulf of Mexico coast.

Table 7.2 Examples of beach nourishment schemes from around the world (For full references, see References at the back of the book).

Country	Authors	Location
Australia	Bird (1990)	Port Phillip Bay
	Bourman (1990)	Rapid Bay
Belgium	Charlier and de Meyer (1989)	North Sea coast
	Charlier and de Meyer (1995a+b)	North Sea coast
	Kerckaert *et al.* (1986)	Ostend
	de Meyer (1989)	Zeebrugge
Cuba	Marti *et al.* (1995)	Varadero Beach
	Schwartz *et al.* (1991)	Varadero Beach
Denmark	Moller (1990)	North Sea coast
	Thyme (1990)	Jutland
Egypt	Fanos *et al.* (1995)	Alexandria
Europe (general)	Verhagen (1996)	North-west Europe
France	Anthony and Cohen (1995)	Riviera coast
	Hallégouët and Guilcher (1990)	Brittany
Georgia	Kiknadze *et al.* (1990)	Black Sea coast
	Zenkovich and Schwartz (1987)	Black Sea coast
Germany	Eitner and Ragutzki (1994)	Norderney Island
	Kelletat (1992)	North Sea coast
Indonesia	de Meyer (1989)	Bali
Italy	Cipriani *et al.* (1992)	Monte Circeo
	Evangelista *et al.* (1992)	Marina di Cecina
Japan	Kàdomatsu *et al.* (1991)	Toban Coast
	Koike (1990)	Tokyo Bay
Kuwait	Al-Obaid and Al-Sarawi (1995)	Kuwait water front
Netherlands	van de Graaff *et al.* (1991)	general
	Verhagen (1990)	general
New Zealand	Foster *et al.* (1996)	Mt Maunganui Beach
	Healy *et al.* (1990)	Various
	de Lange and Healey (1990)	Tauranga Harbour
Nigeria	Ibe *et al.* (1991)	Lagos
Poland	Rotnicki (1994)	Baltic coast
Portugal	Psuty and Moreira (1992)	Praia de Rocha
UK	Child (1996)	Lincolnshire
	Lelliott (1989)	Bournemouth
	McFarland *et al.* (1994)	Whitstable, Kent
	May (1990)	Bournemouth and Christchurch
	Whitcombe (1996)	Hayling Island
Ukraine	Shuisky (1994)	Odessa
USA	Bagley and Whitson (1982)	Oceanside, California
	Bocamazo (1991)	New Jersey
	Campbell and Spadoni (1987)	Florida
	Chill *et al.* (1989)	California
	Dixon and Pilkey (1991)	Gulf of Mexico
	Domurat (1987)	various
	Galster and Schwartz (1990)	Ediz Hook
	Giardino *et al.* (1987)	San Luis Beach
	Griggs (1990)	Monterey Bay, California
	Herron (1987)	southern California
	Leonard *et al.* (1990a1b)	Various
	Marsh and Tubeville (1981)	Florida
	Walton and Purpura (1977)	Various
	Weggell and Sorenson (1991)	New Jersey
Various	Charlier and de Meyer (1995b)	Various
	Davison *et al.* (1992)	Various

attention has been given to developing methodologies, and there are many examples where the technique has been used. Bearing this explosion in usage in mind, and the quantities of material involved, a key question about the availability of source material starts to become clear. Where is it all coming from?

In the following sections we will explore the theory behind beach feeding, before developing some of the issues involved in introducing large volumes of sediment into the intertidal zone.

The importance of beaches for wave attenuation

When waves break and run up a beach, they lose energy; as different profile shapes and gradients will interact with waves to differing extents, the shape of a beach will affect this ability to attenuate wave energy. A dissipative beach, one which 'dissipates' a lot of wave energy, is wide and shallow, while a reflective beach, often occurring on eroding coastlines, is steep and narrow and achieves little attenuation. Hence, the logic behind beach feeding is to turn an eroding reflective beach into a wider dissipative beach, thus increasing wave energy attenuation. Figure 2.1 demonstrated the process by which waves lose energy as they cross the intertidal zone and, in this respect, the wave base is critical. The wider and shallower the beach, the greater the lateral extent of the water/sediment interaction above wave base, and the greater amount of wave energy is lost (Figure 7.1). In a steep beach profile (Figure 7.1a), the depth of normal

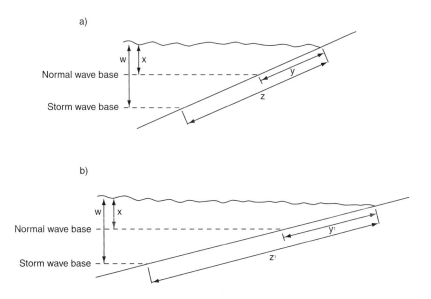

Figure 7.1 Result of reducing beach angle on area of sediment/water interaction. As beach angle is reduced, so greater areas are in contact with moving water, thus increasing wave attenuation.

and storm wave base, measured at 'w' and 'x' metres respectively, result in a lateral distance of sediment/water interaction of 'y' and 'z' metres. In a shallow profile (Figure 7.1b), the same depth of wave base will result in greater lateral distances of water/sediment interaction, measured here as 'y¹' and 'z¹'. Clearly, given that waves lose energy per unit distance as they flow over the sediment surface, then greater wave energy attenuation will occur in the longer profile shown in Figure 7.1b. This then, in basic terms, underlies the principle of using beach feeding for coastal defence purposes.

Beach recharge techniques

As previously stated, beach feeding, in essence, involves dumping sediment onto a beach. However, how this is done varies according to sediment type and source areas. The most common approach is to dredge material from offshore and transfer it onto the beach. Variations on this include: piling the material at high water mark to feed dunes (see Chapter 8), or at low water mark or as an offshore berm to allow waves to move material onto the beach and create a natural profile; placing at one end of a beach to allow longshore currents to move material along the coast; or delivery from land-based sources by truck. Each approach will be discussed in detail later, but we should first make a more fundamental distinction, that of cohesive and non-cohesive sediments. We saw in Chapter 2 how different coastal environments are characterised by different sediment regimes. Wave-dominated environments, such as open coasts, tend to have beaches made up of relatively coarse sand or shingle (non-cohesive sediments), whereas beaches in tide-dominated environments are composed of relatively finer grained sediments, containing silts and clays (cohesive sediments). These two types of grain size characteristics behave in different ways and so need different methodologies when it comes to recharge.

The main difference lies in the particle size and their respective settling velocities. In order to pump sediment onto a beach, it needs to be mixed with water to increase its fluidity. If fine clay and silts are pumped onto the beach in this way, the fine particles will be kept in suspension and drain off the beach as the water runs back to sea. With coarser sediments, this may still occur to a certain degree. In order to prevent this, it is common to construct a system of bunds or barriers to pond water up and allow sediment to settle. This approach is similar to the system of brushwood groynes used to trap fine sediments discussed in Chapter 9, although here we are artificially introducing the sediment rather than trapping sediment naturally available. Perhaps the most important difference, however, is the behaviour of sediment once placed on the beach. The dewatering process is much more complex in the case of mud-dominated environments (clays and silts), so the recharge of these environments is a much more complex process because it cannot be done rapidly. Normally in this case, the techniques for recharge are to increase the accumulation of natural sediments by entrapment; these techniques are outlined in Chapter 9. However, pumped recharge has been carried out in mud-dominated environments, although this has often been with sediments coarser than those naturally occurring in the estuary. Pethick and Burd (1993) discuss an example from Hamford Water in Essex, where sediment was delivered by a technique known as rainbow pumping (Figure 7.2) from an offshore barge. While it did increase the elevation of the mudflat, this was with non-cohesive material, with much of the finer sediment washing off the mudflats as the sediment dewatered. Because of these complications and the difference in approach needed, we will concentrate largely on non-cohesive sediments in this chapter. The recharge of cohesive sediments will be discussed in Chapter 9.

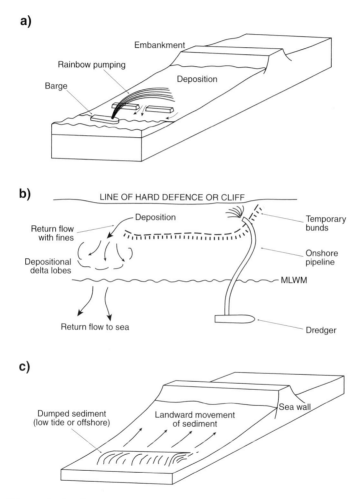

Figure 7.2 Methods of applying recharge sediment to intertidal profiles
  a) Rainbow recharging by spraying of water onto beach. Bunds prevent rapid seaward flow
  b) Pumping direct onto beaches from dredger
  c) Trickle charging. Sediment placed offshore is spread over beach by wave action.

There are a range of methods by which non-cohesive sediment can be added to a degraded beach profile. The primary method is to place sediment directly onto the beach, via pumping from offshore or by tipping from trucks. This is the main category which we will be dealing with here as it represents the most common approach to beach feeding. Other approaches, however, also need consideration. Bird (1996) presents an overview of beach nourishment techniques and includes discussion on the use of longshore drifting to supply sediment from an area of recharge up drift. This method is similar to that of trickle charge (see below), the main difference being that waves are moving the sediment along the coast rather than towards the shore from seaward. Other methods include bypassing, involving the movement of sediment around a drift obstruction, such as a jetty. This approach was discussed in Chapter 4 as a method to overcome some of the difficulties with these structures. The reader should be aware that in some textbooks, recharging beaches will also be discussed in the context of increasing natural sedimentation by using structures, such as offshore breakwaters and groynes. In this book, each of these approaches has been discussed separately (see Chapters 9, 6, and 4 respectively), and so will not be included again here. However, if such a generalisation is required, then the reader should be aware that the majority of the techniques described in this book are aimed at increasing sediment levels on beaches and tidal flats and, as such, all will fit into such a broad outlook.

Direct placement of sediment from offshore sources

The quickest way of increasing the level of beach sediment is to import large quantities from elsewhere. This may be via dredger, or via a land-based source (see below). The most common source of material is from offshore, and is typically obtained from dredging operations. Source areas include offshore shoals, berms, or areas of suitable sea floor sediments. In other situations, dredging of shipping channels provides large quantities of material as a waste product. Charlier and de Meyer (1995b) state that in the United States, the US Army Corps of Engineers regard beach feeding as a suitable method for the disposal of 265.5 million cubic metres of dredged sediment obtained annually, thus avoiding the need for disposal at sea in licensed spoil sites. However, they also indicate that such material is only suitable in 25 per cent of cases due to problems of sediment quality, grain size and transport costs. May (1990) exemplifies this with details of the beach recharge scheme at Bournemouth on the south coast of England (Box 7.1). In this case, beaches were recharged with sediment obtained from maintenance dredging of the navigation channel into the neighbouring Poole Harbour. In addition to the provision of large volumes of suitable material, this joint operation also resulted in substantial financial savings. In Denmark, the local harbour authorities agreed to dump their waste material free of charge offshore at two recharge sites, thus providing a free source of material for recharge operations (Thyme 1990). Other schemes which have relied on harbour dredgings have been carried out in 1978 on the Belgian coast between

Klemskerke and Bredene, where material was obtained from the harbour at Ostend (Bird 1996), and in Hamford Water, Essex, where material from capital dredging in Harwich Harbour has been used to supply a series of recharge schemes in a predominantly mud-dominated environment.

Whatever the source of sediments, where material is pumped from dredger offshore, the method of delivery to the recharge site may be by either the rainbow pumping technique (Figure 7.2a), or via a pipeline linked to the dredger (Figure 7.2b). Either way, in order to pump sediment satisfactorily, it needs to be in a fluid state. To achieve this, it is typically mixed with 10 per cent water, which will then drain off the beach, carrying the fines to seaward, often forming a series of delta lobes (Figure 7.2b). Where recharge is to the mid or upper beach area, temporary bunds may be constructed on the beach to allow a short ponding up of water to allow sediment to settle.

The rainbow method basically involves a dredger obtaining its cargo and then pulling close to the shore and pumping sediment as a spray onto the beach or nearshore area (Figure 7.2a). This method is especially favoured along coast-lines with low tidal range and low wave energy (Bird 1996). In addition, because the dredgers are designed to be able to get close inshore, their shallow drafts mean their capacity and ability to dredge in deep waters is limited, thus making this a technique applicable to relatively small recharge schemes and nearshore sediment sources. Despite the limitations, the method has been successful in recharging many beaches. Along the Danish coast (Thyme 1990), it was used to recharge an offshore sand bar off the west coast of Jutland. In the UK, Riddle and Young (1992) report on its use to recharge beaches at Felix-stowe in Essex, while in Australia, Bird (1996) reports how a converted dredger has successfully recharged beaches at Bribie Island, Brisbane and at Golden Beach, Caloundra.

Where recharge involves large quantities of sediment, the larger dredgers required are unable to get close enough inshore to rainbow charge. In such cases pipelines are used, with booster pumps along their length if required, to pump sediment directly onto beaches where it pours out over the beach (Figure 7.2b). Clearly, having long lengths of pipeline floating on the sea surface restricts the method to calm weather, but when suitable conditions occur, sediment can be transferred rapidly and placed accurately on the recharged profile. This method has been widely used (Table 7.2) and is exemplified by the recharge scheme at Bournemouth, where dredgings were dumped in a temporary store just off the coast, from where a dredger and pipeline pumped the sediment onto the beach (Bird 1996) (Box 7.1). Healy *et al.* (1990) discuss the dredging of sediment from a tidal gorge and pumping it onto adjacent beaches in New Zealand, and Bird (1990) describes the use of dredging and pumping to recharge beaches along Port Philip Bay, Australia. In total, the method was used to recharge 19.3 km of beach around the Bay between 1975 and 1987. It should also be remembered that not all beach recharge is related to the placing of sandy sediments. McFarland *et al.* (1994) and Whitcombe (1996) both deal with the recharge of shingle beaches in the UK. In both cases, material was

Box 7.1

Beach recharge in Bournemouth, UK

Bournemouth and the adjacent coastlines are popular tourist areas and rely heavily on tourist income for sustaining the local economy. This dependence on tourism has arisen partly because of the natural beauty of the area, but also because of the long sandy beaches.

Much of the sediment for these beaches has traditionally been derived from a combination of easterly longshore drift and input from the eroding gravel and sand cliffs backing much of this coastline. However, due to clifftop developments becoming threatened by cliff retreat, a programme of cliff protection was initiated in 1907, combining sea walls and one of the largest vegetation stabilisation exercises in Europe (May 1977) (see also Chapter 5). Fairly soon after the first sections of wall had been completed, beach levels were seen to be falling, and by 1917 it had been decided to construct a series of groynes in order to arrest the longshore disappearance of sand (Lelliott 1989) (compare this with the progressive erosion scenario outlined in Figure 1.2). This was only partly successful, however, and by 1957 local tourism chiefs were expressing concern about the town's golden beach frontage, and the fact that it was not always there! In addition, the sea walls were being undermined in places as falling beach levels exposed their foundations.

Although the construction of the groynes and sea wall had successfully protected the cliff and its cliff top development, it had removed a valuable source of input to the sediment budget, and represents a classic example of solving one problem yet creating another. To compensate for the loss of cliff-derived material and the problems of falling beaches, it was decided to replenish the beach with sand. The initial trial replenishment using $84\,443\,m^3$ of sediment dredged from a nearby licensed dredge site off the coast of the Isle of Wight occurred in 1970 at a cost of £63,600 (Turner, personal communication). This project was carried out in a relatively small area and consisted of a trickle charge scheme, in which material was placed around the low water mark for waves to move it onshore (Lelliott 1989). This was followed by a full recharge in 1974–5 involving $759\,066\,m^3$ of sand from another licensed dredge site in Poole Bay, at a cost of £806,665 (Turner, personal communication). In this scheme, a $1\,800\,m$ long frontage was trickle charged with about $106\,000\,m^3$ being placed offshore to be moved in by waves, and a further $6\,500\,m$ were charged by the direct placing of $658\,000\,m^3$ onto the beach. Beaches were recharged to a typical gradient of 4.5 per cent (1:22) and to a level of +2.0 m O.D. (Ordnance Datum). Clearly, however, this scheme had not treated the *cause* of the beach loss problem, merely the *symptoms* of it. Hence, following recharge, natural processes of coastal erosion continued, in this case longshore drift and offshore movement with no natural replenishment, thus reducing the newly charged beach volume.

Prior to the first full recharge in 1974–5, the beach levels were below the critical beach level (see below) (Cooper 1998) and were measured at 6 million m^3. When the consent was given to recharge in 1974, the UK Department of the Environment made it a condition of funding that a regular monitoring programme involving hydrographic and topographic surveys, along with sediment and volumetric analysis was established to quantify beach response over time (Turner 1994). With data supplied by the monitoring programme, it was clear that a second major recharge would be needed during the early 1990s. By the late 1980s, the critical beach level had been reached again (Cooper 1998) and plans to undertake capital dredging in the entrance channel to Poole Harbour, a few miles to the west of Bournemouth, provided an ideal opportunity to obtain sediment for the recharge operations. Concerns about the grain-size range of the dredged sediment arose in relation to the post-recharge offshore removal of fines by waves, and potential problems for offshore shell fisheries. Following further studies, modelling work predicted that this material would be transported further offshore than the shell fish beds in question and would, therefore, not pose a problem. In addition to this, some of the dredged material was too coarse for the beach regime in question and this was disposed of offshore in a licensed spoil site (Turner 1994). Dredging started in the winter of 1988 and, after a break during the summer tourist season, continued in the winter of 1989; it involved 998 760 m³ of sediment at a cost of, £1,454,501 (Turner, personal communication).

The long-term data set obtained from the monitoring has provided extensive resources by which to model beach behaviour along this coast. Cooper (1998) has used this data to verify predictive models of repeat time. Using the data alone, analysis of post-recharge volumetric decay suggests that the critical beach level, and thus the next repeat will again be necessary in 2003. Using the models of Führböter and Verhagen (see Cooper 1998), the modelling approach predicts repeats to be necessary in 2004 and 2002 respectively. Hence, because of this extensive ongoing monitoring exercise, it has been possible to validate some of the models used in predicting beach fill longevity.

dredged from offshore deposits and transferred onto the beach area, using similar pumping methods to those described above.

Apart from recharging degraded beaches, because of the large volumes of sediment which can be pumped onshore by this method, the technique has also been used to create beaches where none previously existed. In Japan, this method has been used to create completely artificial beaches around Tokyo Bay, where development has been so great that public amenity space is very restricted (Koike 1990). A similar project was carried out to make approximately 20 km of coast accessible for amenity usage along the Kuwait City water front (al-Obaid and al-Sarawi 1995). Similarly, tourist beaches have been created in Ibiza in the Balearic Islands, and Praia da Rocha in Portugal (Psuty and Moreira 1990).

Direct placement of sediment from landward sources

Under certain circumstances, such as when there are no economically viable offshore sources of sediment to dredge, or where beach material is recycled, the delivery of sediment has to be done from landward sources. Principally this involves the use of lorries. Several potential sources are used in this category of recharge: first, material already in the coastal zone which can be used to recharge beaches up drift (such as that which has accumulated up drift of a jetty); and second, material obtained from land sources (quarries, etc.) which is delivered to beaches.

On coastlines where longshore currents are strong, or where sediment movement is held up by shore normal structures (see Chapter 4), then sediment may be recycled from down drift and fed back onto the beach at the up-drift end, or artificially moved around the obstacle. Using 'local' sediments is ideal in the context of beach formation and processes (see below) because the recharge is using material which matches the natural beach in terms of mineralogy, particle size, and particle shape. This will result in a greater stability to the recharged profile.

In essence, recycling material involves physical removal with a digger, and transfer by lorry to the point of recharge. In some instances, the up-drift end of the beach is marked by a harbour, whose protective jetty prevents further down-drift movement. In such cases, it is necessary either to bypass the structure or, if sediment input up-drift is inadequate, to transfer sediment back up drift. At Pett, in Sussex, around 40 000 tonnes of shingle are removed annually from the up-drift side of Rye Harbour and transported back up drift to Pett. Similarly, at Seaford, also in Sussex, 160 000 tonnes of shingle are moved from east to west (up drift) each year, while at Shoreham, shingle trapped up drift of the harbour jetty is used to resupply the up-drift end of this beach. These three examples make an interesting combined study which typifies the battle of coastal engineers and natural processes along the south-east coast of England. A strong west to east drift moves shingle along this coast towards the cuspate foreland of Dungeness, where it meets an opposing drift moving around the coast of Kent towards the west. The continuous drift of sediment has been interrupted by a series of jetties constructed at the mouth of the many ports along this coast. As a result, the natural balance of sediment movement has been damaged, resulting in a series of artificial sediment circulation 'cells' where material moves naturally eastwards by longshore drift, and is transported by lorry back westwards. The result for Dungeness has been serious, as net loss of sediment input has resulted in a situation of net erosion, particularly serious considering the foreland is also the home to two nuclear power stations. The negative sediment budget has necessitated a major shingle recharge scheme here which has recycled 30 000 cubic metres of shingle per year from the eastern side of the foreland and placed it fronting the power station (Townend and Fleming 1991).

Sediment recycling is not restricted to shingle beaches in the UK. In Adelaide, Australia, the strong northerly drift of sand was not compensated for naturally, resulting in a sediment deficit. To remedy the situation, sediment was transported by lorry from the accreting northern section of the beach back to the southern. Although this successfully overcame the problem, pressure from conservationists to preserve the accreting northern beach, and pressure from residents regarding the volume of traffic involved in the recycling scheme, has meant that the scheme has been stopped, to be replaced with using dredged material from harbour dredging (Bird 1996). In Malaysia, recycling by lorries has also been used to recharge beaches although in this case the sediment was rapidly lost from the beaches. This was due to the presence of a sea wall which induced rapid beach lowering through wave reflection (see Chapter 3). This represents another aspect of the recharge problem which will be returned to later. All recharge schemes treat beach erosion; however, this is only the symptom, not the problem, which is often wave activity.

It is not always possible to obtain littoral sediment to recharge beaches and, in such cases, alternatives have to be sought from land-based sources and transported to the site, normally by road. Clearly, no land-based source is going to be as good as a littoral source because the sediment will not have been subject to marine sorting. However, where no suitable littoral sediment source is available, other alternatives need to be used. In Monte Circeo, Italy, the beach was recharged with crushed limestone. In this case, the rock was crushed to a grade coarser than the natural beach material, and provided a matrix in which the finer natural beach sand could become fixed. Following placement between 1980–3, accretion of the beach has occurred (Evangelista *et al.* 1992). Other examples where beaches have been recharged using crushed rock occur in Florida, Ohio, Monaco and Japan (Wiegel 1993). At Smather's Beach, Florida, an artificial beach was constructed using crushed limestone, while on the shores of Lake Erie in Ohio, the recharge material consisted of crushed reef limestone. At Monte Carlo, Monaco, an artificial beach was constructed using dolomite chippings, and in Japan, crushed rock and marble were used to construct a beach in Osaka Bay. In Port Philip Bay, Australia, quarried sand was used to recharge the beach. In this case, the sand was poorly sorted and so many fines were rapidly transported by currents. This reflects the problems of using non-littoral sediment: because the particle size range was not typical of the environment in which it was placed, large-scale post-recharge changes occurred which resulted in some major volume changes.

Finally, land-based recharge also involves the consideration of quarry waste disposal. In several of the examples discussed above, material was quarried in order to recharge beaches. Another aspect of quarrying is that it generates much waste material which, in some cases, has been disposed of at the coast in order to build beaches. From one point of view it could be argued that what is occurring here is not beach creation or beach recharge, but merely waste disposal, and is thus not part of coastal defence. However, it could be claimed that post-mining, the area concerned will either have an artificial beach where none existed before, or will have a much wider beach than pre-quarry. In Chañaral Bay, Chile, around 4.4 million cubic metres of copper mine tailings were discharged into the bay (Paskoff and Petiot 1990). Much of the sand fraction remained at the coast, leading to a seaward beach progradation of 900 m. The finer material was washed out of the sands and deposited offshore. Just along the coast, a beach has been created on a previous rocky shore using tailings from the same operation. In Rapid Bay, South Australia, coastal progradation in the order of 230 m occurred due to quarry waste disposal from a nearby lime-stone quarry (Bourman 1990). Large-scale retreat occurred post-closure in 1982, although a quasi-equilibrium beach has now formed. Further examples include the use of mine waste to recharge beaches in County Durham, UK (Hydraulics Research 1970), tailings from a steel works in south Australia (Bird 1996), quarry waste at Odesia on the Ukranian Black Sea (Shuisky and Schwartz 1979), and old ship ballast in Oriental Bay, New Zealand (Bird 1996).

Trickle charge

Trickle charge is the process whereby beaches are recharged slowly by placing material at a single or series of points on the beach and allowing waves and currents to move sediment onto the beach (Figure 7.2c). Typical areas for sediment emplacement are offshore or low water, where waves will wash sediment landward, or if rates of longshore sediment movement are strong, up drift, where longshore currents will move sediment along the coast. The advantage of this method is that the resulting profile will be natural because it is formed by wave and tidal currents.

Dumping sediment, either as a mound or a longshore parallel ridge, in the nearshore or low water area is often used to recharge mudflats, where rapid accumulation of sediment can be detrimental to infauna (see Chapter 9). The method can also be used to recharge sandy beaches where knowledge of onshore and offshore currents is available. In addition, the placing of such a mound offshore will benefit the coast in other ways, most notably because it will also act as a submerged offshore breakwater (see Chapter 6), and will thus increase wave attenuation. It is important here, however, as reference to Figure 2.1 will show, that any sediment placement needs to be above normal wave base, to enable waves to mobilise sediment and transport it landwards. This was exemplified at New River Inlet in North Carolina, USA, where two loads of sediment were deposited at different distances from the shore. The shallow deposit (depth <4 m) moved onto the beach, while that in deeper water showed no such movement and actually moved seawards (Schwartz and Musialawski 1977). Other examples of trickle charge include many beaches on the Pacific coast of the USA (Clayton 1989c), at Durban in South Africa (Zwamborn *et al.* 1970), Copacabana Beach, Brazil (Vera-Cruz 1972), Denmark (Mikkelson 1977), and the Gold Coast of Australia (Bird 1996).

There is some debate regarding the usefulness of placing sediment offshore, arising because of the apparent success of some schemes and the failure of others. Research into this problem brings us back to the statement made earlier in connection with where the recharge sediment is placed with respect to wave base. Hands and Allison (1991) compared the depth of dumping and the closure depth for a number of cases and found that where material is dumped above closure depth, sediments tended to disperse and move rapidly onto the beach. Where deposited below closure depth, material would move shorewards under certain conditions, but this was not always the case and considerable uncertainty occurred in respect of sediment mobility. Put in the context of Figure 2.1, the circular/elliptical motion of water molecules is responsible for moving sediments when in contact with the sea floor. When these orbital velocities are high enough to produce sediment movement, the deposited mound will move. Clearly, these velocities are not going to occur below wave base because water is not being 'moved' by the wave action. In some cases however, the placement of sediment offshore is not intended to feed the beach, but to add sediment to the nearshore zone, thus reducing the nearshore gradient. Many schemes in the USA (Charlier and de Meyer 2000) involve placing sediment some distance offshore for this reason, so that the sediment can be reworked and spread to build up the sub-tidal profile and push wave base seaward.

The method discussed so far involves placing sediment offshore for waves to move onto the beach or around the nearshore zone. Where a beach has strong longshore currents, it is also possible to use these currents to move sediment along the beach. In effect, trickle charging beaches by placing sediment up drift to slowly recharge beaches is the principle involved in sediment recycling, discussed above. However, the approach can also be used as a primary recharge method, where new sediment is used to prime the beach, rather than recycled material. In essence, sediment is piled up at the up-drift end and moves down drift as a sediment lobe under wave and longshore currents, gradually diminishing in size as sediment recharges the beaches. Grove *et al.* (1987) describe such a case at San Onofre, California. A lobe containing 200 000 cubic metres of sand was placed up-drift and this moved down the coast at around 2 metres per day, diminishing in size by half every 300 days. Other schemes have also shown successful beach regeneration using longshore currents. At Sylt, Germany, 770 000 cubic metres were placed up-drift of a recharge site, with over 60 per cent of the sediment moving along the beach, and on the Gold Coast in Australia, 2.5 million cubic metres of sediment have been used to recharge eroding beaches (Bird 1996).

Important considerations for successful beach recharge

With the range of techniques available and the apparent simplicity of the approach, it is tempting to think that beach recharge is the answer to all beach erosion problems. This is not the case because, as with any coastal defence scheme, we are introducing management into a natural environment. Although introducing sediment to a beach is only replacing a 'natural' material into a 'natural' environment, the process is more complicated than this. Any beach changes shape in relation to the wave and current conditions at different times of year. Any two different beaches are likely to have different grain-size populations, different beach angles, or different profiles. Hence, placing sediment onto the beach is just the first stage, as to be successful, this needs to become 'naturalised' to the beach environment. This naturalisation includes not only waves, tides and wind, but also other structures and usage.

Given this complexity, it is possible to see how recharge schemes may fail because they do not produce a new beach with any degree of permanence. Before any recharge scheme is begun, it is important for the coastal manager to understand one thing: why was the original beach eroding? This piece of information gives the manager the cause of the problem. If the cause is sediment starvation due to impairment of longshore drift, then the justification for adding sediment is obvious. If loss is due to some other factor, such as wave reflection from a sea wall, then the same will happen again unless the sea wall aspect of the problem is cured. In Port Phillip Bay, Australia, a sea wall was built to stabilise a cliff. Wave reflection from this wall, coupled with reduced sediment input due to cliff protection, caused the fronting beach to erode. In 1969, 5 000 cubic metres of sand was added to the beach but the following winter, it was removed by waves. In this case, the scheme had failed to take into account the role of the sea wall in beach loss, as well as the strength of the longshore drift in the area. Similar problems with sea walls occurred on the Klang Delta, Malaysia, where high wave reflection removed 12 000 lorry loads of recycled sediment. In North Carolina, the importance of beach profile was exemplified when recharge material was lost by longshore drift. Construction of a terminal groyne to prevent downdrift loss cured the problem but caused a realignment of the beach to become swash aligned, rather than drift aligned. In this case, the project again underestimated not only the significance of longshore drift, but also the angle of wave approach.

It is clear, therefore, that not only is it important to have sediment available, but also to understand local coastal processes and how these interact with structures already in place. Only with a full understanding can a successful recharge take place. In addition to these local conditions, it is also important to understand two other aspects of recharge, beach profile and sediment grain size.

Beach profile

Beach profile varies according to sediment grain size, wave conditions and storminess, and so estimating this in recharge is difficult. However, once the beach has adjusted to a profile which is in dynamic equilibrium with the ambient conditions, it will become 'stable', and thus present few further problems. In reality, a beach will never become stable in the strictest sense of the word because a profile which is stable in mild conditions is not stable in stormy conditions. However, as far as profile estimation is concerned in schemes which nourish the whole beach, the recharged profile should approximate that of the natural beach, allowing the waves to rework the recharged sediments into the natural concave profiles. A beach which is too steep will become reflective and encourage draw-down and possibly, if over-steep, erosion (Figure 7.3); one which is too shallow will necessitate excessive amounts of feed sediment which will increase cost. One method sometimes used to prevent too much offshore movement, or to reduce the volume of sediment necessary to build an equilibrium profile, is to construct an offshore breakwater and fill up the area to land-

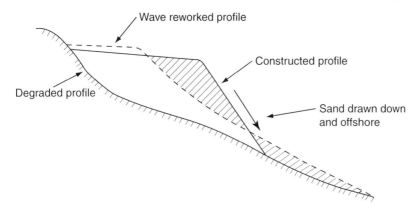

Figure 7.3 Natural adjustment of over-steep recharge beach by waves. Waves cut back the beach face, transferring sediment into the offshore area, thus producing a shallower profile.

ward, forming a perched beach. This process clearly links the techniques of beach fill and offshore breakwaters (see Figure 6.4 and Chapter 6)

Each of the techniques for sediment emplacement described above handle the problem of profile in a different way. Recharge of the whole beach involves an estimation of the natural profile and construction of the beach to this. Trickle charge allows the waves to construct its own profile, while recharge of the upper beach only forms an over-steep beach which can be drawn down by backwash.

In order to estimate the natural profile, the most logical approach is to study existing beaches and model the recharge profile on those. However, the fact that recharge is needed suggests that the existing beach is erosional, and therefore it will not exhibit the 'stable' profile required. Fortunately, it is possible to model the required profile, using variables covering wave height, periodicity and direction, water depth, and sediment properties. Dean (1991) discusses the modelling of profiles and presents a simple mathematical equation:

$$h(y) = Ay^{2/3}$$

where:

h is water depth at a seaward distance y
y is seaward distance from fixed point
A is scale parameter related to sediment characteristics.

Given that the elevation of high water is known, then this will allow water depths to be calculated along a shore normal profile. Although simplistic, such an approach has been shown to compare well with actual profiles, thus supporting its potential to provide reliable data to guide engineers in the reconstruction of degraded beaches (Dean 1991). In addition to this, once the new profile has been estimated, comparison with the existing profile will allow volumes to be calculated, and where it is most urgently needed. Pilkey (1990) argues against this approach, however, and claims the method of using mathematical models to predict volumes is dangerous and leads to failure in many examples. Instead, he favours the Dutch approach which estimates recharge volumes by calculating how much has been lost through erosion, and assumes that these will remain constant post-recharge. Post-recharge adjustments can be compensated for by adding a surcharge to the recharge volume (see sediment grain-size and sediment volume, below)

Given the above, engineers are in a position to estimate the shape of the final recharged profile. This will be achieved partly by directional sediment pumping, and partly by the remodelling of sediment once on the beach by land-based plant. As with any beach, however, this beach angle is dependant both on wave activity, and on sediment. In general terms, beaches get shallower as sediment grain size becomes smaller, hence shingle beaches are steep, sandy beaches are shallower, and muddy beaches (mudflats) can be almost horizontal.

Sediment grain size

In many examples of beach recharge, the importance of using grain-size compatible sediment, i.e. sediment which has the same grain size as the original beach sediment, is consistently underlined. Using basic principles of coastal processes (see Chapter 2), the amount of wave, wind and tidal energy along a coastline will govern the grade of sediment which accumulates on the beach (there are other factors which govern the availability of sediment, such as suitable sources, but these are not pertinent to the current discussion). In its simplest form, lower energy tide-dominated environments are typified by muddy and silty sediments, while higher energy wave-dominated environments are typified by sands and gravels. Where energy levels are particularly high, there may be no sediment at all, the coastal landform being a rocky marine planation surface. Clearly then, the first principle of recharge is to use the right grade of sediment for the depositional environment in question. In addition, engineers have traditionally taken this compatibility a stage further, and tried to match exactly the recharge material to the same grade of sand, shingle or mud, often necessitating extra cost in locating and obtaining the correct recharge material.

Engineers have consistently argued that, given that wave energy and longshore drift are the prime movers of coastal sediment, if sediment of the wrong grade is used, this can have implications for the functioning of the beach. For example, take a beach typified by sediment with a mean grain size of $x\,\mu m$. In sedimentological terms, this indicates that currents on that beach are too strong to allow finer material to settle, and inadequate to transport coarser material. Therefore, recharging this beach with a finer grade sediment will mean that currents can rapidly remove the material. This occurred at Ocean City, New Jersey where fine material was rapidly removed by wave activity (Pilkey and Clayton

1987). Similarly, Berg and Duane (1968) discuss an example of rapid sediment loss from the shore of Lake Erie when too much fine sediment was present in the recharge material, Carter (1988) reports on the loss of 500 000 cubic metres of fill at Rosslare, Ireland because it was too fine, and Eitner and Ragutzki (1994) demonstrate the same problem occurring on the Isle of Nordeney, Germany. It is important to realise, however, that some fine sediment loss will always occur, and is quite acceptable. This process is typically observed in the immediate post-recharge period where a rapid loss of sediment in the first year or so represents the sorting of the recharge material by currents, and the preferential removal of fine material. In contrast, however, if the recharge material used is too coarse, then currents will be insufficient to move the material around. In such cases, the recharge profile (see Figure 7.3) may become fixed, and the subsequent draw down and equilibrium profile not achieved.

Historically, however, engineers have used this grain-size relationship to increase the stability of beaches. Where a beach is eroding, it can be argued that this is because the energy levels of the currents are too great to allow the naturally occurring sediment to accrete on the beach. To increase stability, why not make the recharge sediment coarse enough to withstand the large-scale losses shown by the original? This idea has been discussed by Wiegel (1964) and Dean (1983), and the relationship of increased durability of beaches with a slightly increased grain-size demonstrated, albeit theoretically. Newman (1976) illustrates this point for sandy beaches, recommending a mean grain size for recharge material of 1.5 times that of the original in order to produce a stable beach.

Most of the literature which discusses examples of beach recharge, and text books looking at defence techniques, will underline the importance of grain-size matching. Increasingly through the 1990s, however, researchers have started to question this long-held belief. Following a study of recharge schemes on both the west and east coast of the USA, Leonard *et al.* (1990a and b) claim to have evidence which puts the grain-size theories into doubt. Their study has, albeit on a small population size, shown no relationship between beach durability and grain size of the recharge material, although the beaches used in this study did exhibit a narrow grain size range anyway. Pilkey (1990) and Dean (1991) both support the idea that the importance of grain-size may have been over estimated. Pilkey goes further to say that if further research upholds this change of belief, than considerable money has been wasted in the search for grain-size compatible recharge material.

It is clear that there is a major discrepancy here which is fundamental to the success or failure of the recharge process. Experience has shown that some schemes have succeeded and some have failed. However, because of the established belief that grain size is important, the majority of these schemes have been carried out with grain-size compatible sediment. Hence, we have a series of successes and failures using the traditional methods of grain-size compatible sediment. What has to be remembered, however, is that grain-size is just one of many variables which can lead to success or failure, and it is often difficult to separate one from another.

Sediment volume for use in recharge schemes

In essence, it appears simple to estimate how much sediment is needed to recharge an eroding beach. By taking the volume of the healthy beach, and subtracting the volume of the eroding beach, the difference will represent how much material has eroded. In some cases, however, there is no survey data on the healthy beach, and so this calculation becomes difficult. In the Netherlands (Pilkey 1990), the Dutch overcome this by estimating volume loss by erosion rates, while in most other countries, mathematical modelling is used to predict stable profiles and thus, predicted recharge volumes. There represents a fundamental difference in approach. Modelling will predict the final shape of the beach and allow volume estimation, and at what point on the profile this material is most needed. This approach is fine, except for the fact that the models used are not rigorous enough in that they do not accurately predict and model the functioning of the coastline. In order to predict accurately, it is important to know details of the wave and wind climate, storm frequency and severity, how the sediment moves and how the beaches have behaved historically. In reality, we just do not have this information available for the majority of the world's beaches; modelling may provide a guide, but it should not be used as the definitive method on which to base expensive defence schemes. By using the alternative approach (Verhagen 1992, 1996), it is possible to predict what volume of material has been lost, although even this assumes (not necessarily correctly) that the rates of loss have been constant.

Irrespective of the method used, there is a further factor which needs to be built in to the estimate. It has already been mentioned that the first year or so following beach recharge sees a rapid loss of material. This represents the adjustment of the profile (see Figure 7.3) and the removal of fine material. However, it also represents a loss from the original estimate of sediment volume and so, unless extra precautions are taken, the beach will be in deficit within a year of recharge. To overcome this, excess, or surcharge, volumes are used. Pilkey (1990) indicates a 20 per cent volume surcharge added to the beach, in addition to that predicted, to be necessary. Verhagen (1996) suggests 40 per cent, based on analysis of a series of beach recharge schemes. It should also be made clear, however, that even after the initial losses, further losses are going to occur in the medium- to longer-term; after all, the reason for recharge is to overcome a sediment loss problem, the causes of which are not being treated by beach recharge. The sediment surcharge mentioned here refers purely to the problem of initial sediment loss, not to that caused by continuing erosion and longshore movement. These issues are related to fill longevity and recharge periods, which are discussed below.

While the required sediment volume can be estimated, a further variable needs to be added to the equation. We know that beach feeding is only treating the symptoms, not the cause of the problem. Thus, given that the erosion of the beach is going to continue (unless other steps are taken to treat the cause as well as the symptoms), the Dutch method of estimating volumes using erosion rates can be adopted for longer-term benefit. If we know how much sediment on average is lost per year, then we can predict how long the recharge will last. (This assumes a constant rate of erosion into the future which may also be open to some debate). It is possible, therefore, in addition to the 20 or 40 per cent surcharge, to increase the volume still further in order to increase the longevity of the recharge material. However, before assuming that piling more and more sediment onto a beach will increase its longevity *ad infinitum*, there are the other issues of adequate profiling, cost, and sediment availability to be considered.

Other issues to consider for successful recharge

One prime condition for successful beach recharge is the complete understanding of how the beach in question is functioning with respect to waves, wind, currents, and human interaction. In reality, this is an optimistic requirement given our limited understanding of the complex interactions of coastal processes. Given this, the best that can be achieved is an understanding of the component parts of this holistic view. Issues relating to recharge volume, profile, and grain size, as discussed above, are the key factors involved in the successful recharge of beaches. In addition, there are other factors which need to be considered.

The length of the beach is important with respect to volume losses through longshore drift. Clearly, material moving longshore will remain in the recharge section for greater periods of time if that section is long. To a certain extent, this may be somewhat artificial but it does allow for more effective recharge for other reasons. When recharge occurs, the process of adding sediment to the beach pushes the high and low water marks seaward. This process will clearly affect coastal processes, and may even represent a physical barrier to longshore currents in a similar way to groynes (see Chapter 4). Figure 7.4 demonstrates the problem with small recharges, and the resultant unnatural coastal shape. The idea of using recharge is to maintain a natural coastline and to promote natural processes and, as such, it is important to maintain a realistic shoreline in order to facilitate the maintenance of natural processes. When an area is recharged, the offshore gradient also adjusts as sediment is dragged seawards. This means that the bathymetric contours alter and produce an area of shallow water fronting the recharge site (Figure 7.4a). This leads to wave refraction (see French 1997: 31) and the focusing of wave activity at the ends of the recharge site, increasing the potential for movement of sediment (Work and Rogers 1997). In Figure 7.4b, a series of small recharge schemes have been carried out, leading to a series of coastal perturbations and a crenulate coastline. From the ideas in Figure 7.4a, it is possible to see from the bathymetry how waves will be focused at intervals along this coastline, between the recharge sites. This is typically the arrangement found in popular tourist resorts, where individual hoteliers fund recharge fronting their hotels. In Figure 7.4c, one long scheme provides a frontage parallel to the original coast, leading to a more natural coastal shape.

a)

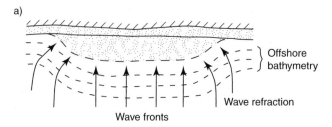

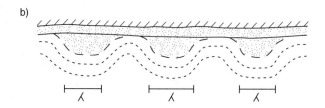

b)

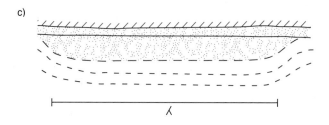

c)

Figure 7.4 Significance of length of recharged area to beach longevity
a) Typical wave refraction pattern, and offshore bathymetry associated with recharged beach
b) Short, individual recharges (λ) produce complex offshore bathymetry and wave refraction. This approach, typical of individually, privately funded recharges, has short longevity
c) Smoother offshore topography fronting single, large recharge scheme.

Komar (1998) looks at this issue from the shore evolution perspective. Using mathematical approaches, it is possible to predict the longevity of recharge schemes as a function of their length. The results of this work are that the prediction of recharge longevity is proportional to the square of the recharged length. Doubling the length of recharge will increase the retention time of sediment by a factor of 4, while tripling it will increase it by 9. Hence, from this simple relationship, it can be seen that short schemes will result in rapid sediment losses, requiring frequent top-up fills (see below), while longer schemes will have greater longevity. However, as if to emphasise the complex interaction of coastal processes, the same equation also predicts that a doubling of wave height reduces longevity to just 17.7 per cent. Leonard *et al.* (1990a) in their study from the East coast of the USA refute this length:longevity relationship, having found no statistical relationship between the two. However, in their study, the longest beaches have lasted the greatest period of time, and it appears that the study, being restricted to the one coastline, may be influenced by the interaction between a series of other variables which cannot be isolated.

If losses from fills is a problem, then one solution is to modify the recharge technique to provide a feeder site (trickle charge) up drift, to allow the constant replenishment of the recharged profile. This will further extend the site's longevity as it provides a mimic for continued longshore input.

Another factor which can affect recharge longevity is the sediment density, i.e. the number of cubic metres of recharge sediment per metre of beach. The density of sediment will be reflected in sorting and grain packing, and thus the effective porosity and the ability of the beach sediment to provide suitable routes for through-flow of water as opposed to surface flow (backwash). This will affect the ability of swash to drain into the beach sediment (see below). The greater degree of infiltration during swash, then the less energy there will be on the backwash, and the less draw down and losses offshore. Work by Stauble and Hoel (1986) has shown that the density of recharged beaches controls the amount of material remaining after a year on Florida beaches, and that a 60 per cent retention will occur if fill is placed at $150 \, \mathrm{m}^3 \mathrm{m}^{-1}$. Clearly, the density of the recharge is also a function of the grain-size distribution, with a wide grain-size range being capable of greater densities than a narrower one (density is higher because the pore spaces are filled with the finer sediment). In contrast, however, such a wide range of particle sizes will be subject to rapid initial losses due to the removal of fines.

Two examples in southern England (McFarland *et al.* 1994) illustrate the problems associated with sediment density. Following nourishment of shingle beaches, the sediment compacted to a density beyond that predicted, largely due to the quantity of fine grained material washing into the pore spaces between the shingle. This dramatically reduced beach permeability and swash infiltration, with the result that swash and backwash became predominantly surface flow leading to seaward movement of sediment. In addition, the combination of shingle and fine matrix effectively gave the beach a concrete-like appearance, which proved resistant to reprofiling by waves. As a result, the beach was not able to adjust to the ambient wave climate and maintained its steep recharge profile, leading to wave reflection and beach cliffing.

Given that the degree of compaction (density) and profile can affect longevity, the method used to place the sediment on the beach must also be important, as this will also affect these variables. Sediment delivered by mixing with water (rainbow pumping) will wash onto the beach and achieve a degree of sorting. Similar effects will occur with sediment delivered by trickle charging, with waves carrying out the sorting prior to emplacement. That dumped onto beaches directly from lorries to form an artificial profile, however, will not be subject to such sediment size grading and so will need to be reworked by waves after emplacement. This again can account for large post-depositional losses. An example from Sandy Hook, New Jersey (Nordstrom *et al.* 1979) suggested that sediment delivered from landward sources via truck lasted for a shorter period of time than that delivered by pumping from offshore. The reasons given were that the initial packing (grain-size distribution) affected through-flow and the ability of waves to rework the beach sediment.

Finally, in some cases, it is not just the sediment size that can be important, but also the sediment roundness. Angular particles are harder to erode than rounded ones, due to their ability to interlock (Bird 1996). Cunningham (1966) reports on an experiment from Florida where imported angular sand (oolitic sand from offshore) was compared with well rounded (local) beach sand. The results showed that the more angular oolitic sand was less mobile and, therefore, more likely to remain for longer periods of time. It should be noted, however, that although the need for lower mobility may be desirable, the material should not become so immobile that no profile adjustment can occur.

*   *   *

From the discussion above, there appears to be a range of factors which affect the longevity of a beach recharged with sediment. Clearly, all of these factors interlink to form an outcome which is difficult to predict, even more so considering the inclusion of other factors, such as variations in wave activity and storminess. What is important, however, is to realise that there are many problems in building beaches, a factor easy to overlook given the plethora of beach recharge examples. What should be remembered is that understanding the limitations and the problems can produce a beach with increased longevity, but any practitioner involved in beach recharge will tell you that no one has yet created a beach as good as a natural one!

What has been discussed above relates to the short- to medium-term. We have mentioned the fact that beach recharge is only treating the symptoms, not the cause. Because of this, any scheme will see a decrease in remaining material over a period of years. This means that if protection is to be maintained, the process will need to be repeated at intervals to 'top up' beach sediment levels. This issue is now discussed.

The issue of permanence

As with any soft coastal defence method which involves encouraging natural processes in the protection of the coastal environment, a scheme's outcome has to be considered in terms of success or failure. Typically, success is measured in terms of how long the scheme remains effective. No coastal defence scheme, of any type, is a permanent solution to a coastal problem because the cause of that problem is constantly being modified (sea level variation, wave activity or sediment supply). In addition, most coastal defence schemes treat the symptoms of the problem, not the cause.

As a result, the permanence of a scheme has to be considered relative to a time frame (how long is it predicted to last against how long it actually lasts).

To this end, there are several approaches to quantify the success of beach recharge schemes. Leonard *et al.* (1990a) claim that beach lifetime can be measured as the time taken for the replenished beach material to be removed. Other approaches do not define a 'lifetime' but use beach 'half-life' estimates, i.e. the time taken for recharge volume to drop by half. Both of these approaches are disputed by other authors, who favour the use of critical beach level to mark the lifetime of the scheme (Figure 7.5). It is important to remember that initial losses can be quite rapid from many schemes as the new profile adjusts to the wave climate of the area. This process has often led to public criticism of recharge because it is easy to perceive the initial loss as a failure of the beach to retain sediment. In the longer term, following this initial phase, losses will continue as natural coastal processes dominate, and longshore and offshore sediment movements occur. It is the rate of these losses which primarily determine the longevity of a scheme.

In the case of beach recharge, all schemes will need to be repeated at intervals because of the continual movement of sediment from a beach due to longshore and shore normal currents. Figure 7.5 explains this requirement for repeats. In order to remain effective as coastal protection (i.e. a certain level of wave attenuation), the beach needs to remain at a certain minimum volume. This is often referred to as the critical beach level (i.e. the minimum beach level which can effectively provide adequate coastal protection). Because of sea level rise, this critical level will vary in time, the beach needing to grow seawards in order to maintain the same level of wave attenuation. Hence, in Figure 7.5, the line representing critical beach level slopes upwards, with each recharge event

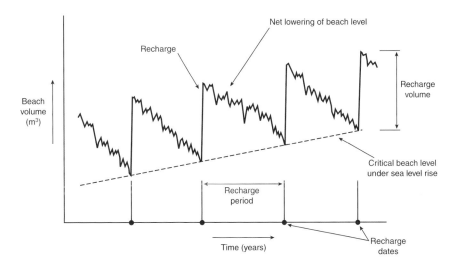

Figure 7.5 Recharge period as function of critical beach volume (CBV) and rising sea level. The upward trend in CBV is caused by the need to maintain effective beach elevation in face of sea level rise.

needing to increase the beach level of the last in order to achieve the same degree of protection.

Given that the rate of sediment loss can be measured, it is possible, therefore, to determine the frequency of recharge in order to maintain adequate beach levels. This can then become a predictive tool for future coastal defence planning. Cooper (1998) has assessed the behaviour of beach replenishment schemes in Bournemouth, and their rate of volume loss. By linking this data to predictive models of Führböter (1991) and Verhagen (1996), it has been possible to predict the rate of material loss with time. Table 7.3 details the repeats of the Bournemouth scheme, and the prediction of the next repeat (see also Box 7.1). As can also be seen from Table 7.3, each repeat involves larger volumes of sediment in order to compensate for sea level rise and to maintain adequate protection. In addition, it can be seen that repeats are necessary approximately every 13 years. In other examples, a period of 13 years is very long. Leonard *et al.* (1990a and b) state that on the east coast of the USA, beach recharge schemes (with a few exceptions) have a life expectancy of less than 5 years, a statement supported by Dixon and Pilkey (1991) for the US Gulf of Mexico shoreline. In Westerland, Germany, the beach was nourished with 1 million cubic metres of sand in 1972, with a further million added (albeit to a wider frontage) just 6 years later, and nearly 2 million cubic meters over an even wider frontage in 1985 (Verhagen 1996). Clearly, in this example we are not comparing exactly the same area but the need for such large quantities of sediment is important. The reason for the difference in time periods between recharge again relates to the dynamics of the beach in question. Different exposure to wave activity, different grain size and different degrees of storm activity will all affect the rate at which sediment is removed from recharge schemes.

Several factors are important here. First, with such large quantities of sediment involved (see also Table 7.1), and with such a short repeat time in many cases, the issue of how to supply such large volumes of sediment becomes important. This is discussed in more detail in the following section. Second is the issue of cost. One argument for beach recharge, apart from the 'soft' versus 'hard' debate, is that it is cheap. While each recharge episode may be cheaper than building hard defences, when this has to be repeated every 5 years or so, costs start to escalate. Engineers often say that beach recharge is like throwing money in the sea, so the question of value for money is pertinent here. Table

Table 7.3 History of Bournemouth recharge scheme. Note gradual increase in recharge volume required (see also Box 7.1).

Date	Volume (m^3)
1970	84 500
1974–5	654 000
1988–90	999 000
2002–3	?

7.4 details some examples of the cost of recharge (all costs converted into US dollars). In some cases, this expenditure occurs intermittently in line with the period of recharge, while in others, such as the German North Sea coast, the costs of recharge are in the order of $100 million a year (Kelletat 1992). This is clearly a huge financial burden on the coastal defence authority. In the 1990s, the US government became concerned about its expenditure on beach recharge (Finkl 1996). In particular, the concern centred around the fact that Florida alone was commanding around a third ($10–$20 million) of the total coastal defence budget (actual cost of defending Florida's beaches equates to $20–$40 million a^{-1}). As a result of this, legislation was proposed to cut this central funding, leaving Florida's authorities with a huge financial problem, and the potentially serious implications concerning the welfare of its beaches, both from a coastal defence point of view, and for tourism. While it is possible to see the case for not spending such large quantities of money, especially considering that the lifetime of recharged beaches in this area is typically no longer than 5 years on average (Leonard *et al.* 1990a and b), it is also important to look at the other side of the argument, i.e. the benefits of spending this money on carrying out the recharge, and the impacts of not recharging. As far as tourism alone is concerned, the income generated as a direct result of Florida's beaches is around $10.2 billion. The industry supports 760 000 people, of whom around 380 000 are fully employed in beach-related tourist jobs (Finkl 1996). Immedi-

Table 7.4 Cost of beach recharge schemes. All prices converted to US dollars for comparison purposes (For full references, see References at the back of the book).

Location	Cost (US$)	Time period	Reference
Australia			
Port Philip Bay	2,930,300	1975–87	Bird (1990)
Denmark			
North Sea coast	c. 500,000	per year	Møller (1990)
Germany			
North Sea coast	3,300,000,000	1955–92	Kelletat (1992)
North Sea coast	100,000,000	per year	Kelletat (1992)
UK			
Bournemouth	2,088,000	1988–9	May (1990)
Seaford	21,200,000	1987	NRA (undated)
USA			
Ediz Hook, Washington	5,600,000	1977–8	Galster and Schwartz (1990)
	970,000	1985	Galster and Schwartz (1990)
Gulf of Mexico	36,403,000	1964–87	Dixon and Pilkey (1991)
Key West, Florida	100,000	1989	Wiegel (1993)
Lake Erie, Ohio	2,100,000	1990–1	Wiegel (1993)
Oceanside, California	3,081,172	1981	Bagley and Whitson (1982)

ately then, we have not only the direct income from beaches which is an order of magnitude higher than the expenditure necessary to maintain them, but the possibility of several hundred thousand people falling unemployed if beaches fail to attract the visitors. These costs relate purely to tourism; the cost of increased coastal protection and security to coastal residents is not calculated but would produce additional ways of offsetting the costs of the recharge programme. Based on these figures, there is a clear case for maintaining capital funding.

This problem of finance is not just one experienced by Florida. Coastal authorities need to assess fully the financial side of beach recharge, both with respect to scheme costs, but also benefits (both protection and tourist) which arise out of it. For example, Hillyer *et al.* (1997) calculate considerable benefits throughout the USA from beach recharge, not just recreational, but also from increased protection. On the negative side, additional concerns which can arise from this increased security is that increased protection (as from any defence scheme) encourages increased coastal development. This in turn further increases the value of the hinterland and, thus, the importance of the coastal defence. This clearly links to factors beyond the purely coastal defence argument, and includes legislation and coastal planning, issues discussed further in Chapter 12.

What this discussion does tell us is that, while beach recharge needs repeating, sometimes on a regular basis, its costs are not only those involved in carrying out the actual project. In some cases, such as at Bournemouth, some costs can be offset with other activities, such as the use of dredging to supply recharge sediment. To obtain a fair balance on which to judge cost effectiveness, we also need to include financial income from tourism and from increased coastal protection.

Where is all the recharge material coming from?

We have seen throughout this chapter how common beach feeding is, and is becoming, as a means of coastal defence. Table 7.1 indicates how important the method is throughout the world and as its popularity spreads and our knowledge of how beach sediments behave increases, the use of the technique will increase. In addition, coastal authorities throughout the world are being encouraged to adopt the 'soft approach', meaning the use of beach feeding (and other soft methods) will increase.

In the UK, Child (1996) highlights the publication of government documents which encourage soft defence usage and also the use of beach feeding (see Pethick and Burd 1993, MAFF *et al.* 1995). Similarly, in the Netherlands, the use of beach recharge is now the main form of coastal defence and Child argues that in the near future, the technique is likely to become the most common method of protecting coasts in many countries. This poses a serious question for coastal authorities: *where is all the recharge sediment going to come from?*

In the UK, around 7 million cubic metres per year are dredged from licensed sites around the coast for the building industry, a figure predicted to increase to

14 million in the first decade of the 21st century. In addition, the UK also uses 2 million cubic metres for beach recharge. World-wide, Bird (1985) states that 70 per cent of the world's sandy beaches are eroding and faced with this loss, coastal communities apply pressure for coastal defence. In areas where tourism is important, there are additional pressures for increased amenity and a new beach can serve both these demands. For many of these coastal problems, beach recharge has become the preferred solution because of its 'naturalness' and practicality in erosion solving. This demand for recharge and its projected increase is likely to have serious implications. Although it is true that offshore deposits naturally accrete, it is unlikely that this will occur at a rate to match removals for human use. As a result, offshore sediment volumes are going to decline. We need to ask whether the supply of sediment for beach recharge from offshore is ultimately sustainable.

If the answer to this is in the negative, then it becomes vital to find and utilise alternative sources of sediment. We have already stated how it is preferable to use littoral sediment for recharge, but it is possible that in the future, these sources will become more costly to use as the economics of supply and demand take over. In the USA, Milton *et al.* (1997) report how dwindling offshore sources of sand have led to increased pressure to find alternative sources of recharge sediment. The use of recycled material partly overcomes the problem of finding new supplies because it constantly reuses what has already been added to the beach, albeit with top-ups where needed. This may be satisfactory in relatively small-scale operations where beach sediment transport is primarily longshore, but is not useful where material is lost to offshore stores. It is a traditional misunderstanding that much offshore movement of sediment goes to form berms — deposits which can be re-acquired to feed beaches at a later date. Although this may happen in some cases, it is more common to find material thinly spread over the subtidal sea floor (Pilkey 1990). There is a further consideration here. If we are using beach recharge to increase beach stability, then why use the material which has been removed from the beach if it is not stable under those beach conditions? The material washed offshore is not 'wanted' by the beach, and so more stable (coarser?) sediment is required.

It is quite possible that engineers will increasingly need to look elsewhere for supplies. We have already discussed some of these earlier, but it is pertinent to restate them in this context. Bird (1996) argues that almost any durable material of suitable grain size can be used to recharge beaches, subject to it being free from pollutants and contaminating material. One relevant issue in consideration of cost is that the source of sediment should be near to its destination; hence, it is not practical to have a few large source areas per country, many small areas around the coast are important. Typical source areas could include:

- quarries (landward transportation);
- reworking quarry/mine waste;

- reworking of 'fossilised' beach deposits;
- reworking from down drift (recycling);
- port and harbour dredging;
- lagoons/barrier island coasts;
- fluvial supplies;
- (offshore).

Each of these sources could potentially help alleviate any shortfall from off-shore sources. As has been discussed earlier, each has related environmental issues which need to be assessed in the overall development of the project. Despite this, however, each method has been used in a variety of scenarios (Table 7.5).

Table 7.5 Examples of beach recharge schemes using sediment supplied from a range of different sources.

Source	Examples	Problems/benefits
Quarries	Monterey Bay, USA	High cost of transport
	Ediz Hook, USA	Reliance on roads
	Michigan, Great Lakes	Competing demands (construction)
	Sidmouth, UK	
	Port Phillip Bay, Australia	
Reworking quarry/ mine waste	Rapid Bay, Australia	Metal contamination
	Chañaral Bay, Chile	Roundness/size
	Torpoint, Plymouth	Transport
	County Durham, UK	
Reworking of 'fossilised' beach deposits	Dungeness, UK	
Reworking from down drift (recycling)	Aberystwyth, Wales	Saves on new sources
	Shoreham, UK	Cheap
	Hvidesande, Denmark	
	Seaford, UK	
	Rye, UK	
Port and harbour dredging	Bournemouth	Uses 'waste' material
	Rockaway Beach, New York	Pollution problems
	Sandy Hook, New Jersey	
	Copacabana, Rio de Janeiro	
	Auckland, New Zealand	
Lagoons/barrier island coasts	Atlantic City, New Jersey	Pollution problems
	Hel Spit, Poland	Habitat stability/ damage
	Lido di Jesolo, Italy	Grain size
	Cairns, Australia	
Fluvial supplies	Black Sea, Georgia	Suitability of source
	Gold Coast, Australia	Delta decay

Beach recharge and tourism

No coastal resort can survive without its beach. Tourist authorities will go to great lengths to protect their prime assets, whether it is by beach recharge or the installation of hard defence structures. Ideally, coastal usage should be flexible enough to allow the relocation of users if the depositional pattern along a coastline changes, i.e. a beach erodes. While such changes in coastal deposition and accretion patterns do happen, a new beach may well start to accrete elsewhere, but the infrastructure of the resort is fixed to its location. As a result, coastal management has to allow for such situations and has to recognise the fact that in some circumstances it is necessary to force a depositional environment to stay where it is. The use of beach feeding in these cases has many advantages, and serves to underline the link between coastal defence and tourist activity.

However, the criteria for recharge tend to change from that used in coastal defence scenarios. We have previously discussed the idea of critical beach levels as marking the point where beaches are not at a level to provide adequate defence and protection. In the case of coastal tourism, we can define critical beach level as the level at which a beach fails to provide adequate tourist amenity. It is clear that this level may be higher than that needed purely for defence purposes, thus initiating recharge at a point which is not justified on coastal defence merits alone.

Because of the ability to create beaches full of sediment the approach has been taken up by some coastal tourism authorities and adapted to support tourism. Along some coasts, where climate and inland attractions encourage tourism, the lack of beaches may be seen as a negative aspect of the region. This problem can be overcome by creating beaches from scratch. The reason for carrying out such creation is purely financial; an incentive for people to visit an area. From a coastal management point of view, placing sand on a beach where no beach deposits formerly existed makes no sense at all. In Chapter 2, we saw how coastal processes, such as wave and tidal currents, govern the type of sediment which accumulates along the coast. If there is no sediment present, this suggests one of two things: either there is no sediment available (such as may happen along some tropical coastlines where calcium carbonate precipitation predominates) or, more commonly, wave activity is too great to allow sediment to accumulate. Despite this logic, there are examples of beaches created primarily for tourist use.

The French Riviera (i.e. from Cannes eastwards to the Italian border) (Anthony 1994) boasts 21.5 per cent of its length as being artificial. Here beaches have been created where they did not originally exist, or where they were insufficient for tourist usage. Between 1974 and 1994, fourteen beaches were created along the 28 km frontage, largely using sand or gravel from inland quarries. Along the Portuguese coast in 1970, around 1.1 million cubic metres of sand were pumped onto the marine plantation surface fronting cliffs at Praia da Rocha. In 1983, this was extended westwards by the addition of a further

350 thousand cubic metres (Psuty and Moreira 1990). This latter recharge had been almost completely lost by the late 1980s because of exposure to wave activity. Rapid coastal land claim for commercial reasons led to the shores of Tokyo Bay losing many of its beaches. Since the early 1970s, nine artificial beaches totalling 13 km in length have been created at a cost of 44.6 billion yen (Koike 1990). Collectively, these beaches now attract around 3.7 million people a year. A series of artificial beaches have also been created along the shorefront of Kuwait City, again largely aimed at the local population, but also at tourism (Al-Obaid and Al-Sarawi 1995).

When considering recharge primarily for tourism, the need for a wide inter-tidal profile, essential for wave attenuation in the coastal protection scenario, is not as great. Because of this, the development of perched beaches has often been used to provide not only a beach, but also a stretch of coastline sheltered from wave activity.

Impacts associated with beach recharge

Beach recharge is perhaps one of the simplest methods of protecting the coast because it involves replacing what the waves have previously removed. Despite this, however, the technique still acts against natural processes because it involves adding material to an environment in which it would not naturally be. The implications for the natural environment of piling large volumes of sediment onto a beach are varied. On the one hand, these impacts are desirable, in that reduction in wave energy and increased sediment availability in the sediment budget are the main objectives of carrying out such schemes. On the other hand, beach recharge can, if not carried out sympathetically, have detrimental impacts on the coastline.

Impacts may be divided into two groups: those caused during the pumping and beach build-up, and those which arise due to the new beach profile. Although it can be argued that any defence scheme will have construction impacts, we consider them separately here because they may actually be pertinent to the success or failure of the scheme. Excessive compaction can lead to reduced beach drainage and increased run-off, for example, thus accelerating beach erosion.

Impacts during the recharge process

When sand is pumped onto beaches, the effects of the excess water can be important, as run-off from the beach can cause scouring in the intertidal zone. We have said that 10 per cent water is added to allow pumping. This excess water has to drain and, if it becomes channelled, can erode and provide a focus for subsequent wave erosion. As this water returns to the sea, it carries with it fine sediments, which will settle out in areas of low wave energy. This fine sediment has the potential to alter the ecology and the profile of the depositional area by swamping communities in sediment, increasing turbidity, or modifying the grain-size regime if deposited in a mud-dominated region. This will have secondary impacts because, as we have seen throughout this chapter, the introduction of a coarse sediment to an environment which cannot move it and shape it into a stable profile can lead to further problems. In addition, the mobilising of fines during dewatering can lead to the new beach material becoming net coarser than required and thus relatively immobile with respect to wave refashioning. Furthermore, if this material is washed into pore spaces, through-flow is impaired and surface backwash increased, thereby increasing surface erosion. Such a problem occurred in shingle recharge in southern England (McFarland *et al.* 1994, Whitcombe 1996), where pore spaces became filled with fine sediment, leading to cliffing of the profile. Other methods of recharge utilise heavy machinery, either to deliver sediment from land-based sources, or to reshape material pumped from offshore. The use of such heavy machinery can lead to compaction of sediment and reduction in pore space. As we have seen, this will decrease water movement through the beach, and increase surface backwash with corresponding increased sediment removal and gulleying. One solution to this is to utilise beach drainage methods to lower the beach water table, thus allowing increased through-flow (see below). However, this will not be effective if water is prevented from infiltration due to surface compaction or pore space filling with finer sediments.

It is also important to consider the methods by which recharge sediment is obtained, and the impacts which this can have on the source area. We have already discussed different sources and some of the associated impacts are readily obvious. Land-based sources will require transportation and heavy plant to obtain the sediment. While the long distance transport from offshore sources is unlikely to have impacts above that normally expected from any shipping movement, dredging pipelines may also cause damage to sea bed fauna or to coral reefs, unless properly monitored and precautions put in place. The primary impacts relating to the source, however, will come from dredging itself. First, from a coastal process point of view, it is important to determine whether the source area is acting as a natural breakwater and thus protecting large sections of the coast from wave activity. The removal of an offshore sediment bank or barrier which protected the coast from waves can lead to increased wave attack at the coast and actually make the original erosion problem worse, or initiate new problems elsewhere. Second, it is also important that dredging does not deepen the nearshore area too much, as this can lead to increased wave exposure, or large-scale profile adjustment.

The actual dredging process increases suspended sediment and turbidity which may be significant in feeding or breeding areas. This can be of critical importance for filter feeders, such as shell fish (Adriaanse and Choosen 1991). Similarly, when this material settles, it may coat vulnerable sea grass beds, reducing their capacity to photosynthesise. A recharge scheme near Marseilles, France used sand from the nearshore zone, the dredging of which led to the destruction of adjacent vegetation (Rouch and Bellessort 1990); in Poland, the recharging of Hel Spit led to vegetation damage in the adjacent lagoon which was being used to source the sediment (Basinski 1994). The area of these impacts can be quite large. Studies by The Rijkswaterstaat (1979) indicated that suspended sediment may impact on an area of several hundred metres radius from the point of dredging.

Because dredging involves the removal of substrate, it will inevitably lead to the loss of any benthic or infauna which occurs in the area being dredged. This will be transferred to the new beach site, although being submarine fauna, it is unlikely to survive. This may sound like common sense, but in some areas, sea bed colonisation can involve communities which are important, either because of their commercial value or rarity. Clearly, it is important to investigate the source area thoroughly before sediment is obtained from it. Adriannse and Coosen (1991) discuss the implications for source area fauna, and state that although recovery may occur, albeit after some considerable time in some examples, it is unlikely to reflect the original population in either size or diversity.

Finally, with the increased pressure to utilise harbour dredgings for recharge projects, it is important to monitor the pollutant levels and toxicity of the source material. Heavily polluted source material may damage environments in the intertidal or nearshore zone. Reilly and Bellis (1983) report on a scheme at Bogue Banks, North Carolina which included material contaminated with high levels of hydrogen sulphide. This led to the modification of the habitat and large-scale mortality in macro-invertebrates. Furthermore, the potential toxicity of dredged sediments should not be underestimated. Sediments dredged from New Bedford Harbour, USA contained excessive levels of zinc, copper, and lead, while PCB levels were the highest ever recorded in an estuary (Fraser 1993). (The levels here were so high that some of the dredged sediments had to be processed according to hazardous waste regulations.)

Impacts of recharge on the coastal environment

Once in place, the recharged beach may also undergo a series of changes with an impact on adjacent areas of the coastline. We have already discussed the post-depositional profile adjustments which typically occur following recharge, leading to the draw down of sediment and transfer offshore. As with the dredging process, this can lead to increased turbidity and the potential to blanket benthic fauna and flora. This problem can be exacerbated if the wrong grain size is used, as this can lead to the rapid removal and transfer of fine sediments.

In Chapter 9, we argue that in mud-dominated environments, the increasing natural sedimentation is a better method than dumping large quantities of sediment onto the foreshore, because it allows the gradual accumulation of sediments and reduces the likelihood of burying infauna. It is also possible for the same problems of infaunal burial to occur in sand-dominated environments. All infauna occupy a vertical position in the sediment column suitable to their lifestyle. Natural erosion and accretion will cause the thickness of sediment above these organisms to vary, but given the time scale involved, it is possible for them to adjust their position and maintain a relatively consistent depth of burial. If sediment deposition is too rapid, however, such as when a metre or so of sediment is added to a beach quickly during recharge operations, infauna cannot move within the profile rapidly enough, and so become buried. Adriaanse and Coosen (1991) indicate that any volume in excess of 0.5 m will cause most of the infauna to die because of this inability for rapid upward movement.

Much is yet to be learned about the impacts on infauna, although some work at Myrtle Beach, South Carolina did reveal post-recharge population declines in species, followed by a subsequent recovery (Baca and Lankford 1988). The rate of recovery is also variable (Pullen and Naqvi 1983) and depends on factors such as sediment type and the time of year at which recharge occurs. Pullen and Naqvi (1983) suggest that the critical factors involved in the survival of infauna include not only depth of burial, but also the length of burial time, time of year, grain size, sediment type, and other specific requirements, including moisture levels and redox potential. One particular group of organisms which contains species particularly susceptible to recharge, however, is coral. Some species of coral polyps are particularly vulnerable to changes in turbidity and have low tolerances of suspended sediment loadings, while others are more tolerant, such as those which are found in the Gulf of Thailand which experiences highly turbid conditions during the monsoon season. Beaches fringed by sediment-sensitive reefs need to be recharged with great care as any rapid offshore movement of sediment could cause mortality and reef decay, leading to increased wave exposure and, potentially, enhanced coastal erosion. Marsh and Tubeville (1981) report damage to coral reefs in south-eastern Florida, both from physical damage during dredging, and also from being swamped in sediment.

Other fauna-related studies have been carried out in respect of nesting turtles. Because some species are relatively rare, and often subject to special conservation measures, it is important, if recharging sensitive nesting sites, to determine any impacts which might impair breeding success. In Florida, the traditional sources of offshore sand for recharge are diminishing, with the result that new sources of sediment are needed. This has raised concern as to whether these new sources are suitable for the successful breeding of loggerhead turtles (Milton *et al.* 1997, Davis *et al.* 1999). The alternative used in some cases has been aragonite sand; although grain-size is similar between this and unrecharged beaches, the grain shape, packing and temperature of the sediment is different. Temperature is consistently lower in aragonite fills, leading to concern that this may have an impact on breeding success, while the difference in grain shape leads to different packing densities (Milton *et al.* 1997) which may affect nest building. Counter to these ideas, Davis *et al.* (1999) claim that difference in packing has no measurable impact on the turtles, and the benefit of a wide beach outweighs any negative impacts of the material on these species.

To a certain extent, many of the problems of infauna and birds/turtles which use the beach for nesting can be overcome with planning, and the avoidance of certain times of the year when nesting is occurring and infaunal bioproductivity high (Domurat 1987).

In other recharge situations, the beach is only one part of the eroding coastal system. Where sand dunes are important, they and their unique ecology need to be maintained via a sediment supply from the recharged beach, and so any supply of new material also needs to conform to aeolian sediment transportation parameters, as well as to the beach environment. The removal of fines by dewatering or wave reworking following recharge can lead to a deficit of certain size classes on the beach, and so can reduce the effectiveness of aeolian transfer, thus leading to dune starvation. The issues relating to the management of dunes will be discussed in Chapter 8.

Benefits of a recharged beach

Despite the potential problems detailed above, beach feeding is the most widely used method for soft coastal protection, and is also seen as a suitable way of disposing of unwanted dredgings. Beach feeding is also a way to protect the coastline while maintaining coastal processes, as well as a degree of 'naturalness', thus increasing amenity value. The technique could also lead to the creation of new habitats such as saline lagoons, thus further benefiting the biodiversity of the area. In addition to this, by placing sediment on a beach, there is no interruption to the longshore movement of sediment, thus overcoming many of the problems encountered with groynes and jetties (Chapter 4) (unless the scheme sits too far seaward and interferes with longshore sediment movement in the swash zone).

Clearly then, beach recharge provides the coastal manager with an ideal way by which to maintain coastal processes and coastal integrity, while providing for defence of the hinterland. As far as amenity value is concerned, the earlier discussion of the Florida coastline and its income generating potential clearly illustrates the ability of the method to act as a stimulus for tourist development and increased income and effectively to pay for itself. It has been adopted in many resort areas to provide what many regard as a basic necessity for successful tourism — a beach.

When it comes to discussing benefits, therefore, we actually have two major areas of consideration: first, that the physical environment is managed in such a way that natural processes of sediment movement are maintained, and second, that the socio-economic environment benefits from increased population security and increased means of income generation. Caution is necessary here, however: as with many techniques, the perceived increased security can lead to increased coastal development. With regards to beach feeding, however, this problem is not just one of 'increased defence equals more security for housing', but that the creation of a healthy beach provides not just this increased defence, but also the resources to develop something much greater — an income stream for the local population. This means that beach recharge schemes are critical for two reasons (defence and local economic stability), not just the one typically attributed to defences.

Beach feeding and sea level rise

Beach nourishment is the most frequently adopted approach to treating eroding beaches because it not only overcomes the problems of decreased amenity potential, but also the opportunity to increase wave attenuation provision. Such trends are likely to become more common under continuous sea level rise. This approach is fine, but is hardly a long-term solution. Once water depths (and wave base) increase by any significant amount, the longevity of nourished sediment will be reduced so much that its cost-effectiveness is going to be brought into question. This brings the additional concern about the source of all this material more to the forefront of any planning discussion. Really, the use of beach nourishment is an emergency stop-gap. Longer time spans are going to need a radical rethink if sea levels reach the heights predicted. This then starts us thinking about whether there are better ways of fighting the problem. There are some which involve pushing the hinterland limit landwards, but these are really only effective on a local to regional scale. We will discuss these ideas in subsequent chapters.

In the short to medium term, beach feeding can be used to offset some of the problems of sea level rise, such as increased wave activity, by building beaches seaward. As has been seen in Figure 7.5, the critical beach level is the crucial point here and the beach can be maintained as long as the cost of maintaining this limit remains tenable. Linked to this is the availability of suitable quantities of material. Hence, as with any erosion problem, beach loss due to sea level rise can be compensated for through building the beach higher. This method is widely used already but, given the potential for loss of beaches through sea level rise, it may become a technique which will expand rapidly in popularity. Nicholls (1998) looked at this issue from the point of view of accelerated sea level rise on beach volumes, and from this was able to indicate likely recharge volumes necessary to maintain an amenity beach under various sea level rise scenarios. Similarly, Ocean City in Maryland, USA, has changed its coastal defence strategy from one of hard defences (sea walls and groynes) to one of beach nourishment because of the threats posed by rising sea levels (Titus 1986). Given that over time the required volumes of sediment will increase, and coupled with the problems of sourcing, there will come a point in all beach feeding cases experiencing sea level rise when it becomes necessary either to change tactics and allow landward migration (according to the Bruun rule), or to combine beach feeding with other methods, such as offshore breakwaters to produce a perched beach. While this would provide increased protection and wave attenuation, it would counter the fundamental issues associated with sea level rise, i.e. the need of the coastline to retreat inland to maintain stability. Although a perched beach would both protect amenity interests and also increase natural dissipative abilities of the beach, it will not form an equilibrium profile and aid coastal stability, and would only serve to promote the artificiality of the coastline.

The problem of sediment dynamics and longevity is more complex. We know that wave and tidal currents move sediment onto, off, and along the coast. We also know that these currents are controlled by wave energy, water depth, and bottom morphology. Rising sea levels will lead to increased water depth which, due to changes in wave base and energy levels, will lead to changes in bottom morphology, and hence, through alterations in current direction, may lead to changes in sediment dynamics to beaches. In a climate where beaches are often experiencing enhanced sediment losses anyway, any factor which promotes additional loss could be a problem. The gravel barrier at Porlock, Somerset, UK is a good example of how varying rates of sea level rise can be shown to have controlled the rate of sediment supply. In this case, as the rate of rise increased, sediment supply increased, and as the rate fell, supply was reduced to the point where net erosion and ridge breakdown was occurring (Jennings *et al.* 1998). Clearly in this scenario, increased sea level rise caused an increased input of sediment which would, in the case of beaches, go some way to offset the losses and the need for increased feeding. However, the opposite may also apply and rises in sea level may also reduce the amount of sediment delivered to beaches, for reasons such as current modification previously mentioned.

Other forms of soft management of beaches

In considering beach feeding, we have touched on two areas of associated concern. First, where recharge involved land-based sources or pumping, it is necessary to profile the beach artificially. This method is also used in cases where no sediment addition has occurred, but where beaches have changed shape due to storms. In such cases, stability is achieved by using a bulldozer to modify the storm profile and to reshape the beach to a more stable angle. Second, where excessive backwash leads to beach erosion, sediment can be lost offshore. Such problems of excessive beach draw down can be dealt with by increasing beach drainage or dewatering schemes, where the water table is permanently lowered and through-flow, rather than surface flow (backwash) is promoted. These two issues can also be addressed by beach management methods.

Beach re-profiling

We have seen many times how waves move sediment around and modify beach profiles, and how this process can lead to large-scale sediment loss from the beach due to draw down. Beach renourishment treats this problem by adding sediment to beaches to increase their volume as a compensatory measure. In some situations actual loss from a beach is small, with sediment being moved to other parts of the profile (such as draw down following a major storm event). In such cases, it is often sufficient just to redistribute the existing beach sediment, and to push it back up the beach to form a mound or berm-type structure, with the intention of protecting hinterland properties. This process is known as beach re-profiling or beach scraping.

The technique itself is relatively simple in concept, purely requiring a bull-dozer to push sediment up the beach. Typically, beach ridges form at the top of the up-rush zone and it is these areas which source the material scraped. It is important to carry this out when these ridge structures are well developed, nor-mally around spring and summer (Bruun 1983). Despite this apparent simplic-ity, however, little research has been undertaken into its scientific validity, and the technique is therefore treated with scepticism in some circles (McNinch and Wells 1992). Bruun (1983) cites arguments that the technique is harmful to adjacent beaches and can interrupt longshore drift. However, providing the method is kept within certain guidelines, these arguments are not justified (Bruun 1983). For example, only a top skim would be taken from the beach, normally between 0.2 and 0.5 m depth, depending on sediment grain size, and if undertaken in a technically responsible way, can benefit adjacent beaches rather than harm them.

The research which has been carried out appears to indicate that the method of obtaining the sediment will play an important role in determining the success of the scheme. At Topsail Beach, USA, a front-end loader was used to pile sedi-ment up to reinforce natural dunes and the use of a control site nearby allowed effectiveness to be determined. In this study, it was found that scraped and filled sections lost smaller volumes of sand, and retreated less than control sites. In addition, two major storms affected the area, with interesting responses from the beach. In one, erosion of the scraped beach was significantly less than the control, while in the other, it was considerably more. Clearly this provides contradictory information but the difference may well be due to the different prevailing wind directions. We have seen how beaches orientate themselves to the prevailing wind, as evidenced by the swash alignment of beaches in groyne compartments, for example; in this case, it is likely that the stability of the beaches varied with the direction of oncoming waves.

From this study, it appears that there are situations where beach scraping can lead to increased protection from storms and beach longevity. Unfortunately, it has been little researched, possibly due to its close links with beach feeding, where similar activities of beach profiling and shape engineering take place, albeit with the addition of sediment.

Beach dewatering and drainage

The deposition of sediment on beaches is a function of swash deposition and erosion, and backwash deposition and erosion (Duncan 1964). Swash erosion mobilises sediment and moves it up the beach, thus causing alteration in beach profile but keeping sediment in the beach environment. Beaches will conversley experience a loss of sediment if backwash is excessive. As waves rush up the beach, their velocity decreases as a function of their momentum, the angle of the beach, the depth of water and the roughness of the sediment surface (Grant 1948). On dry sand, where the beach water table (WT) is low, water can perco-late rapidly into the sand, reducing water depth and, hence, up-rush velocity

(Figure 7.6). As velocity decreases, the flow regime will change from turbulent to laminar, and deposition will occur, often resulting in a thick lens of sand shoreward of the swash zone, such as was observed at Manhatten Beach, California (Duncan 1964). During the backwash period, the change in flow regime is reversed with a change from laminar to turbulent, and hence, erosion (Grant 1948). In order to preserve the lens of deposited sand, this flow regime change on the backwash needs to be delayed for as long as possible. The time of flow regime change is dependant on beach angle and also the depth to the water table. On beaches where the water table is high and water is not able to infiltrate, laminar flow rapidly changes to turbulent and all of the water will return via surface flow, carrying much sediment to seaward. Furthermore, it is possible for the backwash flow velocity to be increased by water emerging from the beach face under a hydraulic head of pressure in what is termed the 'effluent zone' (Figure 7.6).

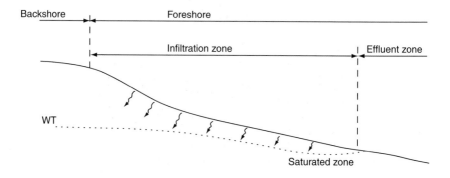

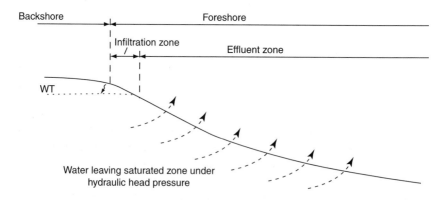

Figure 7.6 Height of beach water table and degree of swash infiltration. A low water table allows greater swash infiltration and reduced backwash surface flow, thus reducing beach draw down.

Plate 7.1 Gabion basket drainage channel to increase throughflow through shingle beach, Worthing, Sussex, UK.

As a result of these links between beach water table, flow regime, and sediment deposition/erosion patterns, it is possible to identify two ways of overcoming problems of excessive backwash erosion. First, the friction and loss of wave energy as the incoming and receding wave, transgress the sediment surface can be altered by increasing beach width (see Chapters 7 and 9). Second, decreased erosion can be facilitated by increasing the period at which backwash occurs under the laminar flow regime. This can be achieved by increasing the degree by which backwash occurs by non-surface routes, and reducing the amount of water emerging from the beach face under hydraulic head pressure in the effluent zone (i.e. by decreasing the level of the water table). Clearly, if infiltration into the sediment surface occurs, then the energetics of backwash are reduced. These basic principles were first proposed by Bagnold (1940) following studies in laboratory situations. Turner and Leatherman (1997) discuss recent advances in this area and link the laboratory studies to real field situations.

In essence, the technique involves lowering the beach water table by means of drains and/or pumps in order to increase infiltration of water on the backwash and increase sediment deposition on the shore face. This approach is supported by Packwood (1983) whose modelling studies have shown that increased permeability will attenuate swash wave energy, and that the unsaturated part of the beach will soak up water on both swash and backwash. The simplest method is to install pipe drains or lines of easier flow. This method is known as gravity dewatering, such as was carried out in Worthing, Sussex (Plate 7.1). The second method is to link a series of pipes to a central drain, from which collected water is pumped.

The first reported field study using a pumped drainage system was carried out in New South Wales, Australia (Chappell *et al.* 1979). Here, two trials were instigated, producing in the order of 65 litres of water per minute from the drainage system. In the first trial, some increase in surface elevation above that observed in the control site was noted, but the second trial produced no increased sedimentation although there was an increase in infiltration. Also in Australia, a non-pumped dewatering system was installed at Dee Why Beach, Sydney (Turner and Leatherman 1992). A series of strip drains were installed at intervals along the coast. This scheme resulted in a drop in beach water table of 0.3 m, although impacts on beach morphology are not clear. It appears, therefore, that in some cases, increased sedimentation may occur during beach drainage schemes, although this is not always the case. What this represents is another example of the complexity of beach processes; it shows that successful enhancement of beach sedimentation involves more than just pumping water from the beach. Chappell *et al.* (1979) argue that in addition, beach slumping, sand liquefaction and the passage of waves (whose amplitudes are not affected by pumping) through the water table can all contribute considerably.

The technique of beach dewatering has also been used for commercial reasons. At Hirtshalls, Denmark in 1981, a PVC pipe was buried 2.5 m down and parallel to the shore along a 200 m stretch of coast with the intention of supplying water to a nearby aquarium and for heat pumps. It was estimated that flow through the pipe would exceed $400 \, m^3 \, hour^{-1}$. Although not primarily for beach dewatering, the situation is directly analogous because of the approach used. In this example, the supply of water fell by 40 per cent after only 6 months due to the increased volume of sediment above the pipe, caused by enhanced sedimentation due to the dewatering process. Other schemes have occurred at various places around the world, such as at Thorsminde and Enoe Strand in Denmark, Nantucket Island, Sailfish Point and Englewood in the USA, and at Towan Bay in the UK (Turner and Leatherman 1992). The scheme at Thorminde, Denmark, used developments from the Hirtshalls scheme. Using a system of pipes and pumps, the typical pre-installation recession of the coast stopped and seawards progradation occurred in some sections. Similarly, at Sailfish Point in Florida, similar trends of deceased erosion and a switch to accretion was observed, along with a drop in water table of around 1 m. Rapid accretion occurred in the Towan Bay installation, but a severe storm removed much of this accumulated material. Regardless of this, however, the impact of the dewatering scheme had increased the beach's capacity to protect the hinterland from the storm.

It appears that the installation of dewatering systems often leads to increased beach levels and also to the possibility of seaward beach advancement. Clearly, though, there are situations where the methodology is more effective than others. Exposed coasts tend not to be as stable in the long term as sheltered coasts, primarily due to the impact of storms. This is because the increased sedimentation during non-storm periods may increase the stability, and reinforce the shape of the calm weather beach profile, but make adjustments to a storm beach profile much greater and more pronounced. It is also true that some of the causes of scheme failure remain uncertain. It is clear that as soon as the beach water table drops, there is a response in the form of changes in beach profile and accretion rate. What is less well understood is the exact nature of these links, such as time lags, the nature of water movement through the sediment, changes of the water table in response to ebb and flood tides, and short-term changes in response to swash and backwash. Developments are ongoing (see Turner and Leatherman 1997); these should give a better understanding of the processes and, therefore, in the methodology by which to manipulate them for the benefit of beach management.

In relation to this, Li *et al.* (1996) have developed a two-dimensional model by which water table changes can be modelled under beach dewatering scenarios. Using this model, they have simulated the impacts of both pumped and non-pumped (gravity) dewatering systems and have obtained results comparable with the field trials discussed above. In addition, they have used this work to attempt to state some guidelines, in that their results demonstrate that the location and size of the drain will affect its efficiency. An increase in elevation and size will increase efficiency, while efficiency decreases the further the drain is moved landwards.

Draining beaches and lowering the water table appears to be yet another method by which to increase beach elevations, without providing a net increase in sediment availability. Therefore, as with groynes, jetties and polder schemes there may well be implications for adjacent beaches in respect of sediment budget deficits. Again therefore we need to adopt any techniques such as this with due regard for the holistic nature of the coastline in question, and not to proceed piecemeal, as has been done in the past.

Summary

This chapter contains extensive information, and is the longest in the book. This is not accidental: it has had to reflect the huge range of material written on the topic of beach feeding. The examples listed in Table 7.2 represent a selection of the case studies available. It is true that many engineers and coastal managers have seen, in beach recharge, the answer to many of their problems. Not only can it provide adequate storm protection, but also the means to attract tourists during the holiday season. This may well prove to be the start of the downfall of this technique. Many authorities are committed to recharge and the required repeats to sustain not only their hinterland protection, but their tourist trades. Many more are jumping on the bandwagon. The problem of sourcing all this material is pertinent here, and very real. These sources are not limitless and we are in danger of developing reliance on a technique which is rapidly becoming unsustainable. Clearly recharge has many advantages over other methods, because by maintaining coastal processes, it preserves the natural functioning of the coastline. The one aspect which it does not have however, and the one which could lead to greater recharge longevity, is to look at the wider coastal cell and what is actually causing the problem of beach loss.

When we recharge a beach, we are treating the problem of erosion. When we recycle sediment back to the beach, we are actually replacing the sediment which the beach 'does not want'. These problems are symptoms, and it is what is causing that erosion which should also be tackled if at all possible. In the earlier example from Morib in Malaysia (Bird 1996), the 12 000 lorry loads of sand deposited on the beach soon disappeared. In this example, the reflective nature of the sea wall had not been investigated as the cause of the problem, hence the cause of the beach erosion remained and led to the removal of the newly imported material. Clearly, if recharge had occurred in conjunction with modifications to the wall, the longevity of the scheme would have been greater. Van de Graaff *et al.* (1991) discuss this problem in terms of treating the disease (the cause of the problem) or the effects (symptom). Given a scenario where beach loss is occurring because the longshore movement of sediment is not being compensated for by up-drift renewal (the symptom), then recharge will allow longshore movement to continue, but the problem of non-renewal will remain. If this symptom is caused by a jetty blocking transport, then bypassing this will treat the cause as well as the symptom, thus leading to greater scheme longevity and integrity.

Perhaps the greatest cause of concern relating to recharge schemes is that their outcomes can be quite variable. Some schemes retain material for long periods of time, others do not. The state of knowledge at the present time is insufficient to explain why, although there are a series of likely reasons. Scheme failure is too often blamed on unexpected storms removing large quantities of sediment, a view also stated by Pilkey (1990). However, beaches are supposed to respond to storms by flattening following sediment draw down. The real issue here is not that sediment is drawn offshore and beaches flatten, but why does the sediment not come back after the stormy period and rebuild the beach like it does on a natural coastline? This underlies one of the main problems with beach recharge: recharged beaches do not behave in the same way as natural ones. There are limitations in their functioning ability, the reasons for which we just do not understand. What is true, however, is that coastal engineers have traditionally operated and planned within this world of non-understanding, putting too much reliance on equations which only go part way to explaining some (not all) of the many variables associated with beach dynamics. It is not my intention to criticise engineers, for there are, after all, many successful recharge schemes. However, it is tempting to surmise how longevity and success could be increased with better understanding of the fundamentals of beach behaviour, and how all the variables which can be measured on a beach, from waves to grain size to drainage, interact to make a 'beach'.

Even without adding sediment to beaches, engineers can still increase their effectiveness and accretion by reprofiling and draining. At its very basic, the moving of sediment around is a simple way of changing the shape of the beach to make it more effective in wave attenuation. Perhaps this approach may be considered as being particularly 'environmentally friendly' because it represents the case of pushing sediment back up the beach from whence it came. Although simple, there are underlying issues here. This represents very much a short-term solution, particularly effective in response to storm draw down, where repair to the high beach is essential before the next storm comes in. However, continuous scraping should be avoided because it will prevent any longer-term adjustment of the beach profile. Beaches, as we have seen, do respond to environmental conditions, both short-term (storms) and longer-term (response to structures or sea level rise). Clearly, to achieve a stable beach, these natural processes of adjustment should be allowed to continue, whereas constant scraping of sediment is hardly conducive to this. One example of this practice occurs in many tourist beaches where scraping is used to 'accelerate' the change from winter to summer profiles for the benefit of amenity usage.

The dewatering of beaches appears a reasonable response to the need to increase beach sedimentation. Clearly, however, from field evidence, the success of a scheme is due to factors beyond just lowering the water table. This is also a short-term solution; while the technique may be suitable to encourage accretion to compensate for beach erosion, if the cause of that erosion is something like sea level rise and coastal squeeze, then there will come a point where the dominance of wave activity will exceed that of beach water table as the fundamental control over beach stability.

This hints at a new line of development which verges on the world of fantasy. In a world where humans are constantly trying to improve on nature, (speaking as a non engineer) if we can master the physics of beach sediment and wave dynamics, then could we, some time in the future, create that perfect (i.e. stable and permanent) beach?

Summary of the benefits of beach recharge schemes

- Increased beach levels provide improved coastal protection
- Better beach development for amenity usage
- Increased socio-economic stability for coastal towns
- Maintenance of natural coastal processes
- Protects coast without altering wave dynamics

Summary of the problems with beach recharge schemes

- Increased turbidity around dredge and backwash areas
- Short longevity of many schemes and need for repeats
- Impacts on infauna
- Inability to function as natural beaches
- Increased amenity usage and encouragement of further development

Recommended usage

- To protect and build-up beaches in areas of sediment starvation
- To increase and protect amenity usage
- Where the maintenance of 'naturalness' is important
- To supplement other defence structures (offshore breakwaters/groynes)

8   Dune building

Introduction

Sand dunes are unique among coastal landforms in that they are formed by the wind rather than moving water, and represent a store of sand in the supratidal zone (i.e. above the landward limit of normal high tides). This is not to say, however, that waves and tides are not important in the dynamics of dunes, because it is these processes which deliver sediment to the intertidal zone, so that wind is able to transport it landwards from the intertidal to the supratidal area. Furthermore, during storms, waves may reach the front of the dune complex and reactivate this sediment store, drawing sediment down onto the beach to form a storm beach profile. These ideas are exemplified by work in the Netherlands which has shown that during a storm, erosion by waves resulted in the seaward transfer of dune sand as far as the outer surf zone (Edelman 1968, 1972). Such studies have allowed the development of models to predict the erosion response of dunes to storms (Edelman 1968, 1972, Kriebel and Dean 1985, Kriebel 1986, 1990). The transfer of sand to dunes has been discussed in Chapter 2, but this process represents a net loss of sediment from the beach sediment budget, and a gain to the sand dune system. This means that for a healthy beach/dune system, the delivery of sand to a beach needs to exceed that lost to the dunes. Illenberger and Rust (1988) detail a study from the Alexandria coast (South Africa) and show that $375\,000\,\mathrm{m}^3\,\mathrm{a}^{-1}$ of sand moves from the beach to the dunes, while only $45\,000\,\mathrm{m}^3\,\mathrm{a}^{-1}$ is returned by wave erosion. Hence, while the dunes grow by $330\,000\,\mathrm{m}^3\,\mathrm{a}^{-1}$, this same figure also represents the net loss from the beach sediment budget. Table 8.1 details other examples of beach/dune sand volume transfers and highlights the extra volumes of sand beaches need to receive above that lost by longshore drift, to keep backing dunes healthy.

As was outlined in Chapter 2, the transfer of this sediment into dune fields is predominantly by wind, but sediment mobility is controlled by other factors, such as wind speed, beach width, wetting area and sediment grain size. The greatest volumes of landward sediment transfer occur on wide, low-angle dissipative beaches, with the lowest volumes on narrow reflective beaches (Viles and Spencer 1995). Because of these contrasting beach/dune relationships, the

Table 8.1 Examples of volumes of sediment transferred from beach to dune systems.

Dune system	Total volume $(m^3\,a^{-1})$	Volume per metre $(m^3\,a^{-1}\,m^{-1})$	Source
Alexandria	375 000	8.3	McLachlan (1990)
Cape Hatteras	528 000	5.3	Pierce (1969)
Assateague Island	462 000	8.0	Bartberger (1976)
Malpeque Barrier system	91 000	2.7	Armon (1979)

extent of dune development will also vary according to beach type, with dune fields on wide, dissipative coasts being generally much larger features, both in respect of size and spatial extent.

Another feature of sand dunes is that their vegetation is not dependent on the inundation of seawater for stability (compare salt marsh). The prime controls are the ability to withstand desiccation and poor soil structure and nutrient levels. Vegetation is vital for the survival of dunes, both with respect to the binding of sediment by root systems and in facilitating the build up of dune sediment by wind baffle. The fact that dunes are not intertidal, and are generally covered by vegetation has, however, been part of the problem in their conservation and protection. Historically, people have seen these habitats as being separate from the sea and so 'available' to be claimed for alternative uses. Typically, these uses have included forestry, housing, sand mining, or amenity uses, particularly golf courses.

As with any coastal landform, sand dunes are dynamic systems, constantly adjusting to variations in wind patterns and sediment supply, and also undergoing a maturation process as they progress inland away from the influence of the coast. Typically, moving inland from the beach into a healthy dune complex with abundant sediment supply, one encounters a distinct morphological and vegetation sequence. First, embryo dunes form around storm high water mark and, once initiated, start to be colonised by salt-tolerant species such as lyme grass. Once vegetation is established, the increased surface roughness caused by leaves, and the greater binding of the sand allows greater build-up of sand, leading to the development of fore dunes (sometimes referred to as yellow dunes). This zone is only rarely under the influence of salt water and so less salt-tolerant plants can colonise, such as marram. The increased vegetation cover further enhances the trapping of sand and the increase in size of the dunes. Marram, for example, can survive sand build up in the order of $1\,m\,a^{-1}$ and can facilitate the development of the fore dunes up to $10\,m$ in height. To landward, sand movement inland is reduced because of the large quantities trapped by vegetation in the fore dune area. However, vertical sand build up can still occur producing dunes up to $30\,m$ high, where they typically become mature (or grey) dunes. Marram will still occur, with additional colonisers being fescue or sand sedge. Further inland, the dunes merge with terrestrial habitats as soil development increases (Figure 8.1).

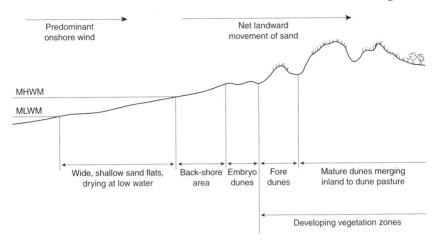

Figure 8.1 Typical landward succession from beach to higher dune communities.

This sequence is perhaps a little idealised, in that it represents the perfect accretion scenario with no limits on sediment supply and no landward restrictions to lateral dune encroachment. It should be mentioned, however, that this scenario is the exception rather than the rule; issues such as sediment availability and hinterland development can modify this sequence. A reference to Table 8.1 will emphasise the quantities of sand needed to supply dune complexes. Given what we know about the factors limiting sand transfer along the coast, it is easy to see why many beach budgets are not adequately sourced to maintain healthy dune complexes. These issues will be developed further below. The important aspects that become clear are adequate sediment supply, and the role of vegetation in trapping sediments and building up the dunes. Clearly, if we are looking for ways to manage dunes, then we can identify a possible management approach. By increasing sediment availability (such as by beach feeding — see Chapter 7), the protection of existing vegetation and the replacing of damaged vegetation can all lead to the rehabilitation of damaged dune areas.

From this introductory section, and the section on aeolian processes in Chapter 2, we have seen how dunes can form, and how, by understanding these processes, humans can utilise these processes to repair areas of degradation and damage. Clearly, it has not been possible to present an in-depth review of how dunes work and the physics of the processes involved. For readers wishing to pursue these lines of enquiry, a good starting point can be found in many coastal morphology textbooks, but more specifically in Nordstrom *et al.* (1990).

There are two other main areas of study before we can look at how dunes are managed and protected: first, why their presence and stability is of fundamental importance in coastal protection; and second, how dunes are being threatened. By understanding these two aspects and by linking this to our existing knowledge on how dunes work, we can better appreciate how we can apply the most suitable management methods for given situations.

Importance of dunes in coastal defence

Dunes represent a vital part of the coastal defence of many sandy coastlines. Not only do they represent a barrier between the sea and the land, similar to a sea wall (Plate 8.1), but the dune face and embryo dunes are also dynamic, being able to supply sand to the beach when it is needed, yet store it when not. Hence, in simple terms, sand dunes represent a form of natural sea wall that can carry out its own beach recharge. Because this dune/beach interaction is so important, it is essential that this capacity for dynamic interaction be protected. During storms, dune faces will experience erosion by waves. This should not be regarded as an immediate management problem in the short term, because it is merely the system's response to increased wave energy. During the summer months, when wave activity is generally lower, much of this sediment is blown back into the dunes. This is exemplified in the earlier example from the Alexandria coast, where 45 000 m^3 of sand is eroded from the dune face each year, yet 375 000 m^3 is returned (Illenberger and Rust 1988). Longer-term trends, however, may be more critical because if annual dune erosion is consistently net greater than the amount of sand returned by the wind, the dunes may enter a phase of retreat. In such situations, the net erosive nature of the dunes means that they may start to decay, and intervention measures may be necessary. Reasons for this scenario are varied, and range from anthropogenic to natural forces, the form of which are described below.

Plate 8.1 Eroding dune face fronted by wide sandy foreshore. Note the wide foreshore area which can source sand for dune formation, Sefton Dunes, north-west England.

Careful management of dunes can provide a high degree of protection to aeolian coastlines. In the Netherlands, dunes form the main line of defence against flooding. Given that many of the country's main industrial and urban areas lie below mean high water mark, the strength and general health of the dunes are vital to the protection of large areas of land (Verhagen 1990). In this example (see Box 8.1 for detailed overview), dunes have, wherever possible, been allowed to function without intervention, with erosion and landward migration permitted. In some cases, however, the dune width did become too narrow and remedial action, such as that described in subsequent sections, became necessary. The result was that many of the dunes along the Dutch coast are artificial, constructed to a high standard to withstand the one in 10 000 year storm event (Komar 1998). Similar importance is put on dunes along barrier island coasts of the USA and some interesting observations regarding the relative functioning of natural versus managed dunes have been made (see below).

From what we have seen so far, the importance of dunes can already be appreciated. There is a further aspect, however, which needs to be considered. When looking at the environmental conditions necessary for dune formation, it was stated that wide sandy beaches are of paramount importance. The problem here is that wide sandy beaches (such as that in Plate 8.1) are also appealing for other coastal uses, and many of these uses can have serious implications for dune health and, inherent in that, for their ability to function along the lines already discussed.

The dune problem

Dunes are highly fragile environments, yet play vital roles in the stability of the beaches fronting them and the protection of the land behind. Their fragility is due to the uncompacted sediment and poor binding by roots which, coupled with their exploitation, makes damage from over-use and misuse easy to cause. Impacts on dunes may be categorised into four types (Tables 8.2 and 8.3). First, 'conversion' includes those methods of exploitation which involve changing vegetation types (i.e. by afforestation, farming, or golf course construction), or the nature of the dunes themselves (development). Effectively, such impacts mean that the dunes will cease to function in their natural manner, either because sediment cannot be transferred inland due to too great a level of vegetation, or because the dunes have been built on. The second type is 'removal' where, as the name suggests, sand is removed for other uses, such as building or mining. Removal can also include examples where dune profiles have been altered in order to facilitate access or to provide 'sea views' for landward development. Thirdly, the 'use' of dunes includes those activities that utilise the current resource. Perhaps the most frequently recognised as impacting on dune health is amenity usage, but also other features such as water extraction and conservation measures may also impact on dune functioning. Finally, we need to include impacts on dunes from beyond the dune environment. Clearly, as we have

Box 8.1

Importance of dunes in the protection of the Dutch coastline

Much of the Dutch coastline is protected by dunes. These structures, covering a total area of 40 000 ha and distance of about 250 km (Arens 1996) are considered to be of 'incalculable' importance to the Netherlands, not only as coastal defence, but also in the provision of drinking water (supplying water to two-thirds of the population, equating to 10 million people (North Holland Water Supply Company 1992)), and as a recreation area of international importance (Helmer *et al.* 1986). The integrity of the dunes is vital for protection and water supply, and so management aims to reduce aeolian erosion and to encourage deposition by erecting sand fences and vegetation planting. Often, such activities are aided by artificially reshaping eroded dunes and feeding beaches to promote dune building (see Chapter 7). While this approach has successfully protected the coast, it has resulted in a largely straight dune coast with no blowouts or other morphological features associated with a coastal dune system. In this respect, the dunes of the Dutch coast cannot be said to be a 'natural' system. However, the approach is understandable given the importance of these dunes in coastal defence, and the system does represent a defence which interacts with the beach to maintain some degree of coastal sediment dynamics. Because, historically, dunes have always been seen as vital defence structures, their management has been largely controlled under the ownership of the Polder Board. Others are owned by water companies, although even here the Polder Board has special controlling powers, emphasising the importance of the role of dunes in defence.

The main 'use' of the dunes is for coastal defence, recreation, nature conservation, and for the supply of drinking water (North Holland Water Supply Company 1992). This means that management has to cater for a wide range of user groups and a wide range of impacts, and to avoid conflicts between these. Louisse and van de Meulen (1991) and Verhagen (1990) identify the essential nature, importance and diversity of dune management along this coast. The Dutch government has spent a lot of money investigating solutions to the management problems, concluding that the dunes should remain as natural as possible, although direct interference will occur in areas of high hinterland value. Where possible, this will be in the form of soft engineering methods.

The paramount importance in the role of dunes is for the safety of the low-lying hinterland. Given that one-third of the Dutch land mass is below sea level (Louisse and van der Meulen 1991), the dunes are a critical area for the protection of both property and life. Their importance is reflected in national policy. If a dune cannot meet required safety requirements (stated as not being able to withstand a 1:10 000 or 1:4 000 year storm, depending on locality), the Polder board is empowered to request national government to fund the improvement of defences (Verhagen 1990). Wherever possible, this improvement will be by the use of 'soft' means, typically beach feeding.

Monitoring has been carried out annually since 1963 via a series of shore normal transects every 250 m along the coast and stretching from below low water to the dunes. Using data from the last 5 to 10 years, dune safety is calculated by reference to the equilibrium profile (Vellinga 1983). Based on this data, the Polder Board determine whether the dunes are adequate to protect the hinterland from flooding and, if erosion trends are evident, in which year this safety limit is likely to be compromised. This approach allows a degree of proactive rather than reactive management, in that potential problem areas can be identified early, and remedial actions, if necessary, taken.

The role in sediment supply is also acknowledged as being critical. The measured transects include offshore bathymetry which is analysed for trends in shore profile adjustment. Volume integration methods allow sediment supply volumes to be predicted and thus likely sediment supply to the beaches and dunes. Given that measurements of beach and dune erosion rates are also ongoing, a sediment budget approach allows a detailed insight into dune behaviour.

Although this approach appears elaborate, it is a basic necessity considering the importance of dunes in protecting the hinterland. Given the likely response of the dune system to sea level rise, this approach takes on greater importance, because only with such a detailed knowledge of historic dune behaviour will it be possible to develop the knowledge to determine when holding the line is no longer feasible, and retreat inland becomes the most viable option.

Disturbing trends are already beginning to emerge which are sounding warnings for the future management of the Dutch coast. Helmer *et al.* (1986) state that many of the dunes are experiencing long-term erosion on their seaward sides. This is due to a variety of reasons, but include dune stabilisation, sediment supply, and sea level rise. The result, however, is that intertidal profiles are becoming steeper, and coupled with this is an increased wave base and erosion potential of storm waves. Important decisions will have to be made. Increasing sediment supply by beach feeding is only a short-term solution here. The main problem is the position of the dunes and sea level rise. Because they have been fixed in position by planting they are, under the influence of sea level rise, in effect getting closer to mean high water. Earlier we saw the importance of the back-shore area for sediment supply to dunes. If this is getting narrower, then sediment supply will be reduced, and the ability of the dunes to increase in size will, correspondingly, also be reduced. Under natural conditions, the dunes would migrate inland, thus maintaining the status quo. In reality, however, development and policies which actively discourage this management strategy have prevented it being adopted as a solution. The time is not too far away when such ideas are going to have to be given serious consideration if adequate protection from the dunes is going to be maintained. Helmer *et al.* (1986) identify areas suitable for this, and explain the benefits which would occur. With the safety of one-third of the land area reliant on healthy dunes, their effective survival is a priority in the Netherlands.

already seen, dunes can only survive with an adequate supply of sediment and such supplies can be high (see Table 8.1); any activity which inhibits this, such as reduced sediment supply due to coastal defences, will also induce net dune decay and potential future defence problems via the beach sediment budget. These issues will now be developed in the context of their impact on dune stability, and on the ability of dunes to function as efficient coastal defence structures.

Whatever the cause of impact, dunes are very susceptible to change. In terms of 'thresholds', these tend to be very low, with only low levels of interference required to induce significant changes. In contrast, however, dune management needs to recognise that there is a balance to be struck, in which use has to be weighed against impact. Although many of the world's coastal dune systems are showing signs of impact and human-induced change, we cannot shut off all these environments from human access. Clearly then we need to operate a balance between the various impacts observed within these environments (discussed below), and the methods employed to protect dunes from both change, and also to increase their defence potential (discussed subsequently).

One key point to remember is that although impacts are being considered as four categories, this does not exclude the possibilities of more than one impact operating on any one dune area at any one time. For the sake of clarity, each category will be considered in turn.

Dune conversion

This category of impacts (Tables 8.2 and 8.3) may be regarded as the most important, as it can involve large scale changes in dune habitats, which can lead to the complete modification of the dune system.

Urbanisation (development)

By developing dunes into housing or holiday accommodation, we are, in effect, sealing all the available sand below tarmac, grass or buildings, with the result that dunes may become static and immobile. This immediately represents the loss of a sediment store. In addition, the remaining dune system is no longer 'natural' and, given that dunes naturally migrate, the fact that a development is now present within the dunes means that the land has increased value; demands for future protection will be increased and any tendency towards inundation by prograding sand resisted. This loss of naturalness may also be noticed in other ways. Habitat fragmentation and loss, introduced species and alteration of local wind regime may all serve to alter the dynamics of the natural system. Perry and D'Miel (1995) show in their study of coastal dunes in Israel that populations of many animal species declined, and plant populations increased following urbanisation of a dune area. Notable in the causes was the loss and fragmentation of habitat, and the introduction of dogs and carrion crows. A further example from Gold Coast City, USA states the main areas of concern with respect to urbanisation as being 1) encroachment onto active dune areas; 2) removal of species to

Table 8.2 Categories of human impacts on dunes. Scales of impact are given in discussion in the text.

Conversion	Removal	Use	External
Urbanisation	Mining	Tourism	Reduced sediment supply
Golf courses	Development	Trampling	Sea defences
Agriculture		Horse riding	Migration prevention
Forestry		Sand yachting	
		Off-road vehicles	
		Water extraction	
		Conservation	
		Military training	

Table 8.3 Estimates of impact severity of different uses of dunes detailed in Table 8.2, with respect to effectiveness of dunes as coastal defence structures.

Category of impact	Implications for effective sea defence
Conversion	
Urbanisation	Major — inability of dunes to remobilise/loss of sediment
Golf courses	Minor — possibility of return to natural system
Arable	Minor/Intermediate — possibility of return to natural system
Forestry	Intermediate/Major — tree felling required
Grazing Land	Minor — ease depends on degree of improvement
Competitor species	Minor — ease depends on degree of invasion and species present
Removal	
Mining	Major — loss of sediment/dune stability
Development	Major — loss of sediment; little possibility of increased mobility
Use	
Trampling	Minor — varies according to degree of vegetation loss and blow-out formation
Horse riding	Minor — as for trampling
Sand yachting	Minor — as for trampling
Off-road vehicles	Minor — as for trampling
Water extraction	Minor/Intermediate — depends on extent of extraction and impacts on water table
Conservation	Minor — problems if rare species cannot be lost
Military	Minor — does not involve change of use/as for trampling
External	
Reduced sediment supply	Intermediate/Major — loss of ability of dunes to build and migrate
Sea defences	Major — removes dunes from dune/beach interaction
Migration prevention	Major — coastal squeeze

create views, thus destabilising dune areas; 3) encroachment of turf and other inappropriate vegetation; 4) uncontrolled access; 5) susceptibility of property to damage during storm events; 6) perceptions that remaining public land is private.

Any development needs services and access; this represents further demands on an already fragile ecosystem. Plate 8.2 shows a dune complex on the Fylde Coast, north-west England. Here, the main coast road runs through active dunes, and fronts a holiday complex. There is, as a result of sand blown inland, constant problems of encroachment onto the road as well as increased access through the dunes from the holiday complex to the beach. Similar examples are cited in the literature. Ranwell and Boar (1986) discuss a caravan park development in dunes at Clachtoll, Scotland where the increase of visitor pressure has caused vegetation loss and breakdown of the thin soils. This has resulted in wind erosion and large bare sand patches developing. Furthermore, they also point out how, being permanent, many sites have no chance of recovery, leading to a gradual worsening of the problem. Carter (1988) also discusses dune damage due to holiday sites, and stresses the degree of damage that can be done in dune systems due to evening barbecues and general vandalism. Some holiday complexes, such as caravan parks are, to some degree, transitory, in that they can be moved with minimal effort and relocated, should the needs of dune protection warrant. In contrast to this are the more permanent developments (see Plate 8.2), such as houses or even hotels and towns.

Plate 8.2 Hinterland development preventing landward dune migration. This holiday camp not only necessitates the stabilisation of fronting dunes, but also increases visitor pressure on the dune area, by encouraging access to the beach (shown in Plate 3.1).

Carter *et al.* (1990) discuss this attraction of dunes to developers. Being sandy and aesthetically pleasing, dunes make ideal locations for hotel developments. In some countries, this development has been taken to extremes. At Matalascanas, near Seville, Spain, for example, a new town housing 250 000 people has been built within a dune complex. Clearly, this development, from a coastal defence point of view is bad enough, but in addition, the remaining dunes to seaward have been lowered and regraded in order to provide a better sea view for occupants of sea front apartments. Putting this in the context of dunes as coastal defence structures, would we seriously consider tearing down a sea wall to give building occupants a better sea view? Such coastal developments are not unique to Spain, however. Island Beach, New Jersey, USA is a further example of development within a dune field (Gares 1990). Here, much of the dune spit has been developed, and as such, has high land values and strong arguments for future defence structures. However, how many people would build a house on a beach?

Golf courses

There are two contrasting arguments regarding golf courses. On the one hand they represent the introduction of alien species and necessitate intensive dune management (such as nutrient addition and watering), thus making the environment unnatural; on the other hand, they provide scrub control and protect dunes from intensive visitor pressure, thus providing necessary management. There is certainly some degree of habitat modification, particularly around the greens, although much of the appeal of the links course is its naturalness and rough vegetation. Because of this, many areas are lightly managed for scrub and sapling encroachment, while still being semi 'natural'. The recent large increased demand for golf facilities has put pressure on dune areas. While the management as a golf course can hardly be considered the same as management as a dune system, it can be argued that the degree of change is low and that the use of dunes for golf could be regarded as a form of conservation. Theoretically, areas of dune used as a golf course could revert to a natural system. However, this may become complicated where dunes have experienced a change of use for long periods because they have not been receiving the same amount of sediment, or interaction with the natural environment. Furthermore, management for greens, etc. would promote the formation of soils and addition of fertilisers would alter natural soil geochemical properties (French 1997). Also, the grasses used in greens are not as drought tolerant as native species, and so watering is often required. Some courses may source this water locally, leading to the risk of lowering the dune water table, which may have implications for dune slack communities (see also Box 8.1).

While the use of dunes for a golf course would stop wind blown sand and sediment migration, it would not necessarily stop the interaction of beach and dune. If major tees and greens are sited away from the dune edge, then normal winter draw down can continue without impinging on the main golfing areas. Clearly, this would require an active zone at the edge of the course.

Agriculture

Many dunes are subject to conversion to arable land, particularly along the landward margins. The nature of the soils prevent this being particularly high-grade land, but cropping of grass may occur, while some root crops are particularly suited to sandy soils. Similarly, the nature of dune soils is such that any ploughing would lead to rapid degradation and nutrient loss. Given that much arable usage is concentrated towards the hinterland, the impact on natural dune functioning and coastal protection is low. However, if not managed correctly, severe breakdown of surface layers can result, leading to increased wind erosion.

A further aspect of agricultural use of dunes does not involve their conversion but utilises natural dunes as resources. Grazing of dune vegetation, particularly yellow dunes, by livestock (and also by wild animals, such as rabbits) can have implications for soil stability. Boorman (1989) reports that a low animal density (greater than 0.5 cattle or 4 sheep ha^{-1} (Ranwell and Boar 1986)) can result in a significant increase in the proportion of bare ground, which can itself lead to an increased risk of erosion and blow-out formation. The main problem cited is rabbits, although cattle trampling is also locally important. Ranwell and Boar (1986) further develop the rabbit issue: not only do rabbits damage vegetation by grazing, but also by their burrowing. This action damages roots both indirectly, and also directly due to nibbling on sugar-rich root fibres. A further aspect of the problem is that grazing reduces the height of the vegetation, thus decreasing its ability to trap and retain sand. Sheep have greater impacts in this respect because they crop the grass much closer to the ground. Band (1979) also reports that grazing reduces rooting depth which can reduce sediment binding ability and also lead to vegetation failure and dieback in drought conditions.

Forestry

The planting of trees on dunes is a long established activity in many areas, although increasingly, there is a growing pressure to remove many of the plantations and allow regeneration and rehabilitation of natural dune swards. Perhaps the main issue here is that trees often lock up sand over large areas of dunes, reducing wind speed, and so prevent landward sand movement. Mac-Donald (1954) has estimated that the interference with wind flow and subsequent sand movement is experienced inland for a distance equal to twenty five times the height of the trees. Similarly, the germination of much forest floor vegetation is impaired, leading to a largely atypical community, with many introduced species. The problems of atypical vegetation become particularly relevant when these afforested dune systems start to erode. Normally, as dunes are cut back, the seaward edge becomes 're-activated' with marram grasses recolonising and increasing the stability of the dune front. With no marram present to seed these areas, erosion of forested dunes may proceed rapidly.

Many of these issues have only recently been realised. Afforestation was once considered desirable in many countries as a means of compensating for the loss of natural woodland inland. Because of this, many dune areas have experienced these problems. In Tentsmuir, eastern Scotland, afforestation begun in 1890 has resulted in 1300 ha of forest, but it is argued that this has produced major changes to vegetation communities and the impoverishment of native species (Ovington 1951). Similarly, the planting of 720 ha of dunes at Newborough, Anglesey by the Forestry Commission between 1947 and 1965 has replaced the natural dune grassland and led to major modifications of the dune ecosystem (Hill and Wallace 1989). Although many disadvantages have been cited, dune afforestation is not a thing of the past. Mailly *et al.* (1994) discuss an example from Senegal where trees are currently being planted in order to stabilise dunes and prevent landward migration onto agricultural land. Here, the programme is being considered a great success because it is preventing ongoing desertification and protecting farmland. The longer-term outcomes, however, are not as clear. In effect, the dunes are being fixed in position and are thus unable to respond to storms or sea level variations.

Removal

This category refers to those practices which result in the removal of dune structure by sand extraction. The implications of removing sand from dunes may appear to be obvious. However, removal need not be a problem in a healthy dune system with a sediment supply in excess of that removed, because although the removal of sand does represent a loss to the sediment budget, abundant inputs to this budget from coastal sources may be more than adequate to offset losses. This may not always be the case, however, and failure to replace material lost through extraction will result in a net loss of sediment to the system. This removal has a series of impacts. First, it results in the lowering of the dune topography which can make the area more vulnerable to storms and subsequent breaching; and second, it may result in an interruption to landward transfer of sediment by wind, and thus result in starvation of the mature grey dunes inland.

Mining

Sand may be extracted from dunes for many reasons. Most obviously for building materials, other uses have included glass making, improvement of soil fertility and the extraction of metal residues (Ranwell and Boar 1986). In addition, sand is increasingly being extracted as a source for beach recharge schemes (see Chapter 7). Many of these uses are well exemplified by Guilcher and Hallégouët (1991) in their study of Brittany sand dunes. Here, uses have included the commercial extraction of sand for building and the taking of calcareous sands by farmers to improve soils. The impacts of these activities have been great, in that mechanised sand removal has out-paced sediment input from aeolian sources which has led to dune degradation and erosion due to a negative sediment budget. Other activities threatening dunes in this area include increases in tourism, afforestation, and the cultivation of asparagus and tulips. Similar problems can be observed at Kenfig dunes in Wales, where 4.5×10^5 tonnes have been removed from the dunes (Jones 1996).

Before ending this discussion about mining, it should also be realised that sand mining need not occur within dunes themselves for an effect to be felt. Mining of the foreshore or even offshore dredging could impact on dune stability if these operations affect the supply of sand to the beach, and thus into the dunes.

Use

Having considered threats caused by the physical alteration of dune areas, we must also consider a major group of activities which do not occur with the intent of altering the physical structure of the dunes, but rather to utilise them as a resource, principally for recreation. Coastal tourism may have a whole range of impacts, but one area where these impacts are most severely felt is in dunes. Although the presence of vegetation helps stabilise dunes against large-scale sediment remobilisation by the wind, it is not adequate to prevent the impacts of trampling caused by people.

One way in which any habitat responds to trampling is by the increase in the abundance of trample-hardy species. In dunes, this change in species is not always possible because the range of suitable species that can tolerate the dune environment is limited. As a result, dunes tend to respond to trampling not by species replacement but by species loss. Boorman and Fuller (1977) in their study of Winterton dunes, Norfolk, UK, have shown a good correlation between areas of intense trampling and bare pathways, with grasses such as marram and lyme being particularly vulnerable. When considering the degree of trampling necessary to cause change, we are not talking about particularly large numbers. Boorman (1976) demonstrated that just ten passes in a month were sufficient to reduce the height of vegetation by 66 per cent. This equates to just five people walking across dunes to the beach and back again. Forty passes per month would reduce height by 75 per cent, eighty passes would start to produce bare ground and 150 passes would lead to a 50 per cent loss in vegetation. Clearly then, with only 150 passes (75 return journeys across the dunes) producing a loss of half the vegetation, the impacts of people can be great, especially in areas where tourism pressure is great. Plate 8.2 shows a clear example of this problem occurring at South Shore, Blackpool. The only access to the beach (apart from walking along the road) is across the dunes, and this is the route taken by most visitors. The resulting problem is the loss of large areas of vegetation and the formation of blowouts.

Apart from trampling, any activity in dunes that exerts similar impacts on vegetation is likely to cause loss. These issues are developed fully in French (1997), but include horse riding, sand yachting, and off-road vehicle use. All of these activities produce stresses on vegetation that ultimately result in loss. Another impact is military training, again with similar effects to off-road vehicles, although one advantage here is that training areas tend to be areas where the public does not go, so some degree of protection may occur here. Despite this, however, military use can have dramatic impacts on dune stability by killing off large areas of vegetation and leaving the dunes bare and open to wind erosion. Such a problem occurred at Gullane, East Lothian where the use of dunes for heavy vehicle training in the 1940s produced extensive dune degradation in part of the dune system, and widespread sand mobilisation (Ranwell and Boar 1986). Similarly, at Studland, Dorset, UK wide-scale dune damage occurred due to military battle training, also in the 1940s. Here, subsequent mine clearance necessitated the burning of large areas of dune vegetation for location purposes, leading to widespread sand mobility.

Water extraction

In some coastal areas, dunes provide the main water supply. One good example of this is in the Netherlands, where coastal dunes provide large sections of the population with drinking water, including Amsterdam. Furthermore, the suitability of the dunes as an aquifer was such that breweries and linen bleaching factories were set up to take advantage of the abundant supplies of water (North Holland Water Supply Company 1992 (see Box 8.1).

As a valued water supplier, dunes clearly have a large potential. There are, however, several implications for water extraction which could have an adverse impact on dune stability. Without sufficient replenishment, continued extraction will lower the water table. Clearly, this will affect any wetland areas which may be located in the vicinity; but of greater importance in this context is the fact that, as we have already seen, wet sand is harder for the wind to move than dry sand. Therefore, the higher the water table, the less sand is freely available for aeolian transfer inland. As the water table is lowered, the zone of dry sand increases, allowing greater volumes to be eroded and, thus, greater lowering of the dunes. Geelan *et al.* (1995) discuss this in the context of dunes adjacent to Amsterdam, where approximately 1200 ha of dunes suffered from excessive water table lowering and desiccation due to continuous water extraction since 1886. Artificial recharge with purified Rhine water has occurred since 1957 but the regeneration of the water table and phreatic communities was not satisfactory because of the excessive nutrients in the river water, changes in direction and volume of flow, and unnatural fluctuations in the water table.

A further impact is that although the water table in dunes is primarily fresh water, as water is extracted, so salt water is drawn in. This saline intrusion may again affect dune slack areas, where dune vegetation may be partly salt tolerant, but not to the continued elevated amounts which may be present with salt water intrusion into the dune aquifer. Examples of such damage have been observed in dunes at Southwold, Suffolk, UK, and in the Netherlands, where the water table has been lowered by up to 5 m in places (van der Maarel 1979).

External

The final group of dune impacts originates outside of the dune system. We have seen earlier how important it is for dunes to receive a constant sediment supply. One reason for dune decay is the stopping or reduction of this supply. The reasons for this are various and are exemplified in many places in this book. Notable, however, is any structure which effects longshore movement of sand, i.e. sea walls, groynes or jetties; or any activity which reduces the amount of sand in the sediment budget, such as beach mining and offshore dredging. As an example, the sand dune system at Kenfig, Wales is currently experiencing a negative sediment budget, in part due to the construction of a jetty at Port Talbot, up drift of the dune complex (Jones 1996).

*   *   *

Throughout this section we have looked at various activities which have impacts on coastal dune systems. In a book concerned with the problems and solutions associated with different coastal defence types, this may at first appear out of place. However, sand dunes are actually a natural sea wall and so any activity which impairs the functioning of that structure must be a cause of concern from a coastal defence point of view. Typically, problems of trampling and tourist pressure on dunes are studied as ecological issues. In our context, however, they are much more than that. We have seen how few people have to walk through dunes to cause the loss of vegetation. This vegetation is, in effect, the 'cement' in our natural sea wall; it is what is holding it together. If that is lost, the strength of the defence is severely reduced. Similarly, because we are dealing with a dynamic system, our natural wall is interacting with the beach, and so needs to be resupplied with raw material. Following a storm, a concrete wall may be damaged, and engineers need to bring along new concrete or blocks to repair the damage. In dune systems, similar damage may occur, but the raw materials for repair (sand) are already present on the beach. In the same way as if the engineers bring insufficient concrete for the repair, without enough sand on the beach, the dunes cannot repair themselves and so lose their ability to protect the coast.

Clearly then, dunes are a good (and cheap) natural defence which need protecting if they are to perform the job that coastal engineers want them to. In this respect, we need to manage and overcome the problems discussed in the last section.

Methods of dune protection and rehabilitation

From the proceeding section, we can identify the cause of much dune degradation. This puts us in a strong position to do something about it. Perhaps the simplest solution would be to prevent any usage of dunes and allow them to function as purely natural systems that are free to migrate and function as governed by the natural forces acting on them. In reality, this ideal can never be achieved; we are already in the position that due to historical uses, many dunes are closely linked to human development and activities, and therefore must be managed and protected in the best way possible. What is of importance, however, is to decide what we are managing them for. There are many cases and points of view to be put in making such a decision, but in our context the importance of dunes in coastal defence is the most important.

Sand dunes provide an ideal coastal defence. This is a statement that has been reinforced again and again. When considering the best approach to management, there are several criteria for successful management:

- *Interaction* The dunes need to experience as little interference as possible and be allowed to interact with natural processes.
- *Vegetation* Vegetation is essential for stability, but over-intensive vegetation can lock up sediment and be a barrier to successful dune regeneration following storms.
- *Migration* Dunes need to have some mobility, so anything that locks sand up is a bad thing.
- *Sediment supply* Dunes go through short-term phases of erosion and accretion. There needs to be enough sediment available in the beach environment to facilitate this.

Given these criteria, and in accordance with the issues raised in Tables 8.2 and 8.3, we can design the best form of dune management for each situation, given that some areas are already heavily used by tourists, or heavily developed for industry, housing, or agriculture.

When considering the management of heavily used dunes, it is important to bear in mind their multiple function of providing a good amenity area which needs to be preserved, a temporary store of sediment to allow for short-term adjustment of the beach during storm conditions, and an effective defence for the hinterland. Consequently, the management and protection of dunes has to be carried out in such a way that they are protected and able to carry out their natural function, while still maintaining public access. As part of this access problem, it should also be remembered that most people who access dunes only do so as a means of getting to the beach beyond. Because of this, the provision of routes through dunes is perhaps of prime importance and the simplest way of preventing widespread dune degradation.

We can recognise three approaches to dune management. First, in heavily degraded areas where the structure and integrity of the dunes has been lost, it may be necessary to consider the complete reconstruction of dunes. This is perhaps the most drastic in respect of coastal management because it has huge implications for the sediment budget due to the large volumes of sediment required (compare the discussion of beach feeding in Chapter 7). It may be that the supply of sand would be naturally available, but at the expense of other habitats down the coast. Despite this, however, artificial nourishment has been carried out in many dune systems, by the placement of sand on the high beach or in the supratidal area. The techniques employed are similar to those used in beach feeding, and have been discussed in this context in Chapter 7. However, in the case of dunes, pumping sand from dredgers into the recharge area is not straightforward. We have seen how many large dune systems are used for water supplies, or can have vegetation communities that are generally intolerant of salt. Hence, the pumping of salt water rich sand into dunes will have major environmental implications, and necessary steps need to be taken to prevent contamination of the dune water table and vegetation communities. Adriaanse and Choosen (1991) discuss the methodology behind dune recharge and also some of the environmental issues. Case studies relating to dune feeding and regrading can be found in Baye (1990) who discusses dune reconstruction in Canada, Cortright (1987) who looks at management of dunes at Nedonna Beach, Oregon, Marsh (1990) who looks at the nourishment of dunes along the shores of the great Lakes, and Nordstrom (1987) who looks at the artificial grading of dunes along the Oregon coast. A further aspect of creating/reconstructing dunes is their 'naturalness' and how they interact with coastal processes. Dolan and Godfrey (1973) discuss the recovery of artificial dunes compared to natural dunes and find that following storms (in this case, Hurricane Ginger), the artificial dunes tended to erode more and take longer to recover. This compares with similar trends observed between recharged and natural beach profiles.

The reconstruction of dunes is the solution to a problem. The key factor to consider in all such cases is the actual cause of this problem. If the reason for dune decay is lack of sediment in the coastal budget, then the introduction of new supplies may well be a good solution. However, if the decay is due to the locking up of sediment due to afforestation or other land use, then this should be the target for management initiatives. The introduction of sediment will not create the perfect morphology or the most environmentally stable regime. For this to occur, the newly deposited sediment needs to be fashioned and shaped by the wind. Large plantations, as we have seen, alter wind patterns, while artificial grasses or crops alter sediment movement dynamics. Clearly, it is important to consider all aspects of recharge before successful schemes can occur.

The second, and most common, management approach occurs in dunes which are not severely degraded, but need restoration and repair to their seaward face. When undertaking dune restoration by whatever method and approach, it is important from a coastal process point of view to bear in mind the method by which sand is transferred to them. As we have seen, sand is derived from the intertidal zone at low water via the wind by saltation. As a result, any structure or vegetation placed on the ground will intercept sediment and promote sediment accretion. The most common way of encouraging the accretion of sand is to impede the wind velocity, typically by installing 1 m high wire mesh, brushwood, or geotextile fences, which interrupt the coarser sand but allow the finer material to pass through and on up into the higher dunes, hence ensuring that these areas do not become starved of sediment (Plate 8.3a). The effect of these fences is felt approximately 8 times the fence height down wind. In other words, a fence of 1 m height would allow up to 8 m of land behind the barrier to undergo some degree of sedimentation. Janin (1987) noticed distinct patterns to sedimentation around dune fences. Up wind of the fence, sediment accumulation occurred by surface creep, with accumulation against the fence. Down wind, sedimentation occurs due to reduction in shear velocity under saltation or suspension thresholds, and sediment spreads out over wider areas (Plate 8.3b). The exact form of sand deposit varies with fence height, as detailed earlier, and also with fence porosity, angle of fencing in relation to wind direction, and sediment grain size. Hotta *et al.* (1987) tested different forms of fencing in an attempt to understand the role of porosity. Using four types, fern frond fencing (unspecified porosity), split bamboo (50 per cent porosity), 3–4 cm diameter bamboo poles (50 per cent porosity) and 15 cm wide by 1 cm thick wooden slat fencing (40 per cent porosity), they found that the most effective with respect sediment accumulation was the fern fencing, and the least effective was the wooden slats. Overall they concluded that any fence with less than 50 per cent porosity would diminish accumulation rate, and the higher the fence porosity, the higher and longer the resulting dunes. This work followed on from earlier studies by Kimura (1957), who had investigated the impact of the angle of the fencing relative to prevailing wind. This earlier work found that the two criteria interacted and were both important, and that as the wind veered from being normal to the structure (i.e. head on at 90°), the porosity of the fencing needed to change for maximum accretion rates to be achieved. When wind was blowing at 90° (directly at the fence), maximum accretion occurred with 67–71 per cent porosity, at 80° to the structure, maximum sediment accretion occurred in fences with 50 per cent porosity, and at 55° to the structure, at 60 per cent porosity (Kimura 1957). A stabilisation scheme at Camber Sands, UK utilised four rows of galvanised wire fencing and demonstrates the effect of fence porosity (Metcalfe 1977). While the trapping of sand proceeded well, a reduction in fence porosity from 69 to 42 per cent resulted in a dramatic increase in accretion. In addition, the use of 1.2 m high spur fences at 90° to the main fence, spaced every 11 m also increased sand trapping.

An important consideration, in addition to that of prevailing wind direction, is the positioning of fences with respect to mean high water mark. A placement too far to seaward would increase the chances of them being eroded out during storms, but too close to the dune edge would reduce the volume of sand trapped and reduce the spatial extent of any developing dunes. Viles and Spencer (1995) also claim a distinct seasonal erosion/accretion cyclicity if fences are placed too far seaward. With these issues in mind, there are a range of suggestions for minimum/maximum distances from the dune face/sea when placing fences. Woodhouse *et al.* (1976a) claim a distance of 100 m from mean high water mark (MHWM) as being the ideal in the case of storm-damaged barrier island dune complexes in North Carolina. Other practitioners quote

Plate 8.3 Use of artificial obstructions to encourage sand accumulation in dune areas
    a) Slatted fences to aid the build-up of foredunes. Two rows of fencing have allowed sediment to accumulate in front of previously eroding dunes, Formby coast, Lancashire
    b) The use of fencing to intercept longshore sand drift, Sefton Coast, Lancashire
        (Both plates reproduced with permission of Prof. Ken Pye, Geology Department, Royal Holloway, University of London).

vertical distances above MHWM as a guide, rather than horizontal distances. Adriani and Terwindt (1974) cite a figure of 1 m above MHWM for most European coastlines, while Brooks (1979) favours 2 m (data cited in Ranwell and Boar 1986).

By using such fences, sediment deposition can be encouraged, but this will only operate until the depth of sand equals the height of the fence. The planting of vegetation can stabilise this sand once it has buried the fence and also, by upward growth, continuously increase the height of the obstruction to wind flow, thus allowing continued accretion. The planting of marram in combination with sand fences and mesh matting have all been successful in regenerating dunes by natural processes of sediment entrapment and vegetation colonisation. In the Mediterranean, where tourist pressure can be particularly high, management of dune systems is of great importance to maintain tourist interest in the area (van der Meulen and Salman 1996). Dunes here have been successfully stabilised by many techniques, including renourishment, sediment entrapment, and afforestation. Clearly it is important to use species native to the dune area in question to promote the formation of 'natural' communities and transitions. Typically, in embryo dunes, marram and sand couch grass are commonly used. The rapid growth and penetrating root system serves not only to bind sand together, thus making it more resistant to wind erosion, but also to act as a wind baffle, thus promoting further sediment trapping.

Because the use of marram is well established in many countries, the method is well researched. The planting design and spacing between plants will vary according to site conditions, bearing in mind the importance of not over-planting and 'fossilising' dunes. Hobbs *et al.* (1983) claim that where accretion is rapid (i.e. up to $0.3 \, \mathrm{m \, a^{-1}}$), plant rhizomes should be planted vertically, but on sites where erosion predominates (such as where rapid blow-out formation is occurring), the rhizomes should be planted at an angle and spaced further apart to increase surface protection.

General planting rules, however, include planting to a depth of 0.15–0.2 m in a 'domino-5' pattern (Ranwell and Boar 1986), with a spacing of 0.3–0.9 m between plants. The exact spacing depends on the degree of site degradation, slope angle, and the degree of wind exposure. As to the question of which plant species to use, this will again depend on site and should be governed by species already present. Marram, sand couch grass, and lyme grass are all commonly used species. Mixed plantings are also used especially where the sand budget is in severe deficit and aeolian transfer from the beach limited. In areas of severe erosion, marram is the most prolific growing, and so is often used with one other species which spreads between to trap further sediment. Work by Adriani and Terwindt (1974) shows that favourable results can be obtained by interplanting lyme with marram or sand couch. Planting combinations were successfully used at Gullane in East Lothian, Scotland. Sand fencing made of buckthorn and wood slats accreted 3.6 m high dunes in 6 years. Following this, planting with lyme on the seaward side, and marram on the landward side successfully stabilised the newly formed dunes. Also in this example, severely degraded blow-out areas were successfully reclaimed using mechanical contouring of imported sand (example cited in Ranwell and Boar 1986). While the planting of grasses represents the most common form of vegetation stabilisation, other species can also be used. Most mature dunes need to be protected with plants that reflect their maturity. This often requires woody perennials, such as buckthorn or white poplar (Ranwell and Boar 1986).

It is clear that a wide range of methods and species is suitable for dune protection. There are extensive guidelines available, often compiled by nature conservation organisations, covering planting patterns, species, and planting times. One suitable starting place, for the interested reader, is Ranwell and Boar (1986). A series of field examples, covering a range of different approaches, locations, and species, is also given in Table 8.4.

Reckendorf *et al.* (1985) illustrate many of the issues raised above, when they discuss the stabilisation of sand dunes in Oregon, USA. In 1885, jetty construction at the mouth of the Columbia River interrupted the sediment supply down drift of the structure, and promoted the accretion of 1214 ha of sand up drift by the early 1930s. This sand started to develop embryo dunes although wind erosion caused rapid degradation. The effective stabilisation of these dunes occurred in three stages: first, by placing fencing to establish effective foredunes; second, planting of these with grass cover; and finally, permanently stabilising with herbaceous and woody vegetation. Within 8–9 months of the fencing being built, dunes up to 1 m in height had developed for 3–5 m either side of the fences. Similarly, in California, the excessive use of off-road vehicles was treated by vegetation planting (Trent *et al.* 1983). As an alternative to vegetation and fencing, the stabilisation of eroding dune faces with some form of mesh, such as that discussed by Ghiassian *et al.* (1997) can be useful, but tends to prevent any further process, such as transfer to the dunes inland.

If the management of dunes is confined to repair and reconstruction without tackling the cause of the problem, then the problems will re-occur. For example, the building of a jetty at Port Talbot in south Wales is blamed for the reduction of sediment supply to the beaches and dunes at Kenfig (Jones 1996). This, coupled with extraction from the dunes and dredging from nearshore means that the dune/beach system has a negative sediment budget. In addition, therefore, it is important to develop holistic management to tackle the general causes of dune degradation. Often dune construction techniques can be supplemented by simple, low cost methods. Given that one of the main causes of dune degradation is trampling, basic visitor management by itself can be effective. One simple method is the construction of wooden walkways. This is a useful way of reducing the effects of trampling; the vegetation on either side of the walkway remains healthy because of the tendency to concentrate walking on these pathways, since the sand itself is very difficult to walk through. Other techniques can include excluding visitors from certain areas of the dunes by fences and keeping out grazing livestock. The problem here is that the wooden fences which tend to be used for this purpose also make very good firewood for dune parties and, as such, many get damaged. Better success has been observed

Table 8.4 Examples of dune protection and rehabilitation studies from around the world (For full references, see References at the back of the book).

Country	Authors	Location
Canada	Baye (1990)	North-east
	Kellman and Kading (1992)	Ontario
Egypt	El Raey (1997)	Nile Delta
	Misak and Draz (1997)	General
France	Bressolier-Bousquet (1991)	Archachon Bay
	Favannec (1996)	Aquitaine
	Guilcher and Hallégouët (1991)	Brittany
Israel	Perry and D'Miel (1995)	General
Mediterranean	van der Meulen and Salman (1996)	General
Netherlands	Adriaanse and Choosen (1991)	General
	van Dijk (1989)	General
	Louisse and van der Meulen (1991)	General
	van der Maarel (1979)	General
	North Holland Water Supply Co. (1992)	North Holland dune Reserve
	Rutin (1992)	Deblink dunes
	Verhagen (1990)	General
	Watson and Finkl (1990)	General
New Zealand	Gagil and Ede (1998)	General
	Partridge (1992)	Kaitorete Spit
Senegal	Mailly *et al.* (1994)	Northern coast
South Africa	van Aarde *et al.* (1996)	Kwa Zulu-Natal
	van Aarde *et al.* (1998)	Kwa Zulu-Natal
Taiwan	Lin (1996)	South-west
UK	Boorman (1989)	General
	Carter (1980)	Northern Ireland
	Carter and Stone (1989)	Magilligan, Northern Ireland
	Hill and Wallace (1989)	Newborough, Anglesey
	Houston and Jones (1987)	Sefton coast
	Jones (1996)	Kenfig, Wales
	Page *et al.* (1985)	Ynyslas dunes, Cardigan Bay
	Pye (1990)	Sefton coast
	Whatmough (1995)	Murlough, Northern Ireland
USA	Cortright (1987)	Nedonna Beach, Oregon
	Dahl *et al.* (1975)	Texas
	Gares (1990)	Island Beach, New Jersey
	Jagschitz and Wakefield (1971)	General
	Lancaster and Baas (1998)	Owen's Lake, California
	Marsh (1990)	Great Lakes
	Nordstrom (1987)	Oregon
	Nordstrom and Psuty (1980)	General
	Psuty (1993)	Perdito Key, Florida
	Reckendorf *et al.* (1985)	Warrenton, Oregon
	Trent *et al.* (1983)	California

in areas where these forms of dune management have been coupled with a programme of public education in the form of information boards and displays. If the public can see why some course of action is being taken, then they are greater respecters of it.

Impacts associated with dune protection

The successful management of dunes is environmentally beneficial in a number of ways. First, it provides an important, self-sustaining coastal habitat which acts as an effective line of soft coastal defence; second, healthy dunes provide a temporary store of beach sand which can be used during stormy periods to supply and maintain beach levels; and third, dunes provide important conservation and amenity areas. These three issues mean that careful management of dunes can have significant environmental benefits for coastal defence and beach stability providing that it occurs in a way which maintains the ability for land-sea interactions, as well as local socio-economic factors.

Because dunes are so fragile, there is a great tendency to over-protect them. Such over-protection, however, is itself a problem. We have already seen how important it is to maintain a constant exchange of sediment between beach and dunes. If the dune becomes over-stabilised, then its store of sediment will become locked up and become unavailable to beach processes, or for landward transfer. This is often a risk when erosion problems are being treated. In an attempt to prevent excess loss of sediment, movement is stopped completely, rather than reduced. The result will, in effect, be a coast which behaves in a very similar way to one backed by a sea wall because it cannot adapt in order to absorb wave energy efficiently, creating further implications for long-term beach stability. This over-management can result from the overuse of fencing and other sediment entrapment methods which may be appropriate for trapping the sediment to build up the dunes, but will not allow the movement of sediment seawards again. While such methods will enhance sediment accretion in the parts of the dunes protected by them, it will, in addition, prevent sediment reaching the back dunes area. The overall effect is the same as when a sea wall is built along the face of a cliff: the hinterland and beach are separated and a potential supply of sediment to the coastal sediment budget is removed.

Clearly then, any of the management approaches discussed earlier need to be carried out with a thorough understanding of the environment in question. Too little management will result in an ineffective protection policy that will fail to achieve its targets, while over-protection will lock up too much sediment and hinder environmental dynamics. However, even given the best dune protection strategy devisable, the fact that dune environments are so fragile and vulnerable to change means that any subsequent amenity usage needs to be coupled with additional adequate management. This may simply be the construction of footpaths or walkways to keep people away from vulnerable areas, or it may be a much grander, holistic approach that tackles interconnected issues of sediment supply, public pressure and extraction. The example of Kenfig Dunes cited earlier (Jones 1996) is one such example where management has had to deal with such a multi-faceted approach to a dune complex. Management planning in this area is adapted to overcoming these problems. Perhaps ironically, given what has already been said about formation of bare areas, erosion hollows created by sheep are considered invaluable in this area, not only as an area of mobile sand, but also as a habitat to attract species which were becoming rare. Little can be done about the jetty but extraction has now ceased. Vegetation planting in certain areas, while leaving others to erode is helping to manage the sediment budget within the dune system and preventing 'fossilisation' of the system. There will, however, be implications for the fore-dune environments as without a new supply from the beach, these will, in effect, roll over landwards.

The internationally important Sefton dunes in north-west England have many of the common dune management problems but, in addition, have the further problem of being located within an urban area, and of acting as a recreation area for the nearby cities of Liverpool and Manchester. Hence, the dune areas need to be managed in an holistic way to provide for amenity usage, production, conservation, and coastal defence. The Sefton Coast Management Scheme (Houston and Jones 1987) aims to mix these multiple demands and encompass environmental improvement, recreation and education development, creation of jobs, and also the conservation and coastal protection uses of the dune area. Much of the coast has conservation designations, while other areas are owned or managed by local authorities or the conservation bodies themselves which, through countryside management and stewardship schemes, have permitted lease holders to farm and manage their land in environmentally sensitive and sustainable ways.

There is no basic blueprint for managing human and physical aspects of dune environments, and what may be applicable for one area may not work in another. However, because dunes are so important for coastal defence, and also because they have great appeal for recreation and production of crops and water, there is always a constant trade-off between usage and protection. Successful implementation of management schemes should allow both to exist in harmony.

Sand dunes and sea level rise

There is much evidence from the recent geological record that rises in sea level promote a rapid landward movement of dunes. The post-glacial marine transgression which occurred along many of the world's coastlines can be shown to be responsible for many landward dune migrations (Hesp and Thom 1990). Providing that adequate sediment is available, extensive transgressive dune fields can result (see Pye and Bowman 1984, Shepard 1987, and Thom *et al.* 1981).

Such observations, albeit from the Holocene, provide a useful analogue by which to consider current sea level rise and the future of dune coasts. Under 'natural' conditions, dunes migrate landwards, with new embryo dune growth on the shoreface supported by new sediment deposition. On many contemporary dune coasts, however, landward migration is no longer possible due to development, and the supply of new sand is limited by budget restrictions, Hence, these are two major constraints for dune survival under contemporary sea level rise. Furthermore, Cooper (1958) argues that dune migration under sea level rise is also driven by increased storminess. This phenomenon is also something which typically occurs under conditions of climate change and can be seen to be happening in several of the world's seas and oceans. Hence, we have observational evidence as to how dunes will react to sea level rise, and the processes which will drive this reaction. Due to increased human-induced constraints on coastlines, however, we also know that such processes cannot be allowed to occur due to lack of hinterland space, or lack of sediment.

In this chapter we have seen the importance of sand dunes as a natural sea defence. Not only do they provide a barrier to flood waters, but also a sediment supplier or store. In the Netherlands, coastal dunes are the major line of defence for prevention of flooding. Many sand dune communities are still 'natural' in that they are free to interact with the sea, although many are showing signs of instability, with cliffed seaward edges indicating that the sea is cutting back into them, removing traces of embryo dunes. According to the Bruun rule, much of this sand will be available for landward transfer and the natural dune succession will reform inland by a process of new dune formation and the reversion of mature dunes to more immature types. By this process, sea level rise will not have a detrimental impact on the dunes system overall, although it will involve some loss of hinterland to newly forming sand dune communities. Nordstrom *et al.* (1990) claim that the predicted rate of sea level rise will cause such processes to become widespread.

The critical issues, however, relate to any infrastructure landward of the dunes, because the presence of high-value land will mean that the dunes cannot move landwards, and so will, therefore, experience coastal squeeze. Along the Fylde coast, for example, the dunes south of Blackpool are undergoing landward migration, and threaten to inundate the coastal road, a holiday resort, and a local airport behind (see Plate 8.2). Because of this, the local authority has a major management issue to decide, i.e. whether to hold the dunes, or relocate the infrastructure.

A further issue particularly relevant to dunes is their vegetation. Variations in water table caused by sea level rise could result in saline intrusion and the alteration of fresh or brackish communities. Although the main stabilising species are not threatened by this because they are saline and arid tolerant, dune slack communities could be affected. While this does not really have direct defence implications, it could impact on conservation value, which if damaged, may lead to changes in conservation status and potential management strategies. Ultimately, such management strategy changes could indirectly affect defence management. Regarding water, some of the larger dune fields, such as those in the Netherlands, do supply large volumes of drinking water, and this supply could be threatened.

Dunes represent a major coastal system which can readily respond to sea level rise, but which, in many places, cannot do so due to increased anthropogenic influences over them. For these dunes, the common approach is to maintain them where they are, trying to protect them in the best way possible without inundating the hinterland. When we considered dune stability earlier in this chapter, we saw how mobile sand can be protected — but have we always been preventing erosion in the right places? Dune faces need to erode under sea level rise as part of the landward migration process (Bruun rule); in doing so they will provide an increased supply of sediment to beaches and thus contribute to the sediment budget. (This idea was suggested by Carter (1988) when he suggested that sea level rise could increase available coastal sediment due to erosion.) By stabilising them, we are trying to 'hold the line' as far as position is concerned, and so with continuing sea level rise, the erosion problem will, in time, become more difficult to manage. There must come a point when managers need to consider whether the continued protection of dunes is sustainable or whether, as with cliffs, there will come a time when there needs to be consideration of letting the coast 'go', and allowing the roll over of dunes landwards. This will involve some difficult decisions with respect to hinterland use.

Perhaps more important here are the barrier island dunes. In the USA, this issue is receiving a lot of attention from the Environment Protection Agency (EPA) because it poses a real threat to large areas of development. Leaving aside the arguments as to whether building on barrier islands is itself a sensible course of action, Titus (1990) has demonstrated the large volumes of sand needed to protect these features under various sea level rise scenarios. Protection may be provided by either increasing the height, or extending them landwards and relocating development (engineered retreat) (either way involves the relocation of property, so why not relocate to safer ground and abandon the threatened barrier islands?). Under the worst case scenario of a 120 cm rise in sea level by 2100, 47.9 million m^3 of sand will be required for Long Beach Island alone. With this in mind, we can revisit the question discussed earlier in the context of beach feeding — where is all this sediment going to come from?

We cannot provide an answer to these issues here, but nevertheless, they are very real. In essence, the sea level rise issue with respect to dunes can be summarised as follows.

- Dunes naturally migrate landwards under sea level rise (compare the Holocene transgressions).
- Modern dunes are often backed by development and, therefore, landward movement is restricted.
- Dune migration needs a supply of sediment to feed newly forming embryo dunes. Sediment budget constraints often make this supply difficult to maintain.
- Many dunes have undergone a change of use and so may not be able to

move. If these are defended, they become artificial and isolated from the natural system.

The debate is continuing. There are many elements to this discussion but whatever decisions are made (and these should be site/case specific) it will have implications for coastal stability over a wide area.

Summary

In this section we have seen how a natural deposit of sediment (dunes) plays a key role in coastal protection. In one way, dunes represent all that is right in coastal protection because they provide effective defence for the hinterland, while interacting with the sea to maintain that protection under a range of wave and tidal regimes. This ability to interact lies in the fact that despite being vegetated, the seaward margins are still relatively mobile, and able to provide sediment to beaches when levels are low, yet store it when levels are higher. Put another way, dunes are cheap defence structures. This would represent an ideal situation for the coastal manager were it not for external pressures put on them. A dune's ability to provide or store sand is ideal for beach levels, but it also represents their 'Achilles heel' in respect of recreation or exploitation, because they remain very fragile and vulnerable to change under minimal pressure and as such, are readily degraded and eroded.

Because research has shown how dunes work, and how sediment is moved into the fore-dune environment, we are able to develop methods by which we can encourage dune growth. In many cases this has been successful, yet if dune systems are to continue to function and provide effective coastal defence, the prime outcome of dune management has to be a system which is protected but still able to interact naturally with coastal processes. This is particularly the case when we consider sea level rise, and how, if they are to remain 'stable', dunes will need to migrate landward. Historically, human interference has been such that as management continues, this potential for mobility has been lost. Over-protection will have a similar effect to that of a sea wall while under-protection will allow the problems to continue. To conclude this chapter we can recognise the importance of dunes, but they can only operate successfully where they can maintain a supply of sediment, and can migrate inland in response to sea level rise. Around many of the world's coastlines, these two basic prerequisites are becoming less and less common.

Summary of benefits of dune protection strategies

- Dynamic coastal defence which can adjust to natural coastal processes
- Once installed, will enable dune growth to continue while not being visually intrusive
- Provision of important conservation areas

- Protection for locally significant sediment store and future input to coastal sediment budget

Summary of problems with dune protection strategies
- Locking up of too much sediment due to excessive defence construction
- Increased immobility due to plantations or conversion
- Increase in 'false' security, leading to inappropriate coastal development

Recommended usage
- To maintain and protect the main line of coastal defence
- To be constructed in such a way that dune mobility and integrity is maintained and preserved

9   Increasing sedimentation in mudflat environments

Introduction

We have already seen in Chapter 7 the importance of having a well established intertidal area as a means of attenuating wave energy and thus reducing the impact of waves on the hinterland. We have also seen how this can be achieved using beach feeding in environments that are experiencing a net sediment loss, and by the use of sediment retention structures.

There are situations, however, where enough sediment is moving along the coast but, due to unsuitable environmental conditions, such as wave exposure or tidal currents, it does not either remain on the foreshore once deposited, or does not get the chance to settle out from suspension. This is particularly the case in low energy environments such as estuaries, where clays and silts predominate. In such situations, it is not necessary to import sediment artificially from other sources, but to encourage it to settle out and remain in the desired area. Thinking laterally, this is something which we have already touched on in other chapters; coastal structures are often put in place to interrupt sediment transport pathways and encourage sediment to settle. Groynes interrupt longshore drift, fencing intercepts wind-blown sediment, and offshore breakwaters reduce wave exposure and allow sediment to settle out in their lee. A chapter therefore, dedicated purely to methods of encouraging sediment to remain on the foreshore might appear unwarranted. To a certain extent this is true, but all the methods discussed in other chapters relate to sandy sediment or coarser (typically referred to as non-cohesive sediments). Such sediments are able to settle out from suspension or other transport mechanism during normal tide and wave conditions. None of the techniques are really suitable for muddy environments in the way previously described, because even under low energy conditions, fine clay particles need time to settle as individual particles, or to flocculate (the process whereby individual clay particles 'stick' together in response to surface charges which attract particles together, formed where saline water meets fresh, effectively increasing their particle size and, hence, particle mass). Methods by which clay and silt particles settle on mudflats, either directly from suspension or in association with creeks, are discussed in the context of tide-dominated environments in Chapter 2. However, as an example, a typical clay particle of 2 μm has a settling velocity of 0.00024 cm s^{-1}. As soon as current velocities fall below this level, the particle will start to fall out of suspension. In reality, such low currents only occur around slack water (as the tide turns from ebb to flood and flood to ebb). Given a depth of water of 1 m over a mudflat, this particle will take nearly 58 hours to settle (Pethick 1984), or in reality, will never settle out. By flocculation, these small particles stick together, increasing their size and, hence, reducing the time required for settling. The time taken will depend on the size of the flocs, which may reach fine sand or coarse silt grade. Fine sand particles between 125–62 μm will settle at velocities up to 0.4 cm s^{-1}. This chapter will look at the additional methods by which the accretion of fine-grained sediments can be achieved in the more sheltered mud-dominated (tidal) environments.

In the last few decades, new advances and techniques have emerged which are specifically designed to encourage deposition in mudflat areas. Because of the major contrast between non-cohesive sands and cohesive clays and muds, the methods need to be different. The principle reason for this is that clays take a long time to settle and, while sands can be trapped by placing structures to intercept drift, finer sediments need a different approach.

Philosophy of encouraging fine sediment deposition

If there is any urgency to a protect an area which needs rapid increases in mudflat elevation, then the most straightforward approach would be to use dredged material (either from offshore or, if suitably clean, from a local harbour) and dump it on the mudflat. This would achieve good results in a short period of time, thus speeding up the benefits from increased wave attenuation. Subsequently, these elevated flats could be artificially vegetated (see below), further increasing their potential for wave attenuation and defence. The key problem, however, is that mudflats are ecologically extremely productive environments, and are claimed to represent one of the most biologically productive habitats in the world (McLusky 1989). Organisms within the mudflats (infauna) adjust to sediment accumulation by moving up the profile, thus maintaining a constant depth of burial. There is a maximum speed at which this can be achieved; if sediment is input too rapidly, this infauna will not be able to adjust its position rapidly enough, and as a result can become buried and die. Recharge methods which introduce a large volume of sediment rapidly could well cause such a problem. Hence, in areas where there is a large infaunal community which supports many other organisms, it is important to maintain the ecological integrity of the mudflats, hence, gradual sediment build-up is necessary. It is possible that sympathetic mudflat feeding can be used in such circumstances, such as trickle charging, where the recharge sediment is piled up in the low shore zone and is then brought in by the waves. Care still needs to be taken, however, not to introduce too much sediment in a short period of time.

Where there is less importance on rapid accretion, and where there is abundant sediment within the water body, it is possible to utilise the natural accretion potential of the water body to deposit the sediment. Clearly, there needs to be some human intervention to promote and encourage sediment deposition. To do this we need to remember that sediment will only settle once the current velocity drops below a deposition threshold (see the earlier example from Pethick 1984). Hence, if we reduce current velocity, we can increase the amount of sediment which settles out. This, however, is only part of the problem, as it is also necessary to keep the sediment on the flats after it has been deposited. For this we can either rely on natural sediment compaction, or utilise other techniques of marsh development (see below).

Methods of increasing natural sedimentation

Ultimately, to increase sediment deposition we need to decrease current velocities to allow fine-grade sediments to settle out. There are several methods of doing this, including some discussed elsewhere. In Chapter 6 the use of an offshore wave break is described for the Essex coast in the UK. Here, rapid accumulation of fine sediments has occurred behind the protection of the structure and has led to significant vertical mudflat accretion. In this chapter, however, we are focusing on methods within the intertidal zone, i.e. actually constructed on the mudflats themselves. There are a range of different approaches to increasing mudflat sedimentation, some of which are more commonly used than others. Each approach will be discussed in turn.

Brushwood fencing

By building permeable brushwood structures on mudflats, either in the form of simple shore normal features (groynes) or a combination of shore normal and shore parallel structures (box groynes), successful wave baffles can be achieved. This can lead to increased sediment deposition due to the ponding up of water on a flood tide, and then allowing water to drain slowly on the ebb through permeable sides. This arrangement produces a sedimentation field and allows sediment to settle out and build up the mudflat level. At Wallasea Island, Essex, brushwood structures (two lines of vertical stakes, between which finer branches or other similar material, such as old Christmas trees, are woven), have been used to successfully build up the mudflat fronting a degraded marsh edge (Plate 9.1a). This example is relatively smallscale and typifies a box groyne system, comprising several sedimentation fields (Plate 9.1b). Other examples also occur around the UK coastline. Pye and French (1993a) discuss the use of box groynes at Rumney Great Wharf, south Wales, where, following construction in 1988, mudflat level had increased by 50 cm by 1991 (Toft and Townend 1991). In the USA, Good (1993) also discusses the use of brushwood fences in the restoration of marshes in Louisiana. Here, sedimentation fields are typically 1.5 m square and 1.2 m high, and constructed of vertical stakes with spaces infilled with old Christmas trees. Again, sediment accumulation within the fields

Plate 9.1 Use of brushwood structures to encourage mud flat accretion
    a) Brushwood groynes being used to encourage sediment accretion fronting an eroding mudflat.
    b) Sediment accumulation within a box groyne system
    Both plates from Wallasea Island, Essex, UK.

has been rapid. There have been many attempts at using both the box-type and the more complicated Schleswig-Holstein arrangements (detailed below). Various degrees of success have been achieved. A summary of the Essex coast experiences is given in Box 9.1.

A more complex arrangement of the brushwood groyne arrangement involves constructing a large box groyne (typical dimensions are around 800 m long by 400 m wide (Hydraulics Research 1987)), and subdividing this with a series of earth groynes in association with a series of shore normal and shore parallel ditches (Figure 9.1). The ditches increase drag on flood tides and allow sediments to settle in the deeper ditch areas. They also increase surface roughness sufficiently to decrease the amount of sediment re-entrained during ebb tides. This technique is referred to as the Schleswig-Holstein method and was first developed on the North Sea coast of Germany (Hydraulics Research 1987). On the Dengie Peninsula in Essex, such a scheme was built in 1981, with a second alongside it in 1983 (Plate 9.2) (Box 9.1). Sedimentation has been rapid and mudflat levels are now considerably higher than those outside the scheme. Along the same coast, Mersea Island to the north had a Schleswig-Holstein system installed in 1988, which has led to net vertical accretion rates of around $10 \, \text{mm} \, \text{a}^{-1}$ (Pye and French 1993).

Box 9.1

Use and impact of sedimentation fields on the Essex coast, UK

The Essex coast, UK has seen extensive use of sedimentation fields in an attempt to enhance mudflat accretion and reduce the impact of increased wave activity on marsh edge erosion. A range of styles has been used, from simple brushwood groynes, to box groynes (Plate 9.1b), offshore breakwaters (Plate 6.3) and the Schleswig-Holstein method (Plate 9.2).

A pilot Schleswig-Holstein scheme was carried out in 1980 at Deal Hall on the Dengie Peninsular (Figure 9.2). The scheme was intended to accelerate sediment accretion in front of rapidly retreating marshes that had occurred over the previous 20–30 years. The scheme, shown in Plate 9.2, was 400 m square with outer edges of brushwood groynes, constructed with double rows of wooded stakes infilled with brushwood. The 400 m square was divided into two by 200 m sedimentation fields by a shore-normal groyne within which shallow ditches were constructed (see Plate 9.2, Figure 9.2). Another scheme, constructed in 1986, involved the formation of an offshore breakwater (Plate 6.3) with two shore-normal groynes at the edges. Similar methods to the above were also employed at Sales Point in 1986, and at Horsey Island in 1989. Details of these schemes, and others, can be seen in Table 9.1, along with some indication of their success or failure.

Table 9.1 Details of sedimentation field schemes from the Essex coast. Locations are given in Figure 9.2, with details in Box 9.1 (Modified from Pye and French 1993a. Original table in Toft and Townend 1991).

Location	Method	Date	Outcome
1. Deal Hall, Dengie Peninsula (Plate 9.2)	Brushwood enclosures (third built in 1989)	1980, 1989	Successful accretion
2. Maydays Creek, Colne Estuary	Brushwood groynes and creek closure	1981–2	Successful accretion
3. Foulton Hall, Hamford Water	Brushwood enclosures	1982	Abandoned
4. Mill Point, Blackwater Estuary	Brushwood groynes and creek closure	1983	Abandoned
5. Marsh House, Dengie Peninsula	Offshore breakwater. Brushwood groynes added 1987	1984, 1987	Successful accretion after 1987
6. Tollesbury, Blackwater Estuary	Brushwood enclosures and groynes	1986	Continued erosion
7. Grapnells Farm, Crouch/Roach Estuary	Brushwood enclosures and groynes	1986, 1989	Successful accretion
8. Brickhouse Farm, Blackwater	Brushwood enclosures and groynes	1986	Minimal impact
9. Sales Point, Dengie Peninsula (Plate 6.3)	Offshore breakwater and brushwood groynes	1986, 1989	Successful
10. Stamfords Farm, Crouch Estuary	Brushwood enclosures and groynes	1986	Minimal impact
11. Wellhouse Farm, Blackwater Estuary	Brushwood enclosures and groynes	1986, 1989	Minimal impact
12. Bohuns Farm, Blackwater Estuary	Brushwood enclosures and groynes	1987	Minimal impact
13. Fleet Hall Farm, Roach Estuary	Brushwood groynes	1987	Successful
14. Mell Farm, Blackwater Estuary	Brushwood enclosures and groynes	1987	Continued erosion
15. Raypits Farm, Crouch	Brushwood enclosures and groynes	1987	Successful
16. Horsey Island, Hamford Water	Offshore breakwater and brushwood groynes	1987, 1989	Minimal impact
17. Cudmore Grove, Colne Estuary	Brushwood enclosures	1988/89	Minimal Impact

From Table 9.1 it can be seen that although methodologies may be similar, scheme success can be variable. However, the definition of success may itself be somewhat variable, in that a successful scheme may be measured in the fact that marsh erosion stops, or that rapid accretion and new marsh is formed. Leaving aside such debates on definitions, it is important from a coastal defence point of view that the marsh is afforded some protection. The schemes on the Crouch and Roach estuaries are generally considered to be some of the most successful (Pye and French 1993a). It is argued that this is primarily due to the abundance of sediment available in the estuarine system which can, once trapped in the sedimentation fields, settle out and compact. The Blackwater estuary has shown much less success, however, and sediment is again thought to be the main reason for this — in this case the lack of it.

The scheme at Deal Hall saw rapid accretion in two of the enclosures, but not the third (built some years later). In this latter field, no ditches or internal embankments had been constructed, emphasising the importance of the internal ditching in the Schleswig-Holstein method. At Marsh House, significant accretion occurred only after an extra groyne was added in 1983. Accretion then progressed so rapidly that pioneer marsh is now starting to form. Similar initial colonisation is also apparent at Deal Hall and Fleet Hall Farm. The offshore breakwater scheme at Sales Point (Plate 6.3) has experienced an average vertical accretion of $5\,\text{mm}\,\text{a}^{-1}$ although considerable seasonal variation does occur, as it does at the Cudmore Grove site. This may suggest a seasonal dynamism to the accreting mudflats, due to the exposure to seasonal variation in wave activity.

It appears from the short discussion above that a range of factors is important in the success of sedimentation field schemes in Essex. Of primary importance here is the availability of sediment within the estuarine water body. A further series of factors, such as wave exposure and the degree of sediment settling and compaction, is also important in sediment retention. The settling of sediment itself is controlled by the presence or absence of a ditch system within the sedimentation field. Hence, a series of rules can be suggested, based on the Essex experience:

- The sediment budget of the system is critical in that adequate supplies of suitable sediment need to be readily available.
- The ditch system constructed within the sedimentation field is a key factor in inducing sediment to settle.
- Suitable wave protection by strategically placed groynes needs to be established in order to preserve the sediment during dewatering and compaction.

The use of these criteria can increase the cost-effectiveness of sedimentation field schemes. However, their use will not reverse the causes of such local erosional trends but will offer suitable wave attenuation and offset marsh erosion rates, and in so doing, provide additional security for the hinterland.

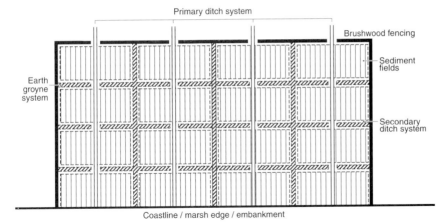

Figure 9.1 Typical Schleswig-Holstein set-up for mud flat accretion. The conjunction of primary and secondary ditches, along with sedimentation fields promotes rapid accretion (See also Plate 7.2).

Plate 9.2 Schleswig-Holstein method of sediment accumulation, Dengie Peninsular, Essex (Reproduced with permission of Cambridge University Air Photograph Library).

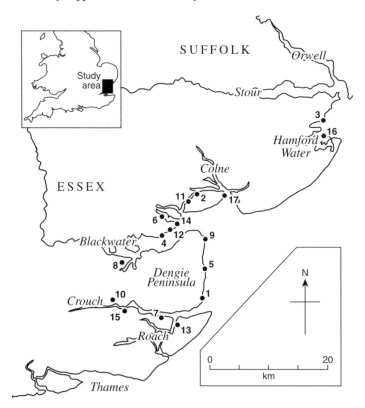

Figure 9.2 Location of sedimentation field schemes along Essex coast (For key to sites, see Table 9.1).

Small scale wave baffles

While the construction of sedimentation fields represents a large-scale approach to mudflat accretion, in recent years there has been an increasing use of small wave baffles to increase sedimentation and encourage vegetation growth on a smaller scale than can be economically achieved with using brushwood fencing. Since their first use in Galveston Bay in the early 1970s (Davis and Landin 1998), the US Army Corps of Engineers has developed techniques of using geotextile tubes or bags filled with sand to construct low breakwaters to encourage ponding of water and subsequent settling of sediment. As with the brushwood groynes, water flows into the protected areas and, as the tide ebbs, is trapped and so sediment settles out under relatively calm conditions.

The idea of using geotextile rolls filled with sand has also been utilised on a smaller scale in the UK. LRDC International (1993) discuss the use of rolls and bags as a means of protecting and restoring eroding marshes at Lymington, Hampshire. Figure 9.3 shows the typical structures used. In each case, the filled bags or rolls provide a wave baffle in a similar way to the brushwood groynes

a)

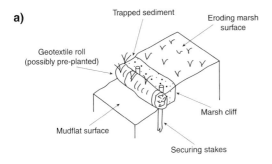

b)

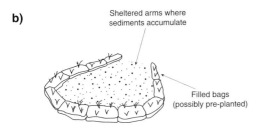

Figure 9.3 Use of geotextile structures in increasing sedimentation rates on inter-tidal flats (Modified from LRDC International 1993).

described earlier, although they are comparatively much smaller. If the structures are planted with pioneer marsh vegetation, then once sediment levels behind the structure have reached a suitable height, vegetation can colonise it and thus produce additional sediment stability (see below).

Recharge and trickle-charging

As the whole rationale behind building up marshes and mudflats is to increase wave attenuation and protect the shoreline from direct wave attack, in areas where available sediment from the water body is low, or where rapid accretion is necessary, some artificial importation of sediment is necessary. The physical principles behind this are to increase the amount of wave energy lost through friction with the underlying sediment surface. We have already seen this principle in Chapter 7, and it also occurs in many other chapters. As mudflats build vertically, the wave base moves further offshore, thus the distance over which waves have to travel increases, and so the amount of energy lost through friction increases. It is not necessary to reiterate details of beach feeding here, but reference is made to the earlier discussion in order to introduce the validity and relevance of this technique in the mudflat environment, accepting that issues relating to the burial of infauna are more acute. Increasingly, the approach involves the use of suitably clean dredged material being pumped onto the mudflat area, followed by a period of dewatering and compaction. After this, the mudflat level is typically at an elevation at which salt marsh vegetation will survive. De Laune *et al.* (1990) describe a fine sediment recharge scheme from the Mississippi, where rapidly deteriorating *Spartina* marshes were causing major concern. Using dredged sediment, the marsh surface was increased by around 10 cm with the result that marsh biomass doubled and new shoots developed, leading to lateral marsh spread. Using this work, the authors suggest that the diversion of sediment-laden Mississippi waters could provide natural sediment to the area which in turn could lead to the rapid regeneration of other areas of degraded marsh.

One approach often used in mudflats is to undertake a method known as 'trickle charging'. This has some similarities to recharge but instead of pumping material directly onto the mudflat, sediment of suitable grade is piled up in the low intertidal area and subsequent tides bring the material inshore (Figure 9.4). The advantages of this approach are considerable. Because the movement of sediment onto the mudflat is gradual, and undertaken by natural processes, the rate of build up is relatively slow, thus preserving the integrity of sediment infauna, and also the morphology of the resulting mudflat.

Increasing wave attenuation with vegetation

At elevations below that which pioneer salt marsh vegetation can become established, the coastal engineer may still increase the roughness of the mudflat surface by the use of either natural vegetation that can tolerate long-term submergence (such as sea grasses or seaweed), or with artificial vegetation. Eleuterius (1975) discusses the use of three species of sea grass (*Thalassia*, *Cymodocea*, and *Diplanthera*) to stabilise low intertidal flat sediments. These plants were transplanted from healthy stands nearby and kept on the sea floor by the use of anchoring devices. In some areas, transplants failed because of

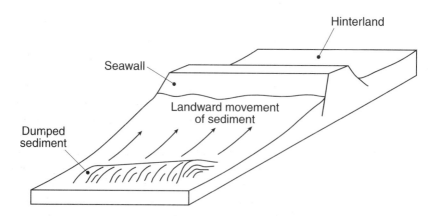

Figure 9.4 Trickle charging of mudflats by placing dredged spoil in low mudflat area.

high wave exposure or particularly unstable bottom sediments. This would indicate that the technique is not appropriate for all areas and certain environmental criteria need to be considered. One of the fundamental issues here is why vegetation is not naturally growing in the area to which it is to be transplanted. If this is due to some physiological reason, i.e. water too warm, too saline, or too rough, then vegetation is not likely to survive when transplanted. Garbisch *et al.* (1975) indicate that successful planting and regeneration can be possible only if certain planning measures are taken. In discussing schemes in Chesapeake Bay, success was achieved by delaying planting until mid-June in high energy areas, and also netting plants to avoid predation from geese. Failure may occur if such criteria, or a range of other factors, such as the injection of large volumes of fresh water from an adjacent estuary, are ignored.

In basic terms, the presence of vegetation in the mudflat environment will reduce the amount of wave energy moving onshore. Work by Fonseca *et al.* (1982) supports such conclusions in their study of the impacts of the sea grass (*Zostera marina L*) on current flow. However, they also found that while current velocity may be reduced within the plant canopy, velocities may increase above the plants. Later studies by Fonseca and Cahalan (1992) compared *Zostera* with three other sea grasses (*Halodule wrightii, Syringodium filiforme,* and *Thalassia testudinium*) under different conditions of water depth and plant density. Results showed that sea grasses can achieve the same degree of attenuation to salt marshes per unit distance, reaching 40 per cent energy loss where plant height approximates to water depth. They also concluded that broad, shallow meadows would also have significant impacts on wave energy. All species tested, apart from *Zostera marina*, showed increased wave energy reduction with increased plant density and water depth.

In essence, what these data showed is that vegetation in the low tidal/subtidal area can facilitate wave energy reduction, and hence suggest a mechanism by which additional protection could be provided for exposed mud-dominated coastlines. Clearly, if the environmental conditions do not favour the growth of natural species, then it is either necessary to adopt a different technique, or use an artificial method to simulate vegetation. Artificial seaweed has been a popular alternative to natural vegetation and may be constructed in several ways, including polypropylene tape strips joined together to form a continuous row, bundles of tapes to produce tufts, or fibre filaments bound together. In each case, the idea is that the artificial seaweed will behave in a similar manner to natural vegetation. Although artificial seaweed has been heralded in some sectors as a means of stabilising bottom sediments and increasing wave attenuation, Rogers (1987) provides a critique of the technique's effectiveness and discusses examples of its use. In both laboratory and field trials, the impacts of artificial seaweed are somewhat below expectation. Work by the UK Hydraulics Research Station (Price *et al.* 1968) reported that the material was only useful in unidirectional flow, while the US Army Corps of Engineers (Ahrens 1976) concluded that it was not effective in wave attenuation for common ocean wave periods. Following design changes aimed at overcoming these criticisms, field trials in North Carolina are reported to have achieved a seaward movement of mean high water of about 58 m, although the role of other structures and natural beach cyclicity may also be responsible for some (or all) of this increase in width. In contrast to this example, other field trials produce opinions which vary on the success of the method. Schemes in Delaware and New Jersey showed no measurable accumulation of sediment (Rogers 1987), although Snyder (1978) suggests that a series of fibres and fronds woven into a basal fabric has been successful in reducing wave activity at an estuarine locality in Florida. Perhaps this indicates the major difference: success may occur in estuarine conditions, while on open coasts it does not, reflecting the fact that the method may be suited to the relatively lower energy conditions of an estuary as opposed to the higher energy conditions of the open coastline. Despite the apparent lack of conclusive field evidence regarding wave energy reduction and sediment accumulation, the method is considered useful in the reduction of current scour in certain conditions, although the majority of monitored field trials have shown the method to be ineffective in controlling wave erosion.

Importance of salt marsh vegetation in wave attenuation

Once the upper mudflat surface has accreted vertically, whether by natural or artificial means, above that at which salt marsh vegetation can become established, greater wave attenuation can be achieved by the increased surface roughness which this provides. This has three further advantages. First, the roughness produces greater energy loss from waves; second, vegetation binds the sediment, thus increasing its stability; and third the leaves and stems increase the rate of vertical accretion by acting as wave baffles.

As waves move across a vegetated surface, the energy levels and wave height decay exponentially. The decay of the wave is a function of wave height approaching the marsh, distance travelled through the marsh, depth of water, and diameter and spacing of the plants (Knutson *et al.* 1982). In suitably wide areas of vegetation, this decay may be sufficient to completely remove the impacts of wave activity under certain conditions. Knutson *et al.* (1982) demonstrates the effectiveness of *Spartina* in dampening waves in Gulf coast situations, and Knutson (1988) demonstrated attenuation with reference to Chesapeake Bay. Figure 9.5 illustrates the increasing reduction in wave energy and height with distance travelled across the marsh. In this example, it can be seen that under the study conditions, all wave energy had been lost within 30 m of the marsh edge. Although many eroding marshes around the world may not actually be 30 m wide, it can also be seen that more than half the wave energy is lost within the first 2.5 m of the marsh edge, and up to 40 per cent of the wave height. In the Yangtze estuary (Yang 1998) average and maximum wave heights on vegetated flats were found to be only 40 per cent and 43 per cent respectively of those on unvegetated tidal flats. Average and maximum wave energy was 16 per cent and 19 per cent respectively. Moeller *et al.* (1996) also found supporting evidence in wave attenuation studies in north Norfolk marshes.

Waves across a series of marsh transects showed a decrease in wave energy of between 47.4 and 100 per cent, with a strong correlation between degrees of energy loss and water depth. They concluded that the marsh vegetation was most effective at energy attenuation at low to intermediate water depths. At greater depths, it is likely that waves would be able to move across the top of the vegetation, in a similar way to that found by Fonseca (1982) for sea grasses; Knutson *et al.* (1981, 1982) were able to quantify this attenuation when they showed that wave energy decreased from $7.2\,\mathrm{J\,m^{-2}\,m^{-1}}$ at a point on the seaward edge of a marsh, to $1.35\,\mathrm{J\,m^{-2}\,m^{-1}}$ to a point 30 m into a *Spartina* marsh sward. Shi *et al.* (1995), using flume experiments, studied the reduction in current velocity with height within the marsh canopy. These studies showed a decrease in velocity with depth, thus suggesting reduced potential for scouring of surface sediments, while Ginsburg and Lowenstam (1958) and Muus (1967) both showed how, under certain conditions, an almost still body of water can form in dense vegetation growth, thus allowing fine sediments to settle. All of these studies emphasise the importance of marshes in coastal defence, as marshes get wider, and marsh growth gets higher and denser, their ability to reduce wave height and energy increases, thus reducing investment required for coastal defence structures. King and Lester (1995) indicated that this relationship feeds directly into sea defence expenditure, because required sea wall height increases almost linearly as marsh width decreases. Complete loss of the final strip of marsh produces a sudden, exponential increase. This situation is

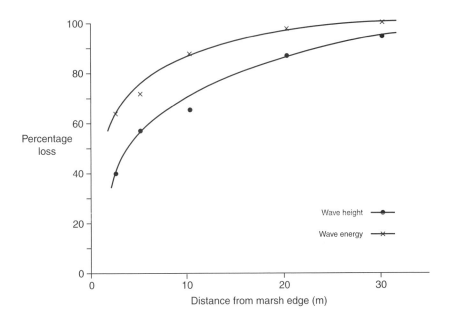

Figure 9.5 Wave height and wave energy reduction with distance travelled across salt marsh surface (Compiled using data from Knutson 1988).

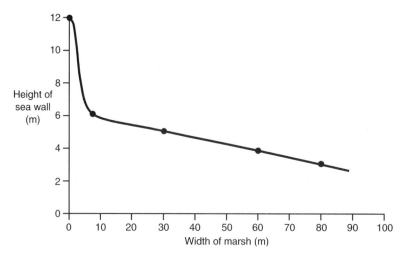

Figure 9.6 Relationship between required height of sea wall and width of fronting salt marsh (Compiled using data from MAFF).

also exemplified using data from the UK Ministry of Agriculture, Fisheries and Food (Figure 9.6) which shows that the required height of sea wall needs to be of the order of 3 m when fronted by a marsh of 80 m width, but 12 m if this marsh were to be completely eroded.

This situation demands consideration of the economic importance of the marsh. Traditionally, the financial 'worth' of a salt marsh has been considered in respect of its value as potential agricultural land, or for wildfowling. As an example, in 1989, a 73 ha area of salt marsh in Essex, UK, was sold to wild-fowlers at a cost of £1,096 per hectare (£0.11 m^{-2}) (King and Lester 1995). Now, however, we need to consider its value in terms of coastal protection pro-vision. In this respect, we can cost 'worth' in terms of money saved on sea wall construction. King and Lester (1995) suggest that the 6 m of marsh fronting a sea wall saves between £250–£600 m^{-2} on wall construction costs. The next 24 m is valued at £30 m^{-2}, while the next 30 m is worth £10 m^{-2}. Clearly, a healthy marsh will equate to money saved in not having to build such a large sea wall, and adds up to far more than the £0.11 m^{-2} paid for by wildfowlers for the Essex marsh.

From the above, it can be seen that marshes reduce wave activity, and thus increase coastal defence provision. Coupled with a reduction in wave energy caused by the baffle effect of vegetation will be an increased ability for sedi-ments to settle out from suspension. This will allow the marsh to continue to develop and accrete vertically, enhancing its ability to keep pace with sea level rise. Gleason *et al.* (1979) demonstrated a positive, linear correlation between the density of marsh plants and the amount of sediment trapped for a planted marsh, a conclusion also made by Yang (1998) from studies in the Yangtze Estuary, China, and by Woodhouse *et al.* (1974) with measurements of vertical accretion rates of between 6 and $12\,cm\,a^{-1}$ from a planted site in North Carolina. Boorman *et al.* (1998) noted a positive correlation between vertical accretion rates of up to $3.08\,mm\,a^{-1}$ at Stiffkey, Norfolk and vegetation height. Furthermore, Yang (1998) also noted a distinct finer grain size population in the marsh environments when compared to mudflats, with silt/clay content being 1.5 times that of mudflats, further supporting the ideas stated earlier of reduced energy levels to allow finer sediments to settle. Pethick (1993) identified stem density as a main method by which flow is modified. Note, however, the observation by Pethick cited earlier that it is not the deposition rate *per se* which is increased by vegetation but the net deposition rate, in that the same amount of sediment may be deposited, but less material is reworked from marshes due to vegetation presence. Such retention occurs partly due to the density of stems and their interference with tidal flow but also, in the medium to longer term, to the fact that the roots bind sediment together. Such binding increases shear strength and makes sediments harder to erode. Work by Pestrong (1972) contrasted the shear strengths of different vegetation stands with unvegetated channels and tidal flats. This showed that shear strength of sediment in root-bound areas increased by 200–300 per cent compared to unvegetated areas. This finding was also supported by Brown (1998) who showed that *Spartina* was effective at binding sediment in the Humber estuary.

It would appear from the available literature that there are additional complications in successfully encouraging sediment to accrete within salt marsh areas. While the link between stem density and wave energy reduction has been amply demonstrated, the link between comparable rates of deposition between marshes and mudflats has been complicated by the findings of Pethick (1993). He argues that the rate of deposition is directly related to the concentration of sediment within the water column (i.e. the amount of available sediment). As a result, schemes that plant marsh vegetation without considering the local sediment budget may fail because of the lack of available sediment. One way of increasing this availability is to trickle charge the planted area, as this would provide a gradual but continuous supply of sediment.

Using vegetation to create salt marshes

The natural succession in any tide-dominated area is for the build-up of fine grained sediments to a level at which vegetation can start to colonise. Initially, the primary colonisers are algae, typically *Enteromorpha* (Plate 9.3a), then primary plants, such as *Spartina* sp. or *Salicornia* sp. (Plate 9.3b). The establishment of a salt marsh takes time but can be speeded up to increase sediment stability by planting pioneer vegetation. It is important, however, to distinguish between two approaches to marsh creation. In this section, we will carry the idea of increasing mudflat sedimentation through to stabilisation via planting. This is, in effect, a method by which the line of salt marsh is pushed seawards. Another method is to move the line of defence inland and create salt marsh from land which has previously been claimed. This practice is known as

Plate 9.3 Increasing the stability of mud flats
  a) Enteromorpha development helps to bind the sediment and resist re-
     suspension by wave action. Foulness Island, Essex.
  b) Primary colonisers (*Spartina*) and higher marsh vegetation
     (*Halmione*), Adur Estuary, Sussex.

managed realignment (also referred to as set back or managed retreat), and will be dealt with in the following chapter.

Basic principles (Chapter 2) tell us that increased mudflat elevation and width will increase wave energy attenuation and thus reduce the impact of waves on coastal defences. We have seen how this can be developed in the lower mudflat area to reduce wave activity. Similarly, as we saw in the previous section, wave attenuation can also be achieved in the upper mudflat area if the surface is covered with vegetation because this not only makes the surface rougher, thus increasing wave attenuation and sediment entrapment, but it also binds the sediment together to reduce erosion. Pethick (1993) raises this latter point in his study of accretion under salt marsh plants. This work revealed that deposition rates were independent of vegetation and not statistically different between non-vegetated and vegetated surfaces. The greater accretion rates observed in salt marshes is put down to the greater ability for these environments to retain sediment due to stem and leaf entrapment and root binding when compared to exposed mudflat environments.

Having seen how vegetation can alter currents and wave activity resulting in increasing sedimentation and reducing wave impact on coastal defences, it is clear that the presence of a healthy marsh will be extremely beneficial to the coastal manager. It is important to realise, however, that as with any coastal defence methodology, the development of a saltmarsh by mudflat planting is site specific, and should only be used in situations where conditions are favourable. The most obvious indicator of site suitability is the presence of marsh already in the area. Not only is this proof that marshes can grow there, but will also provide a seed bank for marsh maturation and succession. If a coastline is experiencing rapid erosion, then it may be necessary to opt for a landward movement of the marsh (managed realignment, see Chapter 10), rather than a seaward movement, as the resulting wave climate will be too harsh for prolonged marsh survival. Other factors are important when locating a planting site. These include elevation, slope, tidal range, salinity, substrate and hydrology (see Broome 1990). It is important to plant species relative to their correct elevation on the mudflat. Salt marshes are typically zoned according to elevation, the zones being controlled by the frequency and duration of tidal inundation. Hence, *Spartina* as a pioneer species, is tolerant of more frequent inundation than higher marsh species, and as such, is often used because it can be planted well down the intertidal zone.

The critical nature of mudflat elevation means that this should be the first area of concern in marsh planting schemes. As we have seen, there are two ways of increasing the elevation of the mudflat prior to vegetation colonisation: first, by using one of the techniques of encouraging sedimentation described above, or second by dumping dredged spoil into the intertidal zone. Many of the available examples cited in the literature (see Table 9.2 for some examples) really relate to this second approach, the aim being to increase mudflat elevation and then to stabilise it to prevent subsequent erosion. Clearly in such situations, the need for defence provision outweighs the importance of infauna. The most commonly used plant for this purpose is *Spartina* sp. Hubbard and Stebbings (1967) carried out a study in the UK and catalogued 35 estuaries where *Spartina* sp. had been planted to increase shoreline and mudflat stability, the earliest being in the Beaulieu River, Hampshire in 1898. Despite its popularity, this species has often received a bad press because of its rapid growth and invasive nature, particularly in relation to unwanted invasion into other areas. However, with regard to mudflat stabilisation, these traits make it particularly suitable. Lee and Partridge (1983) measured the lateral spread of *Spartina* plantations in the New River Estuary, New Zealand, at up to $4\,\mathrm{m\,a^{-1}}$, while Christiansen and Miller (1983) indicate an increase in aerial coverage of $44\,000\,\mathrm{m^2}$ between 1966 and 1978, following initial planting in 1959, in Mariager Fjord, Denmark, an average rate of $3\,667\,\mathrm{m^2\,a^{-1}}$. This spread of *Spartina* rapidly provides enhanced energy attenuation. Extensive research in China (Chung 1993) has shown that *Spartina* plantations produce a wave energy and wave height reduction in the order of 88–95 per cent, and reduction in coastal erosion of 67 per cent in places. With such positive results, the use of *Spartina* in China for coastal defence has become extensive, with over 100 coastal sites being planted covering $36\,000\,\mathrm{ha}$ (1981 figures).

When planting *Spartina*, plants may be sourced from existing plantations, or grown specifically for the purpose in nursery beds. Typically, the most practical method of planting seedlings is by hand. Spacing of plants is important because it is necessary to maintain as rapid coverage of an exposed mudflat as possible, but also to keep the cost of the operation as low as possible. Broome (1990) suggests that a spacing of 45–60 cm between plants is adequate, although on sheltered sites, complete coverage of *Spartina* has been achieved after just one year following a 1 m initial plant spacing.

The ultimate intention of marsh planting is to establish a marsh which behaves in a similar manner to a natural system. Once seedlings have become established, wave attenuation will become more effective and sedimentation will increase within the planted areas. What will take longer, however, is the development of proper marsh soils (especially if the marsh was planted on dredged material), and the infauna and wildfowl populations. Ideally, the plantation should be self sustaining and management free. However, as with any coastal defence structure (which, in effect, is what the plantation is), some maintenance may be required, as well as monitoring to check on growth rates and possible dieback.

There has been considerable interest in planting techniques for stabilising muddy shorelines. Seneca *et al.* (1975) present the findings of a series of research projects following planting in North Carolina. Here, the establishment of new marshes occurred both through planting seedlings and also by direct seeding, both of which were found to be viable methods of marsh generation, although seeding was only really successful on the upper parts of the site. Another study by Newcombe *et al.* (1979) in San Francisco Bay showed that seeding was not effective. This appears to raise important questions as to the

Table 9.2 Examples of mudflat sedimentation and marsh creation studies from around the world. (For full references, see References at the back of the book).

Country	Authors	Location
Brazil	Netto and Lana (1997)	Paranaguá Bay
China	Chung (1993)	Various
	Yang (1998)	Yangtze Estuary
Denmark	Brashears and Dartnell (1967)	North Sea coast
	Christiansen and Miller (1983)	Mariager Fjord
Netherlands	Groenendijk (1986)	Oosterscheldt
New Zealand	Lee and Partridge (1983)	New River Estuary
UK	Alizai and McManus (1980)	Tay Estuary
	Holder and Burd (1990)	Essex
	LRDC International (1993)	Lymington, Hampshire
USA	Broome (1990)	South-east USA
	Cammen (1976a+b)	North Carolina
	Garbisch *et al.* (1975a)	Chesapeake Bay
	Garbisch *et al.* (1975b)	Chesapeake Bay, New Jersey
	Good (1993)	Various
	Goodwin and Williams (1992)	California
	Hardisky (1978)	Georgia
	Lewis (1990a)	Florida
	Lewis (1990b)	Puerto Rico/US Virgin Islands
	Newling and Landin (1988)	Georgia
	Sacco *et al.* (1988)	North Carolina
	Seneca (1974)	North Carolina
	Seneca *et al.* (1975)	North Carolina
	Shisler (1990)	North-east USA
	Snyder (1978)	Palm Beach

controlling mechanisms over success or failure. In some cases, wave exposure, water depth or salinity may not be favourable, while in others, the use of dredged material may introduce contaminants, or may not be appropriate for plant growth.

Before leaving this section on increasing sediment stability by vegetation, it could, in some circumstances, also be possible to increase sediment stability using fauna, rather than flora. Meadows *et al.* (1998) showed how critical erosion velocities can be reduced following seeding of intertidal areas with mussels. As may be expected, however, greatest success was obtained in areas where mussels had gravel to attach themselves to. As such, this technique may be somewhat limited, particularly considering the vulnerability of mussels to inundation by fine sediments, an event which may well occur during stormy periods.

Problems associated with vegetation die back

Under some situations, marsh vegetation can die back, thus removing the protective cover and exposing the sediment surface to the full force of the waves. One way in which this can occur is during marsh drowning (see sea level rise in Chapter 2) or coastal squeeze (see Chapter 10), where marshes revert to mudflats due to increased frequency and duration of tidal inundation. Another cause that remains largely unexplained is the natural die-back of *Spartina*. In the 1930s, many areas of *Spartina* along the south coast of the UK started to die, resulting in large areas of marsh reverting to mudflat. While renewed exposure of the marsh surface to wave energy can lead to erosion, in many places the *Spartina* marshes are being replaced by *Zostera* (Adam 1990). Haynes and Coulson (1982) looked at the problem of die-back in Langstone Harbour on the south coast of England; they argue that the replacement of *Spartina* with *Zostera* is advantageous due to increased biodiversity, better nutrient dynamics, better environments for aquatic organisms and better conditions for waders. This view is also supported by Adams (1990), as well as other authors. One key issue, however, as far as coastal defence is concerned, is whether the two species are comparable with respect to wave attenuation and sediment accretion. Earlier discussions in this chapter would tend to suggest that they are not, with *Spartina* being the favoured species for these requirements. Hence, while die-back may be favoured by ecologists and marine biologists, from the coastal engineer's viewpoint, this replacement does not provide such efficient coastal protection and, therefore, will lead to a reduced sediment stability and accretion potential.

Impacts associated with increasing natural sedimentation

The traditional view is to see marsh creation as being successful if the plants remain alive and in position and can demonstrate some success in the protection of the hinterland. Perhaps an alternative would be to judge success on whether or not the resulting marsh area becomes a 'natural' system, in that it possesses all the ecological and hydrological properties of a natural marsh and has no negative impacts on the environment in which it is located. In reality, both of these views could be considered extremes because, in effect, a marsh creation scheme cannot ever be considered as natural, because it is growing in a location where the system dynamics decreed that marshes would not be. Clearly then, we need to assess success in term of the degree of 'naturalness' and environmental impacts.

In essence, because this technique, as others, produces an artificial modification of the intertidal area, it is going to enforce changes that would not happen under natural conditions. As a result of these changes certain detrimental side effects could occur which would detract from the scheme's success. In some cases, such as with brushwood groynes and the Schleswig-Holstein approach, the techniques are relatively new, and so it is difficult to assess impacts over such a short time scale. However, as Box 9.1 indicates, some schemes are showing favourable results, while others are not. There are a variety of views concerning the success of intertidal flat accretion schemes. Many of those relating to vegetation planting are encouraging, and there is much evidence to support the method. The range of environmental impacts which could arise, however, is wide, and concerns the depositional regime, the ecology, and also sediment characteristics. Potential problem areas are discussed below.

Burying of infauna

We have already mentioned the potential problems regarding the rapid accumulation of sediment on mudflats. It may, at first glance, appear rather contradictory to criticise the rate of sediment accumulation, when the aim of the scheme is to increase the elevation of the mudflat as quickly as possible. However, the fact remains that the sediment infauna is very vulnerable to rapid burial and populations may decrease rapidly if sediment builds up too quickly. If we consider some large intertidal areas, such as the Wash and Morecambe Bay in the UK, the Bay of Fundy in Canada, or Chesapeake Bay in the USA, then all are renowned for their wildfowl importance, and have conservation designations applied to them on those grounds. The success of these birds is primarily due to the availability of large stores of food within the mudflats. Any rapid burial would damage this interrelationship and could potentially lead to reductions in bird populations.

Given this issue, there appears to be a good case for the slow introduction of sediment in such areas, such as that which can be achieved by trickle charging. Despite this, however, the most common way to create artificial marsh areas is to stabilise dredge spoil which has been dumped onto the mudflat. This represents one of the most rapid forms of sediment accretion possible, with a corresponding potential for infaunal damage. Many of the papers reporting such schemes herald the success of salt marsh establishment, sediment accumulation, and increased coastal stability, but other issues such as infaunal population dynamics and impacts on wildfowl populations are rarely investigated. In the few cases which do exist the infaunal biomass appears to be impoverished in marshes created through dredged spoil dumping. Cammen (1976a) noted from North Carolina that infaunal biomass in created marshes was only 10 per cent that of natural marshes after 2 years. This same marsh was studied again after 15 years (Sacco *et al.* 1988) and the state of impoverishment had significantly improved. This led to the conclusion that in this system, it had taken around 15 years for the infauna to re-establish itself. In a different North Carolina system Cammen (1976b) again found impoverished infauna in a dredged spoil created marsh. While the direct link between infaunal impoverishment and method of sediment accumulation is not made, it does appear that something in the marsh creation process is having a major impact on infaunal populations. Perhaps one of the most likely causes would be the rapid increase in sediment levels and the population's inability to migrate within the sediment profile rapidly enough.

Are created wetlands as good as the original?

The issue of infauna survival leads on to another issue, perhaps more related to ecological and conservation issues. Given that marsh creation schemes are increasingly seen as means of mitigating losses due to erosion and development in both the USA and UK, the question of how like the original they are is important.

In comparing created and natural systems, the whole system needs to be studied, including sediments, hydrology, and ecology. A study by Hardisky (1978) at Buttermilk Sound, Georgia, found that the above-ground biomass in the planted *Spartina alterniflora* stands, 4 years after planting, were between 1.3–5.5 times less than natural vegetation, while below-ground biomass was between 2.1–2.4 times less. This would suggest that plant growth is not as rapid on planted marshes and that some time may be needed for a planted marsh to 'naturalise'. A further study by Newling and Landin (1985) carried out 8 years after planting revealed that the biomass was still less than natural systems, although not as marked in the above-ground biomass. This may indicate a maturation process, but concern would exist that root development does not appear to be keeping pace with stem and leaf development, putting the long-term viability of the plantation into question. In this example, the substrate was dredged spoil, which may also play a part in the roots' relatively poor performance. Another location studied by Hardisky (1978) in north Carolina showed no significant difference between the planted and natural marshes, while Newlin and Landing (1985) noted from a site at Bolivar Peninsular, Texas that above-ground biomass was equal to or better than natural marshes, although root biomass was still less; and work by Shisler and Charette (1984) at eight sites in New Jersey showed consistently lower biomass in planted marshes. It appears from these, and other studies, that there is no general trend in biomass equality between planted and natural sites, although in the greater number of cases, biomass, particularly root biomass, is less in planted marshes. Perhaps more significant, therefore, are the properties of the dredged spoil itself, particularly with respect to grain size, water content, and possible pollutants. When considering the implications of these studies for other countries, it is important to remember that all of the above are US examples, and so are typified by US east and west coast marsh types. Geographical variation in marsh type means that not all systems behave in the same way. Furthermore, all the above examples come from schemes in which planting has occurred on dredged spoil. Unfortunately, due to the development of brushwood sediment fields being relatively recent, it has not been possible to identify any research comparing biomass parameters on dredged and 'encouraged' accretion areas. This comparison would be useful because it would allow further assessment of other controls, such as infauna, to be investigated. If we assume some damage to infauna due to the dumping of the dredged spoil, then we may also assume a reduction in internal bioturbation and mixing of the sediment layers. In terrestrial systems, this aspect of soil property is important in the growth rate of plants, and a similar case may exist in the formation of salt marsh soils.

When looking at the sediments themselves, it is important to understand any differences that may occur between artificial and natural sites. Where sediments are deposited in sedimentation fields or by trickle charging, then they are being laid down by the natural processes of the system. This means that they will be sorted and stratified in the same way as other sediments, thus their physical properties should remain constant. Where sediments are dumped, then we are effectively introducing a mass of unstratified and unsorted material onto the mudflat. We have already seen the implications of this for infauna, but there are also implications for water and organic content, redox potential, and nutrient availability.

One problem which does occur with the use of sedimentation fields is that the resulting mudflat typically has a higher water content than that around it. The implications of this again apply to infauna, in that a highly saturated environment is not favourable to a healthy infauna, so this may again lead to impoverishment. To a certain extent, the problem of excess water will become less with time as the sediments de-water, but if this does not occur rapidly enough, damage may be done to the infauna. A further problem lies with sediment grain size. The reason for initiating such schemes is because the wave and tidal currents are too high to allow fine sediment to settle out. As such, the sediment that is present in the mudflat will tend to be coarser than that which settles in the sedimentation fields. Care needs to be taken when establishing vegetation, and the impacts on infauna, that the sediment that is settling out is suitable to allow a successful sustainable environment to develop.

Interruption of sediment supply down-drift

Although this problem appears in most chapters, it is worth constant reiteration because it represents one of the most common oversights in coastal planning and management. If a coastal manager establishes a scheme where sediment is to be trapped on the mudflat, then who is going to lose out? It is critical to understand where that sediment used to go, because if material is trapped in a sedimentation field, it cannot be deposited on another part of the mudflat, which may itself develop net erosion conditions (i.e. transferring the erosion problem), thus requiring additional defence in the future.

In many systems, this is not a real issue because the aim of a sedimentation field is to encourage finer material to settle than would normally be the case. Hence, the material which collects in a polder is unlikely to settle out on any adjacent mudflat anyway. It may, however, be important elsewhere in a coastal cell, such as a quiet backwater where fine sediment deposition does occur.

Aesthetic

One final impact of mudflat regeneration schemes is the aesthetic impact. The sight of a healthy salt marsh is not objectionable, but the sight of brushwood fences, covered in green algae, or of thick mud piled up on intertidal flats possibly is. However, in the case of dredged material, we are only thinking in the short-term because following a suitable period for dewatering, planting will occur and over a few growing seasons, a reasonable vegetation cover will occur. This is not the case with sedimentation fields. The brushwood fencing will remain for a considerable period of time and may prove visually intrusive to some sectors of the community. However, while maintaining some degree of sympathy, a few brushwood groynes which prevent coastal erosion and potential land loss is something that can be put up with, and is also considerably better than many of the developments which go on along the shores of many of our major estuaries.

Benefits of increased coastal sedimentation

The basic advantages of such schemes as described in this chapter are that coastal protection is increased by more effective wave attenuation brought about by higher mudflat elevations and coastal vegetation. From a coastal managers' point of view, the fact that this outcome can be achieved relatively easily and cheaply is a bonus, and removes the necessity for building sea walls or rubble revetments which are not only more expensive, but more visually intrusive.

Accepting the problems associated with the actual mudflat development, the vegetating of a mudflat which is well established will not only increase the wave attenuation potential of the system, but will also have increased ecological benefits. Broome (1972) states that *Spartina alterniflora* marshes play an important role in primary production, energy flow, and nutrient cycling in estuaries, and that this can play key roles in the support of predator species within an estuary.

Intertidal mudflats and sea level rise

In many situations, the cutting back of salt marshes by wave activity is the system's response to rising sea level, or linked processes such as increased wave activity or storminess. In effect, we are seeing a landward movement of the mudflat environment under the Bruun rule. In many cases, the salt marsh is reverting to mudflat as this is the newly stable landform under the new sea level conditions. The Severn Estuary, UK is a good example of an intertidal mudflat environment that is migrating landwards under sea level rise (Allen 1990). Large areas of mudflat are only frequently covered with modern sediments and many have bare mud surfaces that reveal extensive scouring. Adjacent to these flats are extensive marsh cliffs, many of which are retreating by metres per year. As the sea level rises, the outer margins of the mudflats become permanently submerged and subtidal. Corresponding movements in the landward direction replace these areas of mudflat. Where such landwards movements cannot occur, either because all marsh areas have already been lost, or the land is rocky, then the mudflat may start to become compressed (coastal squeeze) and eventually become subtidal.

As sea level, and so water depth, increase so wave base will also get deeper, allowing increased wave energy to penetrate further inland. The presence of sea grasses, such as *Zostera*, will be affected by increased turbulence and also, potentially, increased turbidity as more sediment becomes mobile. As with salt marshes (see next chapter), such sea grass communities will need to migrate inland and re-establish themselves closer inshore, where conditions reflect those previously experienced. If this cannot occur, either due to inappropriate substrate or a rapidly increasing water depth, the sea grass community will die out, thus reducing the coastal protection afforded by them. Counter to this is the further possibility (Burd 1995) that under conditions of sea level rise, sea grass communities could expand due to increased available substrate and nutrients, caused by the flooding of coastal lowlands.

In discussing such issues of sea level rise, there appears to be a conflict between much of what has been said in this chapter. In creating marshes by planting or encouraging sediment deposition in the upper mudflat areas, we are actually pushing the line of coastal defence seawards, and so are going against the natural tendency of the coast to move landwards. In such cases, we are actually fighting against the natural processes of the coastline, and thus contradicting the ethos of soft coastal engineering. Clearly, the techniques of marsh building are not suitable for rapidly submerging coastlines, and even where sea level rise is small, the creation of marshes is only a short- to medium-term solution because in order to maintain any long-term stability, the whole coastline needs to relocate inland. Ways of achieving this artificially under sea level rise are discussed in the next chapter.

Summary

We have seen in this chapter means by which coastal engineers can encourage sediment accretion in mudflat environments, and how, subsequently, these areas can be vegetated to form salt marsh. The methodology is very much a soft defence option because it relies on the accumulation of sediments and vegetation to protect coastlines. Early in the chapter, we saw from wave energy attenuation figures that suitably wide marshes can provide high quality defence. Under normal wave conditions, it may not even be necessary for a sea wall to be present, but the occurrence of storms means that they are still a necessity. Despite this, however, the fact that a healthy fronting marsh exists means that the impacts of storms will be much reduced. There are examples in the literature which quote instances of hurricane damage, and how this is less in areas of fronting saltings.

Clearly then, the presence of a belt of salt marsh fronting a sea wall is beneficial. We have seen how advancing the line of defence seawards can develop such features. A range of examples is given in Table 9.2. Clearly, such an approach is adventurous in a climate of rising sea levels that are forcing the high water mark landwards. This may appear somewhat contradictory but is a longer-term issue than that addressed by marsh creation schemes. Sea level rise has many impacts, one of which is to increase water depth and thus, wave energy hitting a sea wall. The methodology presented here is a direct way of counteracting this issue.

A slightly disturbing trend, however, is starting to emerge with regard to this methodology. Increasingly, politicians and planners are seeing the idea of marsh creation as a way of facilitating development along estuarine shores. Nature conservation bodies generally oppose development which includes claiming of intertidal lands due to the loss of valuable habitat, as well as natural defences. Now that the ability to create marshes has been demonstrated, particularly with the use of dredged spoil, planners see this as a way of replacing claimed areas with created marshes. While this may go some way to replacing habitat, created marshes are no substitute for the real thing, as we have seen. Hence, coastal managers need to resist this pressure from the planners to take mature marshes and replace them with pioneer marshes elsewhere. Historically, we have lost so many hectares of marsh from our estuaries that we should be creating new ones without having to do it purely to maintain the status quo.

Summary of benefits of mudflat and marsh development schemes

- Increased wave attenuation and, therefore, coastal protection
- Increased habitat
- Increased wildlife potential for estuaries
- Useful way of using unwanted dredgings for increased coastal defence

Summary of problems of mudflat and marsh development schemes

- Naturalness of created marshes
- Impacts associated with accumulation of sediment in one place at the expense of another
- Aesthetic intrusion of some schemes
- Can lead to modification of infaunal populations, with possible knock-on effects on wildfowl

Recommended usage

- In areas where marshes once existed, and wave erosion is a problem
- In areas where habitat creation is beneficial for conservation interest

10 Managed realignment

Introduction

As we go through each chapter, we consistently realise that what we are trying to achieve is to prevent the sea from invading and flooding the land. In other words, we are trying to hold the existing line of sea defence, or to push it seawards, which in terms of net land loss, amounts to the same thing. However, when we consider what many regard as the greatest threat to the contemporary coastline, sea level rise, we might gain an appreciation of how fruitless this battle may turn out to be in some circumstances. There is growing evidence to suggest that instead of always trying to keep the coastline static by maintaining the existing line of defence, it may be better to allow the sea to erode parts of the coast by abandoning land to the sea. The idea of actively promoting coastal erosion, or the reversion of land to marine habitat, does perhaps run contrary to what many people may accept by the term 'defence'. However, to the coastal manager, 'defence' relates to successfully protecting the hinterland, so managed realignment, as a method of protecting the hinterland from erosion, is important.

To justify this assertion, it will be useful to remind ourselves of the nature of soft defences. If we accept the basic principles of maintaining a soft coastline and working with nature, then it could be argued that we should be watching very closely what nature is doing. Throughout the geological and more recent past, there is abundant evidence to demonstrate what the coastline does in response to changing sea level. In simple terms, the position of the coast moves landwards (sea level rise) or seawards (sea level fall). Hence, if the natural response of the coast is to move landwards under conditions of sea level rise, should the coastal manager not respond similarly?

This approach is what we will study in this chapter. We will refer to the method as 'managed realignment', although within the literature, it may also be referred to as managed retreat or set back. Increasingly, the term managed retreat is going out of favour as it suggests a negativity in coastal management, 'retreating in the face of the enemy (sea)' which many coastal managers and coastal residents find unacceptable. Referring to the technique as managed realignment does more accurately portray the method as it involves realigning the sea defence from its current position to one further inland. It is at this point that we can draw a further distinction between this method and the 'do nothing' approach, previously mentioned with respect to cliffs, but equally relevant here. Managed realignment involves moving the line of defence landwards to a new line of defence or to high land; an example is Northey Island, Essex where a new wall was constructed 75 m inland of the original (Leafe 1992). In this respect, although erosion is encouraged, it occurs under controlled conditions to predetermined lines. In the case of 'do nothing', the process of natural erosion continues, but there is no interference and it continues unchecked.

Managed realignment is one of the most important contemporary issues in coastal management and is becoming increasingly favoured, particularly in estuarine environments where previous areas of land claim are being returned to the intertidal zone. Some schools of thought regard managed realignment as the ultimate saviour of estuarine protection, in the same way as beach recharge was to open coasts. However, as we have seen, beach recharging has its problems and uncertainties. In the same vein, managed realignment also has many unknowns and environmental concerns.

The concept of both managed realignment and doing nothing incite great passion and feeling, especially for people who are directly affected. The scientific reasoning behind the approach is sound. By defending the coastline we induce changes to the sediment budget and coastal processes which lead to sediment deficits, with the potential of making the overall problem considerably worse that it originally was. By mimicking the natural response of coastline to sea level rise, we are providing a proactive and effective management strategy. Furthermore, the adoption of managed realignment can prove to be cheaper than holding the existing line of defence. Given that existing defences are under threat because of increased wave attack, often caused by loss of fronting salt marsh, repair of the original defences is likely to be ongoing and require further attention in the future, unless the coastline fronting it can be forced seawards. At Northey Island, Essex, the financial aspects of the scheme favoured the realignment option over holding the existing line. Leafe (1992) claims the cost of the retreat scheme was £22,000, whereas the cost of holding the line and repairing/upgrading existing defences was between £30,000–£55,000, excluding the additional costs of future repair.

The main complication when employing this technique lies in the fact that coasts have been very popular for human settlement and industrial growth, and there is not much of the coastline which is not used by someone for some thing. In many cases, therefore, the use of managed realignment is both a scientific problem, and a public education and persuasion one.

Managed realignment is becoming more popular and is increasingly being regarded as a suitable way of protecting tide-dominated coasts in a way which maintains a natural coastline. As with many of the methods of defence studied in this book, it is possible to combine this technique with other methods. In the UK, many schemes have been purely a matter of breaching defences and allowing intertidal processes to become established. In the USA, Brooke (1992) reports on several examples which utilise the ideas of managed realignment in conjunction with other techniques already discussed. In North Carolina, dredged material has been used to increase land levels prior to breaching and planting saltmarsh vegetation. In California, many sites have been breached and used for marsh generation with techniques including the pumping of dredged material, and natural sedimentation to increase surface height, followed by either planting or natural colonisation. In Louisiana, marsh has been created by natural processes, while in Texas, Georgia and New York, dredged material has been pumped and planted with marsh species. In the UK marshes have been restored by realignment of defences along the Hampshire, Gloucestershire (Severn Estuary), Devon, and Lancashire coasts. However, the greatest research effort has been concentrated in Essex, where many schemes have been developed, notably in the Blackwater Estuary at Northey Island (0.8 ha) in 1991; Tollesbury Fleet (21 ha) and Orplands (40 ha) in 1995; and Abbotts Hall (20 ha) in 1996. Details of these, and other case studies are given in Box 10.1 and Table 10.1.

By definition, the technique can apply equally to any habitat, such as sand dunes being allowed to retreat inland over a road, the route of which has been relocated, or a beach allowed to migrate landward following the removal of hard defences. All such approaches involve the landward movement of the original line of defence to facilitate the landward development of habitat. Consultation of the literature, however, will soon reveal a marked bias towards tide-dominated, and in particular salt marsh, habitats. This occurs mainly because it is these areas where landward movement is possible. In many open coast beach environments, hinterland development prevents these strategies from being adopted. One example which is cited in the literature is from a shingle location, Dungeness, on the south coast of England (Maddrell 1996). Here, realignment is used as a cost saving measure to reduce the need for beach feeding.

Rationale of giving land back to the sea

Soft engineering claims to manage the coast naturally. We have already stated that allowing land to flood is the natural way of responding to sea level rise. This can be taken further in many estuaries when we look at the history of each area. There are not many estuaries in 'developed' nations where land claim has not occurred. In hydrodynamic terms, land claim removes land from the intertidal zone, thus constraining tidal processes and reducing the volume of the estuary. Throughout history, land claim has accounted for an estimated 25 per cent of intertidal land in estuaries, resulting in the destruction of large areas of the coastal zone, world-wide (French 1997). The continued exploitation of the coast has important implications for habitats and wildlife, as the more land claim and defence works which go on, the greater the pressure on existing marshes becomes. Furthermore, as sea levels continue to rise, there are further impacts on estuaries. Whether due to natural or anthropogenic factors, increasing water

Table 10.1 Summary table of realignment scheme criteria for Essex coast sites (Compiled from various sources cited in the text).

Location	Area (ha)	Elevation (m OD)	Breaches	Creeks	Mean accretion (mm a⁻¹)	Land use	Time reclaimed
Tollesbury Fleet	21.0	1.0–3.0	1	Relict + dug	29.08	Clover meadow Pasture Arable	c. 200 years
Northey Island	0.8	2.5–3.2	1	Relict + dug	41.25	Grazing marsh (improved)	c 150 years
Abbotts Hall	20.0	1.0–2.5	2	2.2 km dug	No data	Arable	No data
Orplands	40.0	–1.0–4.0	2	Dug	No data	Grazing meadow Arable SSSI	No data

Box 10.1

Use of managed realignment in Essex

The Essex coastline has suffered greatly from relative sea level rise, experiencing both rises in sea level and falling of land levels following post-glacial isostatic rebound. Over the last few thousand years, the area of salt marsh has decreased from 40 000 ha to just 4 400 ha, with ongoing losses of 2 per cent a year (Dixon *et al.* 1998).

On such a low-lying coastline, the impact of losing natural wave attenuation ability through marsh loss has serious implications for sea defence stability, which would need to be raised considerably as the marsh edge erodes landwards (see Figure 9.6). Over the past few decades, this concern has been addressed by a series of defence measures, including polder and sedimentation field schemes (see Box 9.1), and also by the establishment of a series of realignment schemes in the Blackwater Estuary:

Orplands	carried out in 1995	covering 40 ha
Abbotts Hall	carried out in 1996	covering 20 ha
Tollesbury	carried out in 1995	covering 21 ha
Northey Island	carried out in 1991	covering 0.8 ha

Each of these schemes represents an experiment in managed realignment, the findings of which have been used to develop the technique further. In addition to the above schemes, there are an additional thirty sites in Essex where natural breaches have occurred (IECS 1993). Totalling over 500 ha, many of these sites are now colonised by marsh species, although their current percentage coverage and accretion/erosion status, is variable.

The four realignment schemes represent the range of design factors discussed in this chapter (see Table 10.1) including relict and artificial drainage systems, single and multiple breaches, elevation, gradient, and land use history. Of these four, two are particularly well researched and monitored, and written up in the literature.

1. Northey Island

This small realignment scheme, covering just 0.8 ha, was the first to occur in the UK in 1991, and involved the setting back of the old sea wall by 75 m, to a new line of defence. The area flooded had originally been reclaimed between 1843 and 1873 and had been managed as a low productivity grazing marsh. Although there were feint traces of a relict creek system, one creek was dug and the material used to infill a borrow ditch (a borrow ditch is a ditch dug to supply material for the construction of the original sea wall). Although vegetation establishment was initially slow, it has now established and a well-developed marsh, showing some degree of zonation, has formed. This reflects the relatively high elevation of the site (between 2.5–3.2 m O.D.; mean = 2.8 m OD; Pye and French 1993a); and its gradient. The elevation means that the site is only flooded infrequently and to shallow depths. In contrast to this, only a crude creek system has formed and ponding of water is common in surface depressions. Accretion initially ranged from 5.5–41.3 m a^{-1} following breaching (the range varies according to elevation and mudflat/marsh areas), falling to between 0.1–19.2 mm a^{-1} after 5 years.

2. Tollesbury Fleet

Covering 21 ha of agricultural land, the scheme at Tollesbury is considerably larger than the Northey site. The site had been reclaimed in the late eighteenth century and had experienced different land-use practices at different points. Having been split into five fields, land-uses included clover production, cereal growth and pasture. The post-breach elevation ranged from 1.0–3.0 m OD (mean = 1.35 m OD; Woodward 1998). Despite recommendations from the consultants for the construction of a shallow creek system along the lines of a relict creek network, this was dismissed on the grounds of cost, and only one channel was dug to connect the drainage system to the breach. Following breaching, there are signs that a creek system is starting to form although this is very much in the early stages.

The original breach was 20 m wide and designed to drain the site in 2 hours. However, drainage remains slow and considerable ponding of water occurs, as well as distinct scouring around the breach site. Variable rates of accretion occur over the site, ranging from −0.25 ± 3.23 mm to 63.55 ± 5.94 mm a^{-1} (mean = 29.08 mm a^{-1}). This compares to rates of 3.05 mm a^{-1} from locations adjacent to, but outside the site.

*   *   *

After a period of 4 to 5 years, vegetation was present on both sites, down to an elevation of 1.5 m OD This point clearly marks the lower limit of vegetation stability, as governed by the frequency, depth and duration of inundation.

depths and periods of inundation lead to plant communities becoming unstable, leading to greater marsh loss and eventual reversion to mudflats (see discussion of coastal squeeze below). Also, the remaining intertidal area of the estuary may become too small to hold the flood tide and the situation arises where, in order to function safely, the estuary 'needs' some of its original intertidal area back. By carrying out managed realignment schemes, this, effectively, is what happens.

In Chapter 9, we saw how important salt marshes are in wave attenuation. One of the main problems in estuaries throughout the world, however, is that salt marshes are disappearing. In some situations, it is possible to replace these with newly created marshes formed by advancing the line of defence, such as with the methods described in Chapter 9. If this is not the case, then it is necessary to create marsh by moving further inland. Managed realignment is a technique which can achieve this effectively, but is also one which serves a more multi-purpose role. First, it can increase the security of coastal land by its sea defence role; second, it can allow the formation of coastal habitats to increase the conservation potential of an area; and third, it can help offset the impacts of sea level rise and the potential for flooding by increasing the intertidal area of an estuary.

The role of coastal squeeze in marsh loss

The reasons for salt marsh loss in estuaries are varied. In part, it is due to human factors, such as land claim and dredging (see French 1997), but it is also due to 'natural' processes such as sea level rise being constrained by the artificial constraints of embankments. This process is known as *coastal squeeze*.

Vegetation communities which develop on salt marshes often show a spatial zonation according to elevation (see Adam 1990). The elevation at which zones occur is controlled by the depth of water on a flood tide, and the period and frequency of tidal inundation (Figure 10.1a). Different plant species can tolerate different periods and frequencies of inundation, thus those which can tolerate the greatest inundation occur on the low marsh, while those which can tolerate only short or periodic inundation, occur on the high marsh. Figure 10.1a presents a schematic section across a marsh with low, mid and high marsh communities. In a simplistic sense, low marsh vegetation is stable between the depth (d) and inundation frequencies (f) depicted by d^I, f^I and d^L, f^L. Similarly, the mid marsh communities are stable between d^L, f^L and d^M, f^M. Landward of this point, higher marsh and transitional communities dominate. As sea level rises (MHWS to MHWS'), so water depths in the estuary increase, meaning that all zones of the marsh get covered more frequently and for longer periods (Figure 10.1a). Clearly this will lead to plants becoming unstable in their new conditions, causing die-back and re-growth further inland where water depths and periods of inundation match those of pre-rise conditions. Hence, the junction between low and mid marsh communities moves inland to the depth at which it was previously prior to the onset of sea level rise. In the case of Figure 10.1a, this position moves from d^I, f^I to $d^{I'}$, $f^{I'}$, from d^L, f^L to $d^{L'}$, $f^{L'}$, and from d^M, f^M to $d^{M'}$, $f^{M'}$. Support for such landward migration can be found in several marsh areas, not least along the shores of the Gulf and Atlantic coasts of the USA (Bird 1993). In Maryland, freshwater forest communities are being replaced by salt marsh (Darmody and Foss 1979), and terrestrial meadows are being invaded by halophytic vegetation in Chesapeake Bay (Bird 1993).

All the time an estuary is in its natural, undefended state, any long- to medium-term rise in sea level can be compensated for by these landward 'shifts' in vegetation community zones (Figure 10.1a). As soon as sea defences are built, however, any landward shift in these zones is restricted by a physical barrier (Figure 10.1b). The process of sea level rise and habitat zone migration still operates in the same way as in Figure 10.1a, but while in the earlier example all zones can migrate inland, where there are embankments the

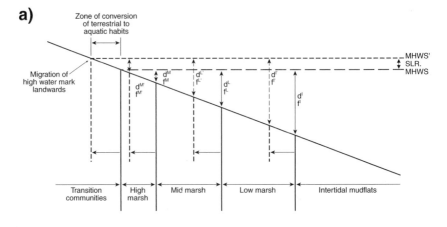

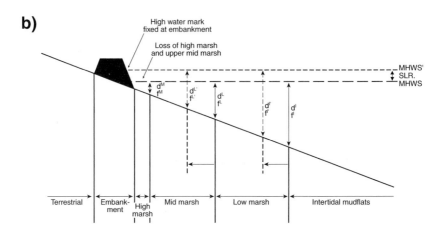

Figure 10.1 Impacts of sea level rise on marsh zonation and the issue of coastal squeeze
   a) No landward limits allow vegetation zones to migrate landwards under rising sea level
   b) Embankments prevent landward movement and lead to loss of higher marsh communities through coastal squeeze.

landward position, i.e. high water mark, is fixed. Therefore, with time, although low and mid marsh communities can shift landwards, there is no space for high marsh, which becomes squeezed against the sea wall and eventually disappears (Figure 10.1b). If sea level continues to rise, the mid marsh may also disappear as inundation and depth become suitable only for low marsh species. As the depth, frequency, and period of inundation increase further, this process will continue with the possibility that a marsh will actually revert to lower marsh species, or even mudflat. In such cases, the increased depth of water and lack of wave attenuation by vegetation will eventually allow greater wave activity and increased threat of overtopping of the defences and subsequent flooding. This would also have implications for the size and cost of the sea wall, with decreasing marsh width resulting in greater wave energy and height received at the wall (see Figures 9.5 and 9.6).

* * *

Given that many estuaries around the world are experiencing sea level rise, and many, as a result, are experiencing marsh loss, there is a strong argument in many areas to give managed realignment serious consideration as a method for the management of estuarine coastlines (see Table 10.2 for a series of examples). It is important to remember, however, that the notion of simply removing existing embankments and allowing the hinterland to flood will not work. There are many aspects of the method which need to be considered before a scheme can be planned in detail. These considerations include what the area is like, but also what it has been used for since being claimed, and implications for the rest of the estuarine system.

Table 10.2 Examples of managed realignment and natural breach studies from around the world (For full references, see References at the back of the book).

Country	Authors	Location
Netherlands	Helmer *et al.* (1986)	North Sea
Puerto Rico	Lewis (1990b)	General
UK	Boorman and Hazelden (1995)	Essex
	Brooke (1992)	General
	Dixon *et al.* (1998)	Essex
	English Nature (1995)	Tollesbury, Essex
	French (1999)	Medway, Kent
	Hutchings (1994)	Essex
	Klein and Bateman (1998)	Norfolk
	Leafe (1992)	Northey Island, Essex
	Maddrell (1996)	Dungeness, Kent
	MAFF (1994, 1995c+d, 1997)	Tollesbury, Essex
	Pye and French (1993a)	Essex
USA	Boesch *et al.* (1994)	Louisiana
	Brooke (1992)	General
	Lewis (1990a)	Florida
	Titus (1991)	General

Methods and environmental factors in managed realignment

Natural estuaries have a typical shape, narrow inland and widening seawards (Figure 10.2) which becomes altered during land claim. This natural shape is also reflected in the bathymetry, which deepens seawards and incorporates distinct channels, on occasions being different for ebb and flood flow. Ideally, a managed realignment scheme should aim at recreating this original shape, bearing in mind that small, piecemeal schemes alone may not achieve this. The basic method of carrying out a scheme is simple: either completely remove an embankment or punch a hole in it. This act will allow the flood tide to flow through the old structure onto the land behind. Obviously, it is important to have some means of containing this water, so new embankments need to be built at the landward limits of the new site, or else there needs to be reliance on naturally rising land. As with any coastal defence scheme, however, it is important to carry out all actions with a full understanding and consideration of other parts of the coast, and the relevant sediment budgets. As most realignment schemes are based in estuaries, it is important to bear in mind the impact that realignments will have on the whole estuarine system.

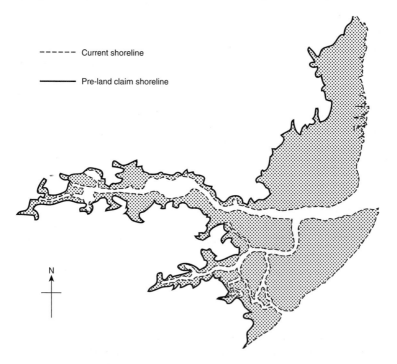

------ Current shoreline

——— Pre-land claim shoreline

N

Figure 10.2 Natural and unnatural estuary shape. Pre-land claim estuary shape (solid line) is considerably larger than modern shoreline (dashed line), reflecting the large areas of land claim (stippled) Crouch Estuary, Essex, (modified from Burd 1995).

Estuaries are highly complex systems which exhibit a series of interactions between the anthropogenic and natural forces, such as tides, rivers, sediment and vegetation. An estuary's shape and depositional patterns are a function of morphology and process. A managed realignment scheme will affect both of these because, first, by opening up a section of previously claimed land, you are actually changing the shape of the estuary; and second, by changing the shape of the estuary, you are liable to change the patterns of flood and ebb flow, which will, in turn, affect the sediment transport pathways and general hydrodynamics of the system. Each of the key parameters will be considered in turn with respect to the impacts of managed retreat.

Tidal prism

The tidal prism is the volume of water exchanged on each tidal cycle (volume at high water minus volume at low water). As a study of any basic textbook on coastal morphology and process will reveal, tidal exchange is one of the fundamental controls over sedimentation and erosion within an estuary. The tidal prism will govern the velocity of flow, which will, subsequently, govern the transport and erosion of sediment. The opening up of a section of sea wall to allow flooding of the hinterland under a managed realignment scenario, increases the size of the intertidal area, and thus, the volume of water which can be held. The volume of water in the estuary after breaching minus the volume before breaching will represent the increase in the tidal volume, and, therefore, the increased volume of water which has to enter and leave the estuary during flood and ebb tides. Such increased volumes can only be accommodated by increased flood and ebb tide velocities. Major breaches, therefore, can represent major increases in tidal prism, which may be coupled with increased tidal scour. This issue was the reason for the rejection of a scheme in the Blackwater Estuary, Essex, where modelling studies predicted that realignment would increase tidal currents resulting in unacceptable scour and erosion at other locations in the estuary. As a result of this study, the location of the scheme was moved and subsequently occurred at the Orplands site in the same estuary (Dixon *et al.* 1998) (see Box 10.1). A further example where a scheme did progress despite problems of scour was in the USA (Goodwin and Williams 1992). In this case, however, the tidal prism was deliberately increased with the prime intention of inducing increased scour to maintain channel openings and prevent siltation.

Estuarine morphology

The ultimate aim of managed realignment is to free up the intertidal zone to respond naturally to coastal processes, with the long-term aim of returning the estuary to a more natural shape (see Figure 10.2). As the natural shape of estuaries is typically funnel-shaped (wider at the mouth and tapering inland), realignments in the lower estuary should, in theory, go back further than those in the upper estuary and hence greater land areas are required. This ideal is complicated by the fact that most modern harbours are located towards the mouths of estuaries, and so these areas are often those where retreat is least likely. In addition to this, it is also important to remember the concerns raised in relation to the tidal prism and increases in flood and ebb current velocities. Realignment schemes upstream from the mouth of an estuary are more likely to have impacts on current velocities than those near the mouth because while the tidal prism is being increased, the width of the estuary mouth may still be confined, thus the increased tidal volumes will need to flow with increased velocity through a relatively small gap in order to fill the estuary.

In reality, the best way to regain a natural estuary shape is not to carry out realignments piecemeal, but to undertake one large-scale retreat covering the whole of the estuary. This way, there will be one major estuarine hydrological adjustment, rather than a series of adjustments in association with individual realignment schemes. Despite this being the best way in theory, in practice it is likely to be impractical in most cases and potentially beyond the capacity of existing predictive models. As such, a series of smaller schemes is often carried out as part of an overall strategy, such as is happening in the Torridge Estuary, Devon, where previously reclaimed marsh on the inside of meanders are being returned to the intertidal zone.

Another important consideration is to carry out schemes in the context of the estuary in question. A wide estuary containing wide intertidal flats with considerable land claim would be evidence of a previous system in which substantial marsh areas may once have existed in harmony with the hydrodynamics of the system. A narrow estuary with steep intertidal gradients and narrow mudflats may not be as suitable for sustaining new marshes, partly due to flood tide frequency and water depth, but also due to the lateral constraints on sustainable marsh. Burd (1995) highlights this issue and cites the Orwell Estuary in Suffolk, UK as an example. In this case, management of the estuary, largely in the form of dredging of navigation channels has resulted in wide channels and correspondingly narrow and steeper mudflats. In this example, Burd argues that there is insufficient mudflat width for marshes to be sustainable in managed realignment schemes due to the steepness of the intertidal profile and the tendency for marshes to slump channelwards under gravity.

Site history

In several instances in the preceding two subsections, hints have been made regarding the form of the estuary prior to the original land claim which removed many of the original marshes. This introduces a further issue, that of site history. When creating a new salt marsh site on claimed land, the most suitable sites are those which were once formerly saltmarsh before being claimed. Such sites provide evidence that the location, soil chemistry and soil structure is capable of supporting the necessary vegetation and infaunal communities, although the degree of alteration and modification since it last did this needs to be determined. If, conversely, the site had never supported marsh growth, then there must be some reason for this, and as such, it may be better to avoid such sites.

Bearing in mind that any successful establishment of vegetation will depend on the soil in which it grows, this aspect of the site needs careful investigation. In the last paragraph we mentioned that sites which were once marshes are the best for realignment because they have once supported marsh flora. The process of land claim does, however, produce major changes in soil structure and chemistry which will have a bearing on its suitability. Since they were last marshes, soils have dried out causing irreversible changes to the soil organic content, soil density, and soil porosity. In addition, the soil salinity would have decreased dramatically and, if used for crop production or intensive grazing, the addition of fertilisers is also likely.

Another important aspect of the site's history is what it has been used for since its original claim. In most situations, marshes are claimed for agriculture and so some land improvement may have occurred and drainage installed. Clearly the degree of improvement will vary according to whether the land has been used for pasture or arable purposes. The Institute of Estuarine and Coastal Sciences (IECS 1993) state that this difference between arable or pasture usage is one of the main controls over the success of subsequent marsh development. The underlying reason is that claimed land used for pasture has had less post-inning modification than one which has been ploughed and fertilised for crop growth. Comparison of a series of breach sites in Essex (Woodward 1998) has revealed evidence to support these statements. A site at South Woodham Ferrers, on the Crouch estuary, Essex, was used for grazing, as had a site at Northey Island, in the Blackwater estuary. Other sites, such as at Fambridge, also on the Crouch, had been improved for arable land and had drainage installed, while Tollesbury, on the Blackwater, had been used for arable farming. In all these cases, the best marsh growth occurred on the sites used for pasture. It is claimed that there are several reasons for this: first, the lack of disturbance of the subsoil and the more 'natural' pedalogical processes operating; second, it is also claimed (Boorman and Hazeldon 1995, Woodward 1998) that the pasture surfaces, containing a layer of dead vegetation caused by seasonal die-back, increased the ability of the surface to attenuate wave energy and to trap sediment, thus, they were able to accrete sediment faster and establish a more dense vegetation surface; finally, comparing the two sites used only as pasture, increased plant diversity was found at one location (South Woodham Ferrers) compared to the other (Northey Island), and it is suggested that this may relate to the fact that the pasture at the former site was frequently limed and manured. It has also been suggested that arable sites may be more prone to erosion, particularly when they have had drainage installed because drainage represents a series of channels (in whatever form) that water can easily flow along. This means that they can also function as lines of exploitation for flood tides, thus increasing the rate of marsh break up and decay. At Fambridge, this process can be seen quite clearly (Plate 10.1). Such comparisons indicate that the degree of colonisation of a realignment site may owe part of its success to the history of the site. When discussing drainage, a further quantification should be made. It has been said (Chapter 2) that a marsh creek system is an effective way of supplying sediment to the back marsh area, and so may be considered a benefit for marsh accretion. This may appear, therefore, to offer a contradiction between artificial drainage and natural creek systems, in that one may be a problem, but the other an advantage. This is a complex area, and one which research does not fully explain. However, there appear to be two main differences. First, the artificial drains are straight and so are not 'stable' channel forms, thus leaving them prone to erosion and adjustments, and second, the frequency of artificial channels, which equates to drainage density in a natural stream system, is very high. This latter point is exemplified in Plate 10.1.

Surface elevation

Partly related to the history of the site because it is a function of the length of time that the area concerned has been claimed (see above), the elevation of the land surface is another critical factor contributing to long-term success in marsh creation. As soon as a marsh is claimed, it is cut off from the sea and receives no further sediment input. Although this might suggest that the surface will remain at the same height, the lack of inundation will mean that the hydrology of the sediment sequence will change. The implications of this can be seen in two ways. We have already seen how the elevation of a marsh

Plate 10.1 Tidal exploitation of installed drainage network, following re-inundation of farmland by sea, North Fambridge, Crouch Estuary, Essex, UK.

will govern the vegetation which grows there, due to the frequency and duration of tidal inundation. When claimed, no new sediment is deposited on the old marsh surface, although those marshes seaward of the sea wall will continue to receive sediment and build-up. Hence, there will be a relative difference (increasing over time) in height between claimed and active marsh. Similarly, the lack of tidal inundation will reduce the water content of the sediments which subsequently undergo a dewatering, which causes the sediments to contract and drop in elevation still further. The combined effect of both these features is that the claimed marsh surfaces fall in elevation relative to the active marsh surfaces. This means that the longer the marsh has been claimed, the greater the height difference is likely to be, and the lower the marsh community which is likely to develop. If this level is below that at which marshes can naturally form, no marshes will grow at all. Research by the UK Institute for Terrestrial Ecology has shown that in Essex, vegetation will only become established on sites above a level of 1.5 m ordnance datum (OD), although density is low. Above 2.0 m OD establishment is much more successful (Woodward 1998). Knutson *et al.* (1990) support this figure and suggest, from their historical study, that elevations above 2.1 m OD on breaching were most successful. This figure is variable, however, and may not be an entirely accurate indicator. Successful schemes have occurred at Orplands, in the Blackwater estuary, where initial surface elevation ranged from −1.0 m to +4.0 m OD and at Abbotts Hall, with elevations of +1.0 m to +2.5 m OD It is much better to look at elevation in terms of frequency of tidal inundation as, after all, this is one of the prime controllers of vegetation development. As an example, in the Blackwater estuary, Essex, an elevation of 2.1 m OD equates to about 380 tidal inundations per year, while the same elevation in the Thames estuary would equate to 490 inundations per year (Burd 1995), suggesting that while pioneer vegetation may be stable in the Blackwater at 2.1 m OD, the inundation frequency at this level is too great in the Thames to allow vegetation to grow.

In essence, the higher the newly flooded surface is, the fewer times that it will be covered by the tide, meaning that more mature marsh vegetation can colonise. In addition, there will be fewer tides to erode sediment from the surface, there will be greater wave attenuation resulting in lower wave energy and greater hinterland protection, and there will be longer time periods between tides to allow compaction of the sediment. The best method or determining whether the proposed realignment scheme is at a suitable level for marsh creation is by relating it to those elevations already present on adjacent marshes. Lower elevations are unlikely to establish vegetation communities until sufficient accretion has allowed the elevation to increase. This situation can be accelerated by artificially raising the surface by sediment pumping or by dumping of dredged sediment onto the surface to increase the elevation sufficiently to allow the more immediate colonisation of flora (see Chapter 9). Another important factor connected with elevation is the slope gradient because this will control the amount of site covered by the tide.

Surface gradient

Natural marshes develop vegetation zones as we have already discussed. This not only increases the floral diversity, but also the ability of the marsh to attenuate wave energy and protect the hinterland. It is possible, however, for marshes to develop on flat land and many of the 'natural' breaches (see below) which occurred naturally around the turn of the century in the UK tend to be on such surfaces. The resulting marshes, as a result are typified by a low species diversity resulting in a rather low-grade marsh.

A sloping site, preferably grading landwards into terrestrial vegetation is ideal because it will develop a range of vegetation types depending on elevation. The slope will encourage a more diverse flora and reduce the amount of post-realignment scheme management needed. Sites with transitional communities, however, are rare in many estuarine situations. Considering the types of locality where realignment is needed, they are rarer still because realignment is most needed in estuaries which have an extensive history of land claim and so are also the very sites where the building of embankments has eliminated all such transitional communities due to their conversion to arable or pasture. Because of these inherent restrictions on the idealised community structure, increased habitat diversity, conservation importance, and sea defence effectiveness is generally achieved if the realignment site has a natural slope which allows a range of marsh communities to colonise, or the surface is artificially graded to produce a range of elevations, to imitate the effects of natural marsh development.

Research by Zedler (1984) in the USA suggests that the slope is fundamental in determining the type of vegetation which develops. This work found that slopes of between 0–2 per cent were generally recommended. Later work by Knutson *et al.* (1990) indicated that the most successful establishment of vegetation occurs on slopes of between 6–7 per cent. Sites in the UK which have undergone realignment following natural breaches have typically been in the region of 0.1 per cent. These have produced successful vegetation colonisation, but it has tended to be monospecific and uniform (Burd 1995). The importance of slope, however, has been shown on the experimental managed retreat site at Northey Island. Here, the site was backed by sloping land, with the result that a range of new plant communities has become established, increasing in diversity towards the higher elevations at the landward portion of the site. Similarly, the slope present at the Tollesbury site has produced a good vegetation diversity.

Sediment characteristics

A study of saltmarshes will reveal that some tend to be siltier than others, but all are predominantly fine grained. Typically, marsh plants perform best in loam to clay soils (Harvey *et al.* 1983). Hence, the sediment present on the site will also exert some control over the success with which plants colonise. There are several aspects of this relevant to managed realignment. First, modification of the soil with coarser sediment may increase its drainage beyond that compatible with marsh vegetation; and second, the level of organic material may also be important. It is also important that the sediment deposited following breaching of the embankment is of the right grade to encourage plant colonisation. This issue is linked to others discussed below but suffice to say here that it is important for fine-grained sediments, similar to those on adjacent marshes to be able to settle out and remain in the breached area. Provided that there is sufficient suspended sediment in the water body to allow accretion, and that this sediment can be retained on the ebb tide, this will provide the basis for increases in surface elevation and future plant establishment.

As with all such accretion scenarios, as long as the rate is not so rapid as to smother the plant communities, the continued deposition of sediment will allow the continued building-up of the sediment surface. The subsequent development of root systems will then bind the sediment to form a more resistant surface. One important criteria, however, is that in order for the marsh to survive in the longer term, it needs to be able to grow vertically to keep pace with sea level rise. Hence, there needs to be a continued supply of sediment to the site from the estuary to allow this to occur. If the sediment availability declines, possibly due to other defence measures, then the marsh surface could be in danger of 'drowning' (see Chapter 2) or reverting to lower marsh communities — the same process which initiated the need for managed realignment in the first place.

Creek networks

In natural salt marshes, creeks develop as an important part of the inundation and drainage dynamics of the marsh, and its associated sediment deposition. It is therefore beneficial to have some form of creek system in marshes created by realignment. Seeing that many of the sites for realignment are on previously claimed land, it may be possible to re-open the original marsh drainage system, which is often still obvious in the form of surface depressions (Figure 10.3). Clearly, this is most likely to occur in land used for grazing as arable land, involving drainage and ploughing, often has any remnants ploughed out at an early stage. Where such relict systems are evident, they may be exploited as part of the new marsh either by allowing natural processes to re-open the natural system, or to artificially open it by excavation.

Particularly where the land has been used for arable purposes, there remain no traces of the former drainage. One option is to open up the sea wall and allow natural erosion to develop its own pattern. This has the advantage of providing the right geometry to the creek network but it is a long-term process and may hinder the establishment of vegetation. Alternatively, an artificial drainage network can be cut, based on equilibrium geometry, which is itself dictated by tidal characteristics, tidal prism, sediments, accretion rates, and vegetation types. Myrick and Leopold (1963) established mathematical relationships between these variables, allowing channel sizes to be predicted. The method has been

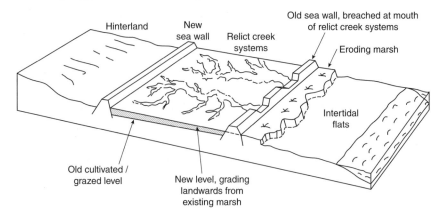

Figure 10.3 Schematic representation of managed realignment site. Note breaches at mouths of relict creek systems.

employed on a series of Californian marshes (Goodwin and Williams 1992), and was also used at Abbotts Hall in the Blackwater Estuary, Essex, where re-establishment of the old creek network was undertaken by digging 2.2 km of artificially meandering channels, into which water was allowed to flow by reversing the existing sluices (Dixon *et al.* 1998). Such prediction is a complicated issue and involves knowing a lot about how the various parameters concerning water and sediment flow across the marsh interact. Constructing creek systems artificially is not a preferred solution, but may be the best given the constraints of a site, and the only possibility where no natural alternative is present. The best way is to allow the site to develop its own system.

Tidal hydraulics

The ultimate factor controlling the delivery of sediment, and its ability to be retained in the realignment site, is the tidal hydraulics. With regard to managed realignment, the tides behave somewhat differently to other marshes because whereas in the normal marsh system, tides rise on the flood and flow over the marsh system from both the front edge and from the creeks, with realignment sites, the tidal floodwaters generally enter the site through one opening, the breach in the sea wall. When undertaking any realignment project, it is important to know what size the breach should be. If it is too wide, the sediment load will be dumped at the mouth and not spread over the whole area, and if it is too narrow, flow velocities will increase and lead to scouring and erosion.

Ultimately, the best method would be to remove the whole of the wall, but where elevation is too low, or the site too exposed to waves, this may not be possible. If not maintained, then this wall will start to breakdown naturally over a period of years and, if of earth embankment type, may also act as a source of

sediment to the newly accreting marsh. If required, wall decay can be accelerated by gradual lowering by mechanical means. The important factor is the
amount of water moved into and out of the site on each tidal cycle, the tidal
prism. Too little will lead to possible water stagnation and lack of sedimentation, while too much will produce scouring. Mathematically, there is a highly
significant linear relationship between the long-transformed breach width and
the tidal prism (Burd 1995). As such, knowing the tidal prism of the realignment site (i.e. the volume of water within it at high tide, minus the volume at
low tide (normally zero)), it is possible to calculate the width of the breach
which will provide the most effective delivery and removal of water, according
to the following equation:

$$W = 37.9e^{0.0000018 \ TP}$$

where:

W is breach width (m)
TP is tidal prism (m^3).

Once the size of the breach is known, it is important to locate it correctly.
Work in the USA (Burd 1995) has demonstrated that there are a series of
important aspects to consider when deciding where to make the hole in the sea
wall. Ideally, the breach should be located at the natural mouth of a relict creek
where these occur (Figure 10.3). Flow within this existing channel will allow its
development and reopening, and provide the movement of material into and
out of the marsh. Second, it is important to assess the impacts of this creek
development. As the creek deepens and develops at the breach, the marsh
outside the opening is in a different hydrologic position than previously.
Originally, prior to the breach, the area just seaward of the breach was at the
headwaters of the creek system. Post-breach it becomes located somewhere
downstream of this point, where flow is greater and faster. As a result, the
potential loss of habitat outside the mouth through ebb tide scour should be
taken into account. Third, in order to encourage marsh development, it is
important to provide as much shelter as possible from waves. This can be partly
achieved by maintaining the walls of the realignment site, but in addition, it will
also be possible to increase the shelter by making sure the breach has sufficient
shelter from wind and waves. Work by MAFF (1995e) demonstrates how the
impacts of breach width and placement can be modelled. Using the case of
Tollesbury, different scenarios were modelled, including variations in size, position, and tidal dynamics to predict the best possible locality of the final breach
(MAFF 1995d); however, post-breach observations may suggest that the eventual placing was not optimal.
   A final point relates to the idea of one or more breaches. Where a relict creek
system occurs, then the breach should be at the head of the creek system
(Figure 10.3). If the realignment site has two clear creek networks, it will be

preferable to locate a breach at the head of each, allowing a 'topographical divide' to develop between the two. If only one such creek system is present, then one wide breach is more suitable than many small ones, because if two or more are present, flow can enter through one and leave via another, thus not transporting sediment onto the marsh.

The realignment scheme at Orplands in Essex, utilised two breaches and an artificially cut creek system (Dixon *et al.* 1998). Being a long, thin realignment (2.2 km long and variable width; total area = 40 ha) a single breach would not have been sufficient to allow suitable flooding of the whole site. The positioning of the creeks was such that openings were sheltered from prevailing wind directions and two drainage networks formed (see also Box 10.1).

Sediment budget

As with any defence structure, the volume of sediment available for deposition is important and there has been consistent reference to the reliance of coastal defence measures upon it. Managed realignment is no different because it can increase not only the area over which sediments are deposited, but also alter the mechanisms by which sediment is eroded and transported (see above). One of the biggest concerns with managed realignment schemes relates to the availability of sediment in the system, and the potential damage caused to existing habitats if this availability was to decrease.

Typically, immediately following a breach, the newly flooded area experiences a rapid increase in sedimentation. Boorman and Hazelden (1995) claim that the natural breach site at Fambridge in 1897 experienced rapid sedimentation, with the old soil surface now up to a metre below the present surface. Similarly, rapid accretion occurred at Northey Island with 41 mm in the first year, and at Tollesbury, rates of 29 mm a^{-1} have been recorded (Woodward 1998). In terms of sediment budget, these accretion rates, when converted to volumes of sediment, become quite significant. At Tollesbury, for example, the site covered 21 hectares, giving an estimate of 6 093 m^3 of average sediment deposition per year. At Northey Island, being a smaller site of only 1 ha, this equates to 410 m^3 per year. While the Northey Scheme is small, at Tollesbury, such a volume could have impacts on local sediment deposition around the site, thus potentially affecting the survival of other marshes.

Storm breaches: a natural analogue to managed realignment

The extensive section above details a series of issues relevant in the planning of a managed realignment scheme. To a certain extent, the issues raised are rather vague in their assessment of impact and importance because, in reality, we just do not have definitive answers to many of the pertinent questions. Because the technique is relatively new (it has only been available for a few decades), we are only just starting to obtain monitoring data to allow the development of the methodology. One avenue open to the research community, however, is to look at natural analogues of the conversion of the hinterland back to intertidal.

In the UK, the last decade of the nineteenth century was particularly stormy, and a series of storm breaches occurred in a number of sea defences. Locations include the Essex Coast (Fambridge, Northey Island (Boorman and Hazelden 1995); the Medway (French 1999), Sussex (French 1991) and the Blyth, Suffolk (Brooke 1991). The 1953 storm surge on the east coast of England also resulted in a series of breaches, as at Titchwell in north Norfolk (Hollis *et al.* 1990). Many of these breaches were never repaired and so present the opportunity to study sites which have experienced renewed inundation by the sea for over a century in many cases. Clearly we are not comparing like with like here, because these events were uncontrolled and no pre-breach monitoring occurred. Boorman and Hazelden (1995) argue that while marshes often develop after these breaches, the longer term results have been unsatisfactory. However, it is clear that some sites have developed healthier marsh than others, raising questions of why this has happened. Plate 10.2 shows a healthy marsh at Pagham Harbour, Sussex. Pagham is a natural harbour and is protected from the sea by a shingle ridge across its mouth. In 1876, the opening in the ridge was sealed and the land divided into grazing land. In 1910 storms breached the ridge, leading to the re-inundation of the land and abandonment of the land claim (French 1991). Since this time, healthy marshes have developed (Plate 10.2).

Plate 10.2 Healthy marsh sward which formed following storm breaching of the outer defence in 1910, Pagham Harbour, Sussex.

In this case, it is evident that the natural breaching of the defence has led to the successful establishment of salt marsh, and may give indications to aid the development of managed realignment methods. The pertinent issues here are the following.

a)   The area had been defended for a relatively short period of time (34 years), meaning that the land level had not fallen too much relative to sea level. Studies from sequential ordnance survey maps indicate that land levels had dropped between 10 and 30 cm during this period. This enabled the rapid re-establishment of marsh vegetation without large scale adjustments to new base levels.

b)   The use of the land for grazing means little modification of soil chemistry and structure had occurred.

c)   The breaching of the ridge allowed a small entrance and relative shelter to the rest of the developing marsh.

In other situations, whilst marshes have developed following breaches, these are currently being degraded by a combination of lateral retreat and internal dissection. This has led some researchers to claim that these natural breaches are not successful because the marshes have not developed the sort of stability necessary to effectively function as defence structures. However, we need to determine the cause of retreat and to assess the longevity of these marshes in relation to those elsewhere in the system. French (1999) looked at a series of storm breaches in the Medway estuary, Kent. In all cases studied, marshes had developed post-breach but all were now retreating. However, far from being a sign of ineffective marsh development and lack of permanence, it can be shown that the marshes which developed as a result of these defence breaches were eroding slower than those which have remained unclaimed. Looking at a comparison of areal loss between marshes, two examples of marshes which had never been claimed have experienced typical average losses of 2.57 and 0.9 ha a^{-1} between 1840 and 1969, while two marshes which formed following breaches have experienced average losses of 0.25 and 0.02 ha a^{-1} (French 1999). Clearly this suggests some factor influencing the break-up and erosion processes in marshes which reform following breaches. This factor may be due to the fact that in unclaimed marshes, erosion is ongoing, while in reformed marshes, certain erosion thresholds needed to be reached before the onset of erosion, or may be a function of marsh position relative to the tidal frame. We do not know the answer to this, but it is clear that marshes which are formed following realignment schemes may behave differently in terms of stability and erodability. The answer may be simple, such as the protection offered by leaving parts of the wall in place, or may be more complex.

As with the earlier section, looking at natural analogues poses more questions, but it does allow us to target our research into more closely defined areas. Evidence from Pagham shows that marshes can form very effectively, and possible environmental factors are suggested. In the Medway, evidence suggests that even when the marshes erode, their rates of loss may be different from marshes which have never been reclaimed. One point worth mentioning, however, is that erosion of a marsh should not be considered a failure. If the estuarine conditions favour net erosion, then marshes will erode anyway because of wider causes, not causes related to the technique of managed realignment. Even in the case of the storm-breached Essex sites discussed earlier, post-breach marshes may be eroding, but they can also be said to have afforded protection to the hinterland for over 100 years, a time period beyond the design life of any sea wall.

Impacts associated with managed realignment

The potential of managed realignment to afford significant increases in coastal protection is clear, in that it conforms to all the ideals concerning increased wave attenuation and habitat provision. However, it is clear that there are many areas where poor planning or failure to understand natural system dynamics has the potential to cause serious problems as a result of realignment schemes.

The most obvious environmental impact is that areas of dry land revert to the intertidal zone. This will produce a loss in terrestrial habitats but as we have seen, this is traditionally farmland. It is unlikely that realignment schemes will occur in areas of valuable natural habitat because it is unlikely that this will fulfil the criteria concerning the history of the realignment site, and the importance that it once supported salt marsh. The exception here could be in some managed grazing land, where rare plant species have become established. In such cases, it may be necessary to offset conservation interests against those of coastal defence and the replacement of freshwater habitats with new intertidal ones. Cley marshes in north Norfolk represent a case in point (Klein and Bateman 1998). This location is identified in the local shoreline management plan as one in which realignment will occur. Cley is a fresh water nature reserve inland of a shingle ridge, and is acknowledged, both nationally and internationally, as a key site for wildfowl breeding. The site is considered to possess important natural value as it is, and it is considered that conversion to a saline environment will destroy this. This subject and its future is currently the topic of intense debate, based around local concerns to maintain existing freshwater areas, and coastal managers' who see it as a way of protecting the coast against future sea level rise. One major issue in this example is that the current habitat is only maintained by the protective influence of the shingle ridge which is clearly showing signs of instability and the need to roll over landwards.

From the detailed list relating to the specific site requirements for successful realignment schemes outlined above, it is clear that some locations favour realignment while others do not. As such, this method of coastal defence must be considered along the same lines as others, as far as environmental suitability is concerned. It is clearly not, as some practitioners claim, a universal solution to estuarine erosion. Considering the limitations and problems cited in the previous section, research has shown that the most successful schemes have occurred where:

1) the elevations of the original site were above the level at which marsh vegetation can grow, and were of an elevation suitable for the desired vegetation communities;
2) the site was adjacent to an existing marsh area which acted as a seed bank for flora and as a migration site for fauna;
3) the hydrological conditions of the site were managed to prevent scouring and allow sufficient water stand to allow sedimentation. In other words, the breach was of the correct size, relict creek networks were present, and the site was sufficiently sheltered from waves;
4) the sites had been well researched and details of the hydrological, sedimentological, and ecological understood.

Clearly, such criteria impart considerable constraints on possible site selection.

Finally, as with any defence strategy, it is important to appreciate the impacts of all actions on both other parts of the coast and the estuarine sediment budget. Given the predominance of these schemes in estuaries, this means that it is important to bear in mind the impact that these actions will have on estuarine processes. The amount of sediment available to an estuary will remain the same whether a realignment scheme goes ahead or not. As such, increased deposition in the realignment area will mean that sediment will not be deposited in other areas. This could trigger off new erosion, or modifications to channel morphology. Such concerns, however, are universal in any coastal defence scheme, but underlie the importance of understanding the nature of sediment movement, and the changes which are likely to occur in response to the environmental modifications brought about by implementation of the defence scheme.

There are many issues raised by managed realignment, and these are significant in terms of scheme success and negating impacts on the adjacent environment. It is true that this approach to estuarine defence is becoming an increasingly common method of protection, but it needs to be underlined that there are many unknowns about the impacts of the method, and the changes in process which arise through its use. We are on a steep learning curve when it comes to understanding hydrology and current activity, and there are still many questions needing answers. What this means, in reality, is that while it is important to understand how each estuary 'works' before a scheme attempts to modify it, we are frequently in the situation that we do not have all the information we need, hence each scheme is an experiment which provides data to improve our scientific understanding of estuaries.

Economics of managed realignment

We have mentioned that when discussing and planning defence strategies, each method is generally subject to a cost benefit analysis. A scheme in the

Blackwater estuary, UK (Pearce 1992) came under considerable criticism locally when sea defences were strengthened at a cost of £4 million, while the value of the land protected was only £400,000. Clearly, this site did not, on a cost-benefit basis alone, warrant such investment. Managed realignment options also need to be judged along the same criteria. There is expense involved with relevant engineering works, especially if new inland defences are required. Similarly, the creation of new habitat is also an expensive undertaking if this has to be artificially encouraged or enhanced because the original site is too low. Offset against these costs, however, is the value of the new habitats, both as wildfowl refuges, and also as coast protection measures. Research in the USA has estimated that coastal wetlands are worth between £18,500–£26,000 per hectare (Brooke 1992), based on the services and functions provided (habitat, defence) (see also Chapter 9). Against this however, is the cost of providing that habitat. According to the US Army Corps of Engineers (Brooke 1992), creation of marshes by realignment costs in the order of £3,200–£44,500, the wide values being due to different site requirements and situations. In the UK the Northey Island scheme cost £22,000, largely due to the necessity to construct a new sea wall at the landward edge of the site. This price compares well with the cost of £66,000 for repairing the original sea wall (Hutchings 1994). Also a small scheme will cost relatively more than a larger one due to economies of scale and area:perimeter ratios. This suggests that the economics are better favoured in some areas than others. In some instances, this type of defence scheme can gain economic justification under the auspices of habitat creation. In some countries, such as the USA and Canada (Brooke 1991) this can attract increased funding, such as in the case of the St. Lawrence River in Canada, where such schemes have attracted £2.4 million of government funding.

Interestingly, there have been examples where economics have been used to argue against managed realignment schemes, The example of Cley on the north Norfolk coast, cited earlier, has used estimates of income from visitors to the fresh water reserve to argue against turning it into a saline one by a realignment scheme. Klein (1998) estimate up to £480,000 per year income based on 100 000 visits, compared to an estimated £20,000 to £30,000 per year to maintain the defence. Although these figures look conclusive, they do not take into account the increasing cost of defence maintenance with increasing sea level rise, and it could be argued that the estimates of income are only that. This latter point becomes more relevant when considering the lower figure in the range of visitor income estimates, just £40,000.

There is one further issue in terms of economics. In most of the defence strategies which we have discussed, we are considering strategies which adopt the 'hold the line' or the 'advance seawards' approach. As a result, we are not considering techniques which actively promote land loss. Clearly, any method which does this (such as managed retreat) is taking land which may well either belong to someone, or is occupied by someone who has considerable capital tied up in it. Any proposal which encourages its loss, therefore, is going to have to concern itself with compensating the occupier of the land. In the UK, there are several schemes which compensate farmers for loss of farmland and, to date, no realignment sites have occurred in places where there is infrastructure or buildings. (Interestingly, in the UK, land flooded by natural storm breaching, such as in the cases previously described, carries no financial obligation on behalf of the defence authorities, nor any obligation to repair the defence structure). In the USA, federal, state or local governments are actively buying coastal land in areas where erosion is threatening the stability of the coastal belt. Here, the situation is more complex because of the promotion of mitigation schemes, whereby any development (such as port or harbour expansion) which claims wetland areas, need to, as part of the development, create new areas to offset the losses (Brooke 1992).

Managed realignment and sea level rise

From earlier discussions, we saw how flood tides deliver sediment to the marsh surface by both direct flooding from the seaward edge and via the creek network. As sea level rise increases, this frequency of flooding will increase, thus producing increased amounts of sediment all the time there is sufficient within the system. By such processes, marsh surfaces can accrete vertically at a rate which is equal to, or greater than that of sea level rise (please note that we are not talking about the short-term movements of sea level associated with the tidal cycle, but the longer-term trends in sea level). The vertical accretion of a marsh is primarily a function of sediment supply, but it is also due to currents, waves and vegetation (Reed 1995).

How marshes respond to sea level rise depends on how these variables interact, and in particular the relationship between organic and inorganic (minerogenic) soil components in respect of the position of the water table (Reed 1995). Acknowledging the importance of the organic and inorganic components of the marsh soils, Allen (1997) has produced a series of models with which to assess their relative importance. This balance is expressed simply as:

$$\Delta E = \Delta S_{min} + \Delta S_{org} - \Delta M$$

where:

ΔE is change in marsh surface elevation
ΔS_{min} is thickness of minerogenic (inorganic) sediment added to marsh
ΔS_{org} is thickness of organic sediment added to marsh
ΔM is change in relative sea level

By integrating over time, it is possible to investigate marsh behaviour in response to ongoing sea level rise scenarios (Allen 1997). French (1993) has also shown the impact of sea level rise on marshes albeit along the north Norfolk coast rather than the Severn Estuary. This study determined the impacts on vertical marsh growth of a range of sea level rise scenarios. One scenario involved a rising then static sea level where vertical accretion continued to a position of stability as marshes reach the level of highest astronomical tides (HAT); while another used a higher and continuing rate of sea level rise under which marsh growth could not keep pace and 'drowned' (see Chapter 2). This point occurred when the marsh surface lay 2.0 m below HAT. This work revealed that, given adequate sediment input, sea level rise rates of $4 \, \mathrm{mm \, a^{-1}}$ would be paralleled by upwards marsh growth, while rates of $15 \, \mathrm{mm \, a^{-1}}$ would lead to drowning. This, therefore, helps to put into context the significance of marsh vertical growth and sea level rise, with extreme rates needed for drowning to occur. French supported this by indicating that lateral marsh erosion (marsh cliff retreat, see below) was by far the more significant process in marsh loss under sea level rise.

Although the precise response of marshes to sea level rise is important, and the subject of much work by a range of authors, from our point of view, it is the general response of marshes to rises in sea level which are important, particularly the ability of the marsh to 'keep pace' with increasing water depth. We have seen how important marshes are in wave attenuation, and in particular, the role of vegetation in both wave attenuation and sediment accretion. As sea level rises, the nature of flooding and the water table changes, and unless the marsh surface can increase vertically, these changes will cause more prolonged satura-tion and higher water tables which may induce retrograde changes in the vege-tation succession. Clearly, such changes will reduce the effectiveness in wave attenuation, and the ability to trap sediment, meaning the effectiveness of the marsh as a coastal defence structure will become significantly impaired.

Many marshes do reveal evidence that they can respond to increased water depth with increased sedimentation. Work in the Severn estuary (French 1996) has shown that three distinct marsh surfaces which experience different water depths on inundation are accreting at $12.1 \, \mathrm{mm \, a^{-1}}$ for the lowest marsh, $6.4 \, \mathrm{mm \, a^{-1}}$ for the mid elevation marsh, and $2.3 \, \mathrm{mm \, a^{-1}}$ for the highest (Figure 10.4). Although this is not directly related to sea level rise, it can be used as a model of stochastic reasoning. The study shows that within a system at a single time, increased water depth results in increased sediment accretion on the marsh surface. It can, therefore, also be used to indicate that as sea levels rise over time, the response of the marsh to increased water depth will be to accrete faster in order to 'catch up' with it. These ideas are also supported by Shaw and Ceman (1999) who have demonstrated the ability of marsh accretion to speed up or slow down as sea level varies. In a study from Nova Scotia, Canada, they demonstrated periods of rapid accretion interspersed with slower accretion rates reflecting eusta-tic sea level changes superimposed on a background of crustal subsidence.

Given this proven ability of marshes to keep up with sea level rises over long periods of time, it would appear that there is no real issue as far as marsh sur-vival goes under sea level rise scenarios, providing the extreme situation, as dis-cussed by French (1993) is not exceeded. However, while it is important for vegetation survival that marsh surfaces keep pace with sea level, such increases in water depth, as we have already seen, are coupled with increased wave energy which, even in an estuary, will be sufficient to cause lateral marsh retreat. Such

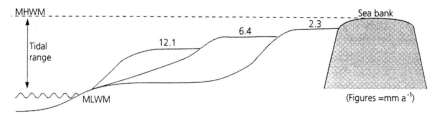

Figure 10.4 Variation in marsh elevation and surface accretion rates (Data from the Severn Estuary) (After French 1997).

processes will have a much greater potential impact on marsh loss and really underlie the point that marshes need to be behind a shallow intertidal mudflat which can attenuate wave energy. There is no point in marshes keeping pace with sea level by growing vertically, if they are being eroded laterally at a rapid rate. Lateral marsh erosion involves a series of marsh changes. First, a marsh cliff is formed which moves inland; and second, the creek system will experience widening and deepening in accordance with the changing hydraulic regime, as well as headward retreat.

Marsh retreat is a common system response to rising sea levels, but in some circumstances accretion of the seaward edge during sea level rise can occur, providing the system is highly charged with sediment (something which can also be managed artificially by trickle charge). In such cases, the marsh edge will remain stable and the marsh could actually increase in width due to landward migration of upper marsh communities, providing there are no constraining sea defences.

In salt marshes, therefore, there are two components to survival: first, that marshes are able to keep pace through vertical growth with sea level rise in order to reduce the risk of marsh drowning; and second, that they can either receive enough sediment to maintain their seaward margins, or migrate inland to maintain an adequate intertidal mudflat frontage to help protect against wave attack.

Summary

Managed realignment makes good sense in areas where hinterland usage or development does not get in the way, as it provides a way in which we can compensate for the impacts of sea level rise, and encourage the natural environmental processes to continue. No coastal manager will argue against such logic, but many might debate the basis on which schemes are sometimes carried out. Realignment is not just a question of punching a hole in a sea wall. We have seen here the complicated set of environmental variables which can influence the success of realignment schemes, everything from estuary size and shape, wave exposure, tidal regime, down to what the land was used for and what was put on it. A series of case study references are given in Table 10.2.

One of the key problems with managed realignment is that it does appear to be an obvious solution. Leaving aside the scientific issues which are yet to be resolved, there is also the socio-economic aspects of giving land back to the sea. From a public perception aspect, this is 'giving-in' to the sea, yet in reality, all that is happening is that engineers are returning land to the sea which has previously been reclaimed.

Clearly then, managed realignment is one of the most complicated methods of coastal defence, not least because it involves so many variables which are notoriously difficult to determine. Perhaps this is where the hub of the problem occurs. We can see the logic to the process, and we can see from existing estuarine locations where the sea has re-invaded claimed land in the last century. What we do not know are the processes which led from the initial breach to the present situation, or the conditions of the site before the breach. Hence, we know the method can work — the example of Pagham Harbour in southern England is proof of that (Plate 10.2) — but it is difficult to recreate the idealised conditions. This is clearly an area in which more research is needed, not just related directly to the technique, but also to basic research into estuary dynamics.

The relative 'newness' of the techniques is also a concern. To date, coastal managers have only really been 'dabbling' with the method. This is not a criticism, but merely a comment on the fact that many are working in the dark and learning as they go along. The bottom line remains that if, under sea level rise, it is necessary to allow the sea to re-invade the land, then the resulting marsh needs to reflect natural conditions as closely as possible. It could be argued that small, isolated realignment schemes cannot do this, and that what is needed is one large realignment scheme which returns an estuary to its 'natural' shape.

Summary of benefits of managed realignment

- Increased intertidal width and wave attenuation capacity
- Increased conservation potential of habitats
- Improved protection against sea level rise
- Increased 'naturalness' of estuaries

Summary of problems with managed realignment

- Danger of modification of tidal prism
- Uncertainties about hydrology and sediment movements
- Novelty of the technique
- Complexity of potential compensation issues

Recommended usage

- To increase intertidal width in areas of marsh retreat
- To increase habitat for flora/fauna
- To increase estuary functioning

Part IV

Coastal defences in a changing environment

We have seen how the coast may be defended and protected from erosion by the use of a range of different techniques, each of which affect the coast in different ways, and each of which tackles the problem of erosion by different means. The decision making process will not be developed here in detail, but we have laid down several 'rules' already, including:

- Do not develop the hinterland near to eroding coasts as defence needs may increase.
- Always try to manage the coast in as environmentally sympathetic a manner as possible, so as to reduce the impact on natural processes.
- Always attempt to understand how the coast 'works', where its sediment supplies are coming from, where they are stored, and where they leave the system.

By keeping to such rules, coastal management will be made easier in the future. In addition, it must be acknowledged that there will be times when we have to interfere with processes and build hard defences, purely because of development. Perhaps more pertinent is that this pressure for hard defences is going to increase. One reason is that our fundamental unit of planning, sea level, is changing, so that many of the processes seen at the coast will also change. Another reason is that coastal processes vary naturally over time anyway; hence, a coastline which was accreting a century ago may be eroding today, while one which is accreting today may not be in 10 years time.

Finally, we must realise that we are going to reach a point when we have to stop building defences and tackle the problem of coastal erosion in a different way. With sea level continuing to rise, we need to ask whether we can manage the coast in a way which avoids using engineering approaches. Some of these ideas are highlighted in this final part which looks at the strengthening of insurance costing or planning laws to make development on unsuitable coastlines cost-prohibitive.

11 Coastal defences revisited

Introduction

Looking back over the various techniques outlined in Parts II and III, we can come up with one of two conclusions. The first is that we can feel assured that coastal defence planning is moving away from the 'hold the line', 'defend at all costs' approach to something softer which preserves the integrity of coastal processes as much as possible. The second is a deep anxiety because of the realisation that whatever we do (hard or soft) there are always impacts (to solve a problem you cannot avoid creating another), and because in addition we also have sea level rise, there appears to be no solution.

In reality, however, we have to be realistic about the problem. We know of the problems with historic coastal development and coastal erosion, and that this is a legacy which we cannot avoid. Because of this, we have to protect high-value hinterland. We also know that we now have new approaches in the form of soft defence techniques, which have provided a range of new methods with which to tackle the problems of coastal erosion, although these still have some impacts, they are better than some of the more traditional methods. New coastal managers entering the profession need to realise that:

a) There are coastlines which will need protecting, yet there are coastlines which can be left alone.
b) There are methods by which varying degrees of land/sea interaction can be facilitated (stopping erosion or reducing it).
c) There are coastlines where development may be acceptable, but there are a greater number where it is not.
d) There are still too many occasions where development is going on along unsuitable coastlines.

It is clear, particularly with reference to points (c) and (d), that the final outcome lies beyond the realm of coastal managers. All too often, the coastal manager is left to pick up the pieces after the actions of others, and to take the blame, but we must consider the following:

- Coastal development needs planning permission. This represents one way of dealing with high-risk coastlines.
- Coastal development needs insurance. Would true-cost insurance be a solution to inappropriate development?

Before we develop these issues further, we can identify some commonalties from the preceding chapters. If we do this, perhaps we will be in a position to develop a new set of guidelines which can be used at the coastline.

Common problems with coastal defences

Regardless of the defence technique used, there always appears to be one common issue arising. If you were to study the sections of each chapter relating to environmental impacts, the most frequently seen topic will be the *interruption of sediment supply*, or problems of defended areas *taking sediment which was originally deposited elsewhere*. This all relates back to sediment availability, the sediment budget and the balance between sediment into a beach environment against sediment out of a beach environment. Therefore, any defence scheme which encourages sediment deposition can, theoretically, impact on the balance of the sediment budget for other parts of the coast, and so cause additional problems which need solving. Despite the rationale of soft engineering and 'holistic' management, this would argue that we can never really solve coastal problems holistically, but only locally and regionally. In counter to this, what it also implies is that for every defence which involves sediment deposition and negative impacts on the sediment budget, we need other schemes which free up part of the coastline and increase sediment provision, i.e. the benefit:loss approach, potentially referred to as the Sediment Budget Model of coastal engineering. By taking the idea of holistic management a stage further, it could be a good management consideration that whenever defences are planned, a fundamental part of the decision should be an estimation of the volume of sediment lost to the sediment budget as a result of the scheme, and ways of compensating for this. For example, the building of groynes will trap sediment through longshore drift, hence this trapped volume would be lost to the sediment budget. A scheme to construct groynes, therefore, could also identify sources for increased sediment provision to compensate for these trapped volumes. This could be by the acknowledgement of increased erosion in safe areas, away from property, or it could be some other form of sediment input, such as beach feeding. If such calculations are incorporated into planning decisions, then many of the secondary issues, which have been discussed under 'impacts' in the preceding chapters, disappear. Operationally, however, this may be difficult because the basic knowledge required to make such volumetric calculations is often lacking. However, as a development on the soft defence policies, it may be workable in some situations, even if the answer to the second part (compensation) is to beach feed.

A second issue pertaining to all defence types is that they interfere with natural processes. Thinking in basic terms, they must do by definition. If this is not clear, think of the following.

Q. Why do we build coastal defences?
A. Because the coast is eroding and threatening development.

Therefore, the natural processes operating on that coastline are erosional (destructive), and by building defences we are changing these basic parameters. In its simplistic form, this is often represented by some secondary reaction to the defence, such as beach lowering or flanking, to name but two. Perhaps it can be argued that alteration of processes from erosional to accretion is good, and so it is, but we should consider a further aspect. One factor which has been consistently highlighted is our lack of understanding of how coastal processes work. This also includes processes relating to post-defence construction. Do we really understand the implications of our actions on many coastlines?

In reality, these ideas are somewhat philosophical, and the stuff for a pint or several on a cold winter evening, but we should be concerned because what our coastal defence endeavours really amount to is interfering with something which we do not fully understand, and the repercussions of which we do not really know. One area in which this readily becomes apparent is where defence techniques involve the creation of new habitats. This has been discussed in the context of salt marsh, mudflats, dunes and beaches, all of which are created and managed as part of soft defence methodologies. However, what is also obvious from each of the relevant chapters is that when these environments are created, none behave in the same way as their natural equivalents.

- Created beaches behave differently from natural ones under storm conditions.
- Artificial dunes do not erode at the same thresholds or maintain the same degree of mobility.
- Created mudflats contain more water and so do not have the same appeal to infauna as their natural equivalents, or have the same shear strength.
- Planted salt marshes do not achieve the same above- or below-ground biomass.

Clearly, we may be able to create coastal environments artificially, but these do not precisely compare to their natural counterparts, and so will respond differently to coastal processes. The reasons for this are not difficult to understand: all involve the artificial importation of sediment which is deposited by artificial, not natural means. As such, deposition is quicker and without the degree of natural transport or sorting which normally occurs. It is in this area that engineers are now starting to realise the problems linked with soft methods. Despite this, we are reaching a point with soft engineering that we did with hard, in that we have to make do with the levels of technology and understanding that we currently have.

Effectively, therefore, we have to operate as we are. This means defending where we have to, not defending where we can get away with it. We must, however, learn our lessons and stop building along unstable coasts where we will have to intervene in the future. Given that we already have a severe problem in this respect, what can we do about it? There are two aspects to the problem:

1) What to do on developed coastlines?
2) How to prevent further problems by new development?

Developed coastlines: a location for 'staged retreat'?

On developed coastlines, the 'do nothing' option is not viable because of the loss of infrastructure and impact on socio-economic functions. Similarly, managed realignment is also a problem because there is no clear line to set back to, as the existing line of defence frequently represents the limit of hinterland investment. This situation is purely short term in respect of the history of a coastline and should not be seen as a situation which cannot change over time. If we consider coastal buildings as being transient features (consider the lifetime of many buildings as being 50 to 100 years), then there will come a time when they will no longer be a feature in defence calculations. Does this not, therefore, provide us with a mechanism of gaining vacant land to give back to the sea in the future?

As buildings reach the end of their lifetime, any land adjacent to the coast should not be redeveloped. In time, this will free up a coastal strip which can facilitate future shoreline retreat. Fletcher (1992), using the US scenario, claims three approaches to future coastal management at the federal level. One of these is:

> Promote the improvement of . . . redevelopment practices in erosion-prone areas.
>
> (Fletcher 1992: 97)

Basically, this is saying that additional interested parties should be brought into the defence management arena. All too often, planners and developers see derelict plots as viable sites for redevelopment, particularly if the land either side is already developed. By leaving these plots, in time more space will become available for coastal retreat. The basic argument against this is economic. Building land, especially in tourist areas, has a high value to builders, but this situation can be remedied if planning laws governing redevelopment in 'at-risk' areas are changed to suit the coastal manager rather than developer. In effect, this process is not managed retreat but 'staged' retreat, i.e. the setting back of the coast in land following the removal of development through sympathetic redevelopment.

Although we could see this as a way of allowing coastal management on developed coasts, many areas where realignment is urgently needed will remain unsuited for many years, if not centuries, purely because of the nature of existing development. Overall, therefore, developed coastlines remain a problem, and will generally require the use of large engineering structures to protect them. However, when redevelopment is appropriate, it should be the role of planners, supported by an active coastal management plan, to make sure that redevelopment does not incur greater defence needs in the future.

Undeveloped coastlines: planning and insurance — true costing in coastal areas?

Despite the fact that planners and developers know about coastal erosion, one of the greatest problems on contemporary coastlines is that development is still going on. The problem is that while development is the responsibility of developers and planners, the provision of coastal defence is not, and so there are no financial disincentives for builders on coastlines. However, prevention of such practices will mean future benefits in not having to extend defences into new areas. The simplest way of achieving this would be to incorporate local planning into coastal management. Coastal managers identify areas of erosion and potential problems, and planners avoid them. Simple? It would appear so, but in many instances, this does not happen.

Given that some unscrupulous developers still build on vulnerable coastlines with the expectation that defence authorities will oblige by building defences when problems occur, perhaps we need to look at ways of forcing them to stop. Planning departments are the obvious way, but these do not always appear to be working well. What, therefore, are the alternatives? Perhaps we can consider two possibilities.

Use of erosion zones in coastal classification

We first mentioned erosion zones (E-zones) and the formation of set-back lines in Chapter 5 (see Figure 5.10). Fletcher (1992) discusses them in connection with US federal law and details how they can be established to allow different types of development according to coastal stability. In essence, by predicting future coastal positions, it is possible to identify how long a piece of land will be there, and so any development upon it would have to have a life expectancy less than the predicted longevity of the land. Where erosion rates are high, temporary structures may be allowed, with no infrastructure provision. On land with a predicted time of about 50 years, some temporary structures may be built, with planning constraints regarding their removal at specified times. Such methods allow some development while accepting the fact that the land is eroding, and will not be defended. This could have particular appeal to the tourist industry, where caravan sites or beach chalets are suitable temporary structures. Having said this, other factors are also important here, not least the fact that the proliferation of unsightly caravan sites along aesthetic coastlines is clearly not a desired outcome.

In the USA, around a third of coastal states have identified set-back lines for

coastal development (National Research Council 1990), while others have strategies which successfully serve to limit development in high-risk areas. States
which do not employ such measures are generally those which do not have an
erosion problem. The criteria used in the establishment of these hazard limits
vary between states. In many cases, an erosion rate of around one foot per year
(about $0.31\,\mathrm{m\,a^{-1}}$) is needed to instigate the formulation of E-zones while in
others, any eroding coastline qualifies because of the potential threat to property. This represents perhaps the most proactive approach to the problem and
offers the best safeguards against increased demands for defence construction.

Experience with this approach has found it to be a useful tool in reducing
the threat to property. While property built pre-policy is still threatened, E-zone
restrictions have provided good management tools for the siting of more recent
buildings. However, as with any coastal management scheme, monitoring and
experience reveal certain areas where they need modification. Such areas
include:

- greater set-back limits for larger buildings;
- improved design (and definition) of movable buildings in the most at-risk
  areas;
- suitable methods of tackling building improvements and extensions;
- re-evaluation of erosion rates to determine whether limits are still viable in
  the light of other coastal protection measures.

Updating is critical in the long-term viability of these schemes because, as we
saw in the earlier discussion on set-back lines in respect of cliffs, these lines are
only as accurate as the data and models used to predict them. In the US case,
Pennsylvania and Indiana monitor their predictions every 2 years; north Carolina every 5 years; Michigan and Texas every 10 years; and Florida every 10 to
12 years (NERC 1990).

House insurance and true costing

A more direct approach to deter developers wishing to build on unsuitable land
would be to make the cost of insuring those properties too great to be viable.
In many countries, the increased insurance risk associated with coastal (plus
other) high risk development is carried by all customers, wherever they live. In
the USA, the National Flood Insurance Program (NFIP) tries to tackle this
problem by putting the onus on local politicians and planners. According to
Annex D of the Act, affordable insurance coverage will be maintained only if
communities act to limit future losses. This doesn't actually mean that houses
cannot be built on vulnerable land, but that if they are, they need to be of des-
ignated standards to reduce flooding (see National Research Council 1990 for
full summary). Amendments to this act included the problem of loss through
erosion. The Upton-Jones amendment to the US National Flood Insurance
Programme encourages retreat from eroding shorelines, not protection of them

(National Research Council 1990). One section (Section 544) provides for compensation to owners for the demolition or relocation of their at-risk properties (110 per cent of their value for demolition, or 40 per cent of their value for relocation). This provides a good alternative to instigating high-cost protection and features in the cost-benefit analysis of any defence schemes as part of the viability of the 'do nothing' approach. (Interesting comparisons can be drawn here with the case of the Holderness coast in the UK.) Insurance policies issued to people in the at-risk areas contain two parts. The first represents the 'normal' cost of insurance a property of that size and type, as would be issued to any property owner. The second additional component reflects the increased risk of being located in a coastal erosion or flood area. This amount decreases proportionately with increased distance from the risk.

This legislation tackles the problem of coastal management in at-risk areas in two ways. First it gives occupants the chance to relocate and leave the area, thus leaving the sea to erode the land. In the context of this book, this represents the abandonment approach. Second, it passes on the inflated cost of insurance to the people most at risk, the owners. While this does not get over the problem of these properties being present in an erosion area, the high cost of insurance will make further development less likely due to the higher costs of protecting the property.

Such methods provide ways of alleviating the problem of coastal defence in developed and developing areas which could provide a useful model for other countries to adopt. Clearly, eroding coastlines do represent high risk areas, and so should attract inflated premiums. This could be said to be unfair to owners — after all, they did not build the properties — but if adopted as a national policy, then insurance companies would promote such premiums, which could come to light in property surveys etc., thus making people avoid purchase and developers rethink their development strategies.

Summary

In this short chapter, we have seen some possible non-structured defence methods for solving coastal problems. In effect, one represents a way in which people can be persuaded to leave their homes and property and move inland to safer ground, thus allowing soft engineering approaches to be implemented, and the second really represents an indirect way of preventing development in inappropriate areas. Clearly, the planning laws should be sufficient but many still are not. In reality, if builders cannot sell properties in an area, they will not build there.

The methods outlined have not been included in any great detail, simply because such detail does not exist. The USA has adopted many of the ideas mentioned, but actually enforcing them can be problematical if challenged in a court of law. The volume by the National Research Council (1990) discusses many of these problems and issues in considerable detail. My reason for including them in this end section is simple. Although they do not really fit into the earlier sections on hard and soft defences, they are methods by which the implementation of such methods can be facilitated. By being aware that there are these developing ideas with which to prevent the problems encountered along our coast, then readers may develop them and think about new coastal strategies for themselves because after all, YOU are the next coastal managers.

12 Coastal defences in the future

Introduction

As we reach the end of this review of coastal defence types and their impacts, it is perhaps pertinent to pause to consider in what areas the subject needs to develop in the future in order to address some of the issues raised. There have been some large-scale changes in attitudes to defences over the past few decades, most notable of which is the development of softer techniques which are being used whenever possible in preference to more traditional methods. The key thing here is that now such techniques have been used for some time, data on scheme performance is available and is starting to feed back into improved design. However, engineers are now realising that even these softer techniques produce their own series of environmental impacts.

Research and development has not only been concerned with the newer methods of defence however. Hard methods, such as sea walls and groynes, have also changed in design over recent decades. Most new groyne construction, for example, is carried out with boulders, rather than the more traditional wood, concrete or metal. Similarly, sea walls now often have sloping, irregular faces which serve to deflect wave energy in many directions, rather than just back onto the beach. Both of these developments, when appropriately designed for the environment in which they are constructed, have produced hard structures with less environmental impact. In summary, therefore, on the one hand we have the traditional structures which are becoming more 'environmentally friendly' and on the other, a new set of methods which work in parallel with coastal processes, rather than against them. Both of these are areas in which development needs to continue. As new designs and methods are used more and more, so more data will become available with which to gain a greater understanding of the interactions they have with the environment.

The increased data availability is also feeding another rapidly expanding area of coastal science, that of modelling. This topic was introduced in Chapter 1, with discussion of various models and their applicability to coastal process understanding and management. Over time, as more verified data becomes available, so these models will increase in predictive ability. Already we are seeing the use of some simple models in predicting future coastline positions, such as in the establishment of E-zones and set back lines. Once this power of prediction becomes more reliable and accurate, so it opens up the greater possibility of leaving the coast alone for as long as possible.

Perhaps one of the greatest advances over the past decades, however, is the greater organisation of coastal defence planning and management as an holistic issue. The new area of *Coastal zone management* (CZM) has developed in response to the need to integrate all aspects of coastal use into one management plan, incorporating land use, defence type and physical process. By doing this, the conflicting interests of groups such as developers, managers, planners, etc., can be overcome and thus inappropriate development can be avoided.

All of these issues point to the way in which the topic of sea defence planning and management is going. In addition, however, there are also a series of important issues which need addressing in the short term. First, there is the problem of sea level rise and the implications which this will have for the use of hard and soft defences; and second there is the important area of public perception. When sea defences are built to protect land from flooding or erosion, there is a need for these defences to fulfil their role, but also a need for people living in the affected areas to feel safe. While on the one hand, a recharged beach may be a perfectly acceptable way of attenuating wave energy, people may feel safer behind a large concrete wall because it is a large, physical structure, and it makes them 'feel' safer, and perceive a greater degree of safety. This again relates to the hard versus soft debate: the physical structures have the greater perception of safety, but the softer structures have the least environmental impact.

Sea level rise and the hard versus soft debate

Of key importance to the successful functioning of many coastal defences is their relationship with the coastal system in which they are constructed. As sea levels rise, high tides become higher, wave base gets deeper and energy received at the coast increases. Furthermore, with ongoing sea level rise, the threat of overtopping of walls will also increase, making more of our coastlines susceptible to flooding. With natural changes in coastal processes superimposed on top of sea level rise, it is easy to see how the coastline can be subject to change over time — in some cases, such as storms, over quite short periods of time. As a result, erosion may increase in some areas, while deposition may increase in others. These changes become critical in relation to defences because, while soft defences may adapt to such changes, hard defended coasts cannot. One thing which is certain, however, is that with ongoing sea level rise, there will be an increased need for adaptation and response to the changing coastline.

We have seen that the ability of different coastal morphological units to survive contemporary processes, as well as changes caused by rising sea levels, lies in their ability to react and respond to changing environmental forcing factors, such as by relocating inland under the Bruun rule. Some of the defence structures used along our coastlines will enable this adaptation to be achieved more easily than others. Although hard defences are still the most common type of defence seen on our coastline, it is clear that they are the least ideal under sea level rise because they fix the coastline in position. Sea walls will stop sea level rise, causing erosion of the hinterland and thus landward relocation, but they may also encourage wave reflection and scour of fronting beaches. In another situation, groynes will not stop shore normal currents but will interfere with changes in longshore currents brought about by current changes induced by rising sea levels and deepening wave base. Here, there is the problem of beaches down drift not responding to sea level due to sediment starvation. The real concern is that although we know of these problems, and we know that rising sea levels will exacerbate them, we just have no option in many cases. High-valued hinterland needs a sea wall to protect it, therefore beaches cannot be allowed to migrate landwards. Similarly, to stop beach loss, groynes may be the most effective method, meaning that areas down drift will have to lose out.

Many coastlines are experiencing these problems. Table 12.1 details a range of studies linking coastal defence structures to sea level rise, and investigates many of the problems caused (see also Chapter 2). In such situations, it is often not possible to adopt softer methods because of the inherited legacy of hinterland development; the coastal manager will therefore have to make the best effort using adaptations to hard techniques.

An alternative approach of setting the line of defence back landwards is a more sensible way of tackling problems of sea level rise, because it is a human response which mirrors what nature will attempt to do anyway. In many estuaries and undeveloped coastlines, such techniques can be easily adopted, with salt marsh formation by managed realignment, sand dune roll-over inland, or cliff abandonment. However, with millions of pounds or dollars (or whatever currency you choose to name) locked up in hinterland development, it is still going to be cheaper in many areas to fight the sea and hold the line using hard methods. In many cases we have no choice but to stick with these hard structures. For example, the predominance of nuclear power stations on soft eroding coastlines as reported by French (1997) means that we are committed to a 'blank cheque' approach, because for the next 200 years at least, these structures will need to be protected no matter what the cost (200 years is the estimated time needed for the radioactive levels in the core area to reduce sufficiently for it to be safe to dismantle).

Through reading preceding chapters which have discussed the response of coastlines to sea level rise, it becomes apparent that the main underlying rationale in defence construction is the adoption, wherever possible, of defence techniques which allow the coast to respond naturally. Given that the natural response of the coastline under sea level rise is to move inland (and, as counter to that, during sea level fall it moves seawards), then the easiest way to manage the problem is to allow the coast to do just that. Hence, managed realignment allows marshes to move inland, the 'do nothing' approach allows erosion to continue unchecked, and dune migration allows sand to migrate landwards. This may all sound simple but there are many complications, most notable of

Table 12.1 Examples of sea level rise studies from around the world (For full references, see References at the back of the book).

Country	Location	Rate (mm a^{-1})	Age range	Source
Argentina	Buenos Aires	1.5	1905–1988	Douglas (1997)
	Quequén	0.8	1918–1982	Dennis *et al.* (1995b)
Bangladesh	Bay of Bengal	3.3	2000–2100	Castro-Ortiz (1994)
Canada	Halifax, NS	3.03	1896–1995	Shaw *et al.* (1998)
	Quebec	*0.75*	*1940–1989*	*Shaw* et al. *(1998)*
China	General	1.2	1900–2000	Wang (1989)
	North Jiangsu	5.0	until 2050	Chen (1997)
	Pearl River delta	8.0	until 2050	Chen (1997)
	Yangtze River delta	9.0	until 2050	Chen (1997)
	Yellow River delta	6.0	until 2050	Chen (1997)
Denmark	Esbjerg	0.82–1.38	1901–1969	Shennan and Woodworth (1992)
Egypt	Nile delta	5.0	Current	Day *et al.* (1995)
France	Brest	1.4	1880–1991	Douglas (1997)
	Ebro delta	1.0 –5.0	Current	Day *et al.* (1995)
	Marseille	1.2	1885–1991	Douglas (1997)
	Rhone delta	1.0–5.0	Current	Day *et al.* (1995)
	Rhone delta	2.1	since 1905	Suanez and Provansal (1996)
Italy	Venice lagoon	8.0	Current	Day *et al.* (1995)
Netherlands	North Sea coast	2.0	1900–1985	den Elzen and Rotmans (1992)
	North Sea coast	1.5–2.0	20th century	van Malde (1991)
	North Sea coast	0.73–2.53	1901–1987	Shennan and Woodworth (1992)
New Zealand	Auckland	1.3	1904–1989	Douglas (1997)
	Wellington	1.7	1901–1988	Douglas (1997)
Norway	*Bergen*	*0.74–1.18*	*1928–1986*	*Shennan and Woodworth (1992)*
	Stavanger	*0–0.42*	*1928–1986*	*Shennan and Woodworth (1992)*
Senegal	Dakar harbour	1.4	1943–1965	Dennis *et al.* (1995a)
UK	Aberdeen	0.54–0.98	1901–1988	Shennan and Woodworth (1992)
	Avonmouth	0.25–1.07	1925–1980	Shennan and Woodworth (1992)
	Dover	1.6–3.0	1961–1987	Shennan and Woodworth (1992)
	East Anglia	3.75–6.25	1990–2030	Clayton (1989b)
	Forth estuary	4.3	1990–2012	Shennan 1993
	Lowestoft	0.45–1.81	1956–1988	Shennan and Woodworth (1992)
	Mersey estuary	5.3	1990–2050	Shennan 1993
	Newlyn	1.7	1915–1991	Douglas (1997)
	Portsmouth	5.0	1962–1987	Woodworth (1987)
	Solent	4.0–5.0	1896–1996	Cundy and Croudace (1996)
	Thames estuary	6.8	1990–2050	Shennan 1993
	Wash	6.3	1990–2050	Shennan 1993
USA	Chesapeake Bay	2.7	since c. 1700	Lisle (1986)
	Gulf of Mexico	2.0–2.5	1938–1993	Warren and Niering (1993)
	Key West	2.2	1913–1991	Douglas (1997)
	Mississippi delta	10.0	current	Day *et al.* (1995)
	North Carolina	1.9	1940s–1980s	Hackney and Cleary (1987)
	San Diego	2.1	1906–1991	Douglas (1997)
	San Francisco	1.5	1880–1991	Douglas (1997)

Note: Italic text signifies a falling sea level.

which is that we are unable, in many cases, to adopt soft methodologies because of the value of the hinterland and the strategic importance of fixing the coastline. In reality, therefore, we know that a soft coastline can adapt to sea level rise, but we also know that in many cases, these are methods which cannot be adopted and used at the present time.

Because we are now aware of the problems associated with hinterland development, we can provide for better planning provision along the coastline in future. These ideas are linked with successful coastal zone management plans, which will be discussed below. While this may help in preventing future coastal defence problems, it does little to solve many of the problems currently experienced along coasts which are already extensively developed. So, in effect, we are stuck with an historic legacy of coastal development which is now causing problems due to sea level rise. Although it may be tempting to do so, we cannot point an accusing finger at historic developers, however, because at the time when many of these developments were originally built, sea level rise was a phenomenon which was hardly known about, if at all. At the present time, there is nothing to do but throw money at these problem. The possible alternatives, such as staged retreat or the construction of effective E-zones, will take time, and government policy, to implement. In many coastal areas, however, these may be the few alternatives left.

Public perception of hard and soft defences

The common theme throughout this book is that the adoption of any coastal defence methodology is based on its suitability to a particular environment, coupled with an acknowledgement of a series of environmental impacts which may occur to differing degrees. One aspect which has been rarely touched on here, however, or in the many papers published on coastal defences, is how the public react to different defence methods. This is due, in the main, to a general lack of such studies. People who live on the coast will have an interest in how their property is to be defended, and will want such defences to be suitable for the job in hand. This suitability, however, will often be a perceived suitability, rather than real because having experienced erosion or flooding, coastal communities will want some visible means of preventing such problems from re-occurring.

Another aspect of public interest is more general than the above. In many countries, where public money is involved in the building of coastal defence structures, it is natural that the public will often have a reaction to the way in which their money is being spent. Such reaction can be grouped into three categories. First, there is the reaction to how effective a scheme is, i.e. does the first storm following new sea wall construction damage the structure? Second there is the suitability of the scheme, such as the use of managed realignment which many people still regard as being negative and 'giving in' to the sea. Third, and perhaps most critical, is the perception of safety. Each of these categories involves human reaction, and so the range of possible ideas and issues differs widely across the broad spectrum of human nature. However, despite the general lack of research on this topic, this is an area which the good coastal manager should be aware of, and should address in public consultations. In discussing the three main categories, it will be necessary to adopt a more 'suggestive' rather than 'absolute' style due to the lack of information available, but readers should be aware of the potential challenge in this, in that the reaction of people to defence planning is one area where increased attention from managers is needed.

Public reaction to scheme effectiveness

The extent to which an individual will judge a scheme to be successful will ultimately depend on their own set of interests, and whether they are upheld by the scheme. For example, a coastal engineer may regard a beach feeding scheme to be successful if it maintains critical beach levels (see Chapter 7) for a period in excess of 5 years, or if it affords hinterland protection during storms. In other words, the effectiveness of the beach in preventing damage is paramount. In contrast, a tourism manager will regard the scheme a success if visitor numbers are maintained. In this case, damage limitation is not the prime concern, but beach attractiveness to the fee-paying visitor is. Finally, a local property owner will only regard it as being successful if the value of that property is protected, likely only if beach width is maintained (Camfield 1993).

One of the greatest problems of beach feeding is its tendency to lose a lot of sediment during the first year after fill. This commonly leads to claims by local populations of scheme failure but, as we have seen in Chapter 7, it is actually the adjustment of something artificial to a natural set of environmental conditions. Similarly, large storms which draw down a lot of sediment can lead to claims of beach loss, whereas in reality it represents a natural adjustment to beach morphology which will generally revert over time. In effect, both these areas of public response and claims of failure relate to reaction to short-term events, and not to the natural cyclical behaviour of coastal systems. However, what we really have is a public education exercise: unless the public realise what will happen to the beach, they cannot be criticised for showing such concern. Frequently, the planning of a beach feeding programme may appear appealing, particularly in a tourist resort, but to see sand disappearing at an apparently alarming rate, will lead to local insecurities and concerns.

Such a public response to beach feeding, as well as other comments arising from other defences, such as the failure of vegetation to colonise realignment schemes, storm damage to sea walls, lack of sediment accretion behind offshore structures, or the failure of marram planting on dunes are all understandable as there are no visible tangible ways of measuring success from such defence schemes. The provision of increased wave attenuation, for example, may be obvious to the coastal scientist with abundant measuring equipment, but not to the casual onlooker. One thing which readily becomes obvious about coastal management is that there is no quick fix, and many soft techniques do not provide an overnight solution. Furthermore, as we have seen time and time again, one solution normally involves the instigation of other problems which also need to be dealt with. These are natural occurrences and not scheme failures.

Public reaction to scheme suitability

There has been much comment made on the increased preference for soft defence schemes over hard given the choice. This often means that any coastal problem will often have a series of suitable techniques with which to solve them, ranging from the hard to soft. Throughout preceding chapters, we have discussed the merits of these techniques and the situations in which they may be used. Such reasoning and rationale may appear quite logical to the scientific community, but many of the salient points of the decision-making process may be lost on members of the public. There is a large section of the public, for example, who hold strong views with regard to maintaining the coastal position (holding the line), and so the idea of managed realignment or 'do nothing' is not something which should receive serious consideration. However, as we have seen, there are many situations in which such a policy should be adopted as the preferred means of defence.

Similar problems relate to the choice between hard and soft defences. To a certain extent, this overlaps with the following section, but it is also pertinent here. To many people, the creation of a salt marsh or sand dune is purely an ecological benefit, and has nothing to do with coastal defence, and could even be regarded as being totally negative due to the loss of land involved. Clearly, such a statement is not borne out by scientific fact but has to be taken into consideration by planners and those involved in the public relations side of the sea defence management.

Linked to this issue is the future of the coastline. Where hard defences are instigated, they promote a much greater security and perception of permanence to local populations. Once this perception is held, it can be very difficult to change, and so could lead to increased problems for a coastline. Burby and Nelson (1991) discuss this problem with respect to the promotion of further development in areas perceived to be safe following new defence construction. So not only will hard defences provide a perceived more appropriate way of defending coasts, they could also provide a greater misperception of coastal stability, and a tendency for increased coastal development which goes counter to the actual risk associated with an erosion or flood-vulnerable coastline.

Public perception of security

One further aspect of hard defences is that of perceived security to coastal inhabitants. Put basically, people feel secure behind a large, concrete sea wall (Plate 12.1). However, as we have seen in our discussions of wave attenuation, a healthy beach or salt marsh can provide adequate protection in itself, necessitating just a small wall for protection against storm waves. Such features as beaches or marshes, however, do not give the impression of being coastal defence structures. Consider the house shown in Plate 12.2. This property, located at Kennack Sands, on the Lizard peninsular in Cornwall, south-west England does, on first inspection, look about to fall into the sea. However, since I first visited this site in 1984, the cliff line in question has not moved, and had not moved in the years preceding then. What this locality has, however, is a wide, shallow fronting beach, with a wide dry beach width, which has proved quite adequate in preventing cliff erosion.

There are important links between perceived safety and the problems associated with coastal development. People will continue to move into an area all the time they perceive it to be safe. Furthermore, we have also seen on many occasions how inappropriate coastal development is often the problem along many coasts. Herein lies a cause and effect scenario. People perceive the coast to be safe and so build. In actuality, the coast is not safe and will require additional defences to protect the new development. Leaving planning permission issues aside, it is tempting to ask the question of why there is this perception of security. Why can people not see that an eroding coast is going to pose a threat to homes and businesses? Burby and Nelson (1991) suggest a possible link when they identify that the main reason for identifying risk is experience. People who have experienced loss or damage through coastal storms are more aware, while those who have not, are less aware of the issues involved, and also, to some

Plate 12.1 A house with a view. The perception of increased security behind hard defences can lead to some losses of coastal views. These houses are below the height of spring high water, Bispham, Fylde Coast, Lancashire, UK (See location map in Box 3.1).

Plate 12.2 A house with a view. No major defences are necessary for this cliff-top house due to the presence of a wide fronting beach which provides adequate wave attenuation, Kennack Sands, Lizard Peninsular, Cornwall, UK.

extent suffer from the 'it won't happen to us' type of belief. Clearly, there is an important element here, and that is the education of people to risk.

These three categories of public interest in coastal defence schemes all contain one key element which although difficult to handle in the planning process, is vital for successful scheme implementation. This element is perception, the belief that something is effective, safe and secure. All the time the public has this belief, they will argue at public meetings for hard sea walls, because these are 'safer'. Once they have them, the planners will argue for increased development, because the hinterland is 'safe' behind the sea wall. And so the spiral continues. One of the key battles in coastal management is not only developing the soft alternatives, a process already well under way, but in convincing the public that they are viable alternatives to hard defences, a process not really looked into. Once this is done, planners will have not only a greater acceptance of soft methods, but also a reduction in the level of inappropriate development along the coastline.

Where is coastal defence going?

Throughout this book we have looked at many aspects of coastal defences, and identified key issues therein. There have been many unanswered questions about some of the methods discussed, such as the response of the natural system to various techniques, and the lack of 'naturalness' to many created environments. Such issues are being addressed as techniques develop. Each time a technique is used, it builds on the lessons learnt from earlier attempts, most notable in this is the use of managed realignment.

Perhaps the greatest step forward in coastal defence planning, however, is the integration of all the aspects of the process into one overall management scheme. We have seen how coastal processes, land use, development, nature conservation, and humans all interact to make the coastline what it is. Integrating all of these into one Coastal Zone Management (CZM) plan can provide the coastal planner with a tool which prevents unsuitable development, allows coastlines to be left free for erosion, and allows most suitable defence policies to be made for other parts of highly used coastlines. There is not time or space to discuss CZM in detail here, but some basic points are relevant. In its purest form, the concept of coastal zone management (CZM) involves an ecosystem approach to the problems of management. Any human activity or development along the coast must be undertaken only with due consideration of the environmental needs of erosion, accretion and habitat creation/stability. Increasingly, however, the term has been applied to the whole field of coastal activity, such as the management of competing land-use activities, fishing, shooting and sailing, for example.

The main underlying principal regarding CZM is that it has to view the coast as a single unit. The realisation of this need has come about because of the major problems coastal engineering structures have caused along the world's coastlines due to their lack of environmental sensitivity and their rather piecemeal approach. Because engineers now realise that the coast is a complex system with processes operating over a large area, not only is the adoption of softer techniques important, but so also is the management of coastal processes over the whole area which they affect. Many of the problems which we have seen exemplified, particularly, although not entirely, with hard defences, have occurred because of the increase in defence construction which has resulted in the isolation from coastal processes of more and more sediment sources, which has, in turn, produced the need for further defences.

Given that we have acknowledged the failings of past approaches to defences in many areas, we are beginning to recognise a new opportunity which could, with decent CZM planning, be seized to correct some of these problems. Many of the existing defences are quite old, especially those located in established tourist areas where defence structures were constructed a hundred or so years ago. Given this, many have experienced damage and also changing sea conditions (sea level rise), and can be said to be loosing their effectiveness, and thus are nearing the end of their useful life. As defences become due for replacement, it is possible to re-assess their viability and usefulness by asking questions such as: does the area in question still need defending? Can it be better defended by a softer approach? If we free-up sediment movement along the coast, can it protect itself by habitat generation? It is by asking such questions that areas have adopted cliff retreat and managed realignment as management strategies, but it is only because the coastline is being looked at as an holistic unit that such decisions can be made.

These areas which are able to adopt a new methodology will also benefit under conditions of sea level rise. The option of a more natural coastline can only lead to future benefits because the more sediment which can be freed up, the healthier the sediment budget and more 'natural' coastlines can develop. This means that these coasts can respond to changes in wind, waves, and sea level more easily. Sensible CZM plans should be able to identify such areas and manage them appropriately, freeing up sediment to help supply beaches where hard defences need to be maintained. However, CZM also needs to identify those areas where sea walls and groynes are still necessary. A major tourist resort depends for its survival on such defences, and so in such cases, it is important to maintain the line of defence. Similarly, the Netherlands would take on a whole new outline were it not for its hard defences (although CZM is adopting an holistic management scenario involving dune and beach feeding to supplement sea walls). Clearly then, coastal management has to take on board all of these aspects, and present an argument and recommendation as to whether the future defence of an area, if new defences should become necessary, hold the existing line, advance seaward, or retreat landward.

We appear to be entering a wide range of issues here which become integral in forming a decision about the most suitable way of defending a coastline (see French 1997 for an overview of this decision-making process). There is general agreement among most coastal managers that because of the increased knowledge of how the coast works and how different human uses can impact upon it, it is necessary to adapt a form of coastal management along the holistic approach. Only by looking at the coast as a whole can we produce a sensible management strategy. Although this may appear an obvious statement, it actually represents a major step forward in coastal defence planning and management, because it takes on board the concept of the interaction of not only natural processes with the coastline, but also human interference.

The concept of coastal zone management (CZM)

Because the world's coastlines are used for many things there has been a proliferation of controlling and interested groups when it comes to coastal management (ranging from national heritage groups, to conservation, and to activity groups, such as yachting associations). This has led to much conflict between different groups as to how best to manage individual bits of the coastline and their usage. Due to the large extent of the coastline, management often needs to be in small sections (an embayment, an estuary). However, while the development of CZM plans are important, they still have the potential problem of activity in one of these smaller areas having a detrimental effect on processes in another. It is essential, therefore, that these more local plans are linked, in a hierarchy, to a broader scheme based on regional and national management strategies. The common denominator in all of this is the formulation of national coastal defence strategies, based on coastal zone management plans.

So, using the modern definition of the term, coastal zone management can be employed as a process which protects the coastline from damage and change from any activity. The topic itself has been the feature of many texts (see French 1997 for overview, or for more detail, texts such as Beatley *et al.* 1994, Clayton 1993, Goldberg 1994, ICE 1989, Kay and Alder 1999, OECD 1993, and Penning-Rowsell *et al.* 1992), and a detailed discussion is beyond the scope of this book, although as a tool with which to moderate the human impacts on the coast, it is important to consider some aspects of the topic. There are many ways in which the CZM process can be summarised, although as a general rule we can say that CZM represents a dynamic process which develops and implements a coordinated strategy to allocate resources to achieve the conservation and sustainable multiple use of the coastal zone.

The methodology behind coastal zone management

We have consistently seen in previous chapters how different defence types can cause knock-on effects for natural processes and coastal stability. Coming out of this is the growing realisation that if left to its own devices, this continued intervention by defence construction will only produce further conflict between managers and the natural environment. Eventually this will impact on the whole viability and stability of the coastline.

Many nations have now realised the problems associated with defending the coast, and have undertaken initiatives to correct them, largely by constructing formal, holistic management policies and the adoption of softer methodologies. This has put CZM into the international arena, resulting in an increased awareness of many transboundary problems, not only sediment supply and the health of sediment budgets, but also pollution and tourism. Barston (1994) highlights the increased potential for holistic defence planning which an international arena provides, but also demonstrates how this can produce a need for improved information systems and the development of international standards. On the one hand, it is this international arena which should be the scale of CZM, after all, consider the significance of sediment from the UK's Holderness coast and its importance for North Sea states, as discussed earlier. It is here, however, that a further, and significant problem arises. Once put on an international scale, coastal managers and other experts tend to loose control of the whole CZM process, with politicians becoming the driving and negotiating force. This remains, however, an important issue and is one on which continuous efforts need to be made to address over the coming years. There are many international organisations already established, such as around the North Sea or the Mediterranean, and perhaps greater emphasis needs to be put on these to take responsibility to oversee CZM. In order to do this, however, such groups also need to be given the necessary power and authority to make decisions. One such step was made in 1992 with the publication of Agenda 21, following the

Earth summit in Rio de Janeiro, Brazil. This meeting led to the blueprint for environmental action. Chapter 17 deals exclusively with oceans and coastlines, and probably represents the strongest international commitment to CZM that currently exists. Agenda 21 commits coastal nations to the implementation of integrated coastal zone management initiatives, and to the sustainable development of coastal areas and marine environment under their jurisdiction. Within this, a number of co-objectives commit signatories to more specific aspects of coastal management (from French 1997).

1) To provide for an integrated coastal policy and decision making process in order to promote compatibility and balance of coastal uses. This also stipulates the inclusion and co-operation of government departments, ministries and agencies which have control over specific aspects of the coast.
2) To apply preventative and precautionary approaches in development, including prior assessment and systematic observation of the impacts of major projects.
3) To promote the development and application of techniques which reflect changes in value resulting from uses in coastal areas. This includes pollution, loss of value due to erosion, loss of natural resources and habitat destruction, on which assigning a cost is difficult; and the increase in value due to development of the hinterland.
4) To liaise with all interested groups, to provide access to relevant information and to provide opportunities for consultation and participation in planning and decision making processes associated with the development of management plans.

It appears, therefore, that all of the guidelines and incentives are in place for countries to develop detailed coastal management plans for their coastlines. There is general agreement that it is essential to get away from the traditional approaches to management which tend to be sectorially orientated and fragmented, and go for the management of the coastal zone as a whole unit. While this may be true for the upper levels of the hierarchy, there are still problems which successful coastal management needs to tackle. Most important in this are questions such as: who does what, particularly at the ground level? Who formulates the plans? Who enforces the plans? Who pays for the plans? This brings us onto the next important consideration for the next generation of coastal managers, actually turning the concept into a workable framework and policy, which represents perhaps the prime area in which the above ideals have faltered in many cases.

Summary

In this final chapter, we have been considering the way forward. There has been extensive detail in preceding chapters as to the issues which relate to particular sea defences, but what is also important is the integration of these different defence types into a wider coastal plan which will protect coastal areas with high hinterland values, allow others to interact with natural processes, and allow others to erode. By doing this we can build up a management plan for a whole section of the coastline which is constructed on the principals of coastal processes. To some extent, this is still artificial because the coastal planner is still being constrained by development and land use, and so even the best coastal management plans have some degree of 'adjustment' to allow defences in some areas that may be considered unsuitable, were decisions to be made purely on coastal process grounds.

A further complication results from the presence of people, and their perceptions of what is suitable, correct, and safe. This is a relatively new area of research but one which is starting to play an important part, particularly when planners and managers hold public meetings to discuss a defence scheme. Perhaps the greatest outcry occurs when a decision is made to move the line of defence landwards, whether by realignment of existing defences, or by just allowing erosion to continue. All the evidence points to reasons why this should be adopted as the preferred solution in many cases, yet the idea of giving in to the sea and retreating inland cannot be comprehended by many members of the public. Perhaps this points to another important issue which needs to be addressed in the coming years and decades: that of public education. For too long people have got used to expecting sea walls and hard, physical structures when faced with a coastal problem. However, times and methods have changed and new approaches are being used which may go counter to long-held ideas and feelings. With time, as these newer methods are seen to work and be effective, so their acceptance might improve, but this process will be accelerated if coastal managers and planners take the time to explain the rationale and the processes which have gone into making these decisions.

In closing, therefore, it is useful to ponder where the subject is going. There are many research establishments the world over investigating the sorts of issues and problems raised in the earlier chapters on different defence techniques. Over recent years, these institutions have provided good value for investment in improved structure design and method improvement which makes our current defence types more efficient and effective. Long may this continue. There are further points, however, which now need attention. CZM is recognised as a good way of facilitating coastal management, yet many countries still cannot decide who should fund it or even manage the process. Developers still build near coastlines because of the perceived security afforded by sea defences, yet is a building with a life expectancy of 70 years going to be safe behind a sea wall with only 20 years of design life remaining? The whole area of coastal development needs addressing because at present, we are stuck in a spiral where defences allow the development, and because of the development, there is a need for better/improved defences. Finally, the relationship between perception and reality needs investigation. People want to live at the coast, yet only really feel safe behind large structures. Is this the ideal for coastal occupancy — views of concrete sea walls? (See Plate 12.1.) This area is complicated and involves a range of different research techniques, entering not only areas of physical geography, but also social and cultural.

These are questions to which there are no answers at present. Throughout this book there are instances where questions have been raised and no answers given. This is because at the present time there are none. Readers of this book are entering a subject area which has, and is still, undergoing rapid development, with new ideas surfacing all the time regarding the methods used, and how these are managed and implemented. There is an increasing need for coastal managers with a wide-ranging background, not just in engineering, ecology, or coastal morphology (although these are clearly important), but also in the social and environmental sciences, so that the applied nature of the discipline can be fully developed.

References

Adam P. (1990) *Saltmarsh Ecology*, Cambridge Studies in Ecology, Cambridge University Press, Cambridge.

Adamus P.R. (1988) 'Criteria for created or restored wetlands', in Hook D.D. *et al.* (eds) *The Ecology and Management of Wetlands, Vol. 2: Management, use, and value of wetlands*, Croom Helm, London: 369–72.

Adriaanse L.A. & Coosen J. (1991) 'Beach and dune nourishment and environmental aspects', *Coastal Engineering*, 16: 129–46.

Adriani M.J. and Terwindt J.M.J. (1974) 'Sand stabilisation and dune building', Rijkswaterstraat Communication No. 19, The Rijkswaterstraat, Netherlands.

Ahrens J.P. (1976) *Wave attenuation by artificial seaweed*, Misc. papers no. MR76–9, Coastal Engineering Research Centre, USACE Fort Belvoir, Virginia.

Ahrens J.P. and Bender T. (1992) 'Evaluating the performance of sea walls', in Institution of Civil Engineers (eds) *Coastal Structures and Breakwaters*, Thomas Telford Press, London: 13–24.

Aldridge, J.N. (1997) 'Hydrodynamic model predictions of tidal symmetry and observed sediment transport paths in Morecambe Bay', *Estuarine, Coastal and Shelf Science*, 44: 39–56.

Alizai S.A.K. and McManus J. (1980) 'The significance of reed beds on siltation in the Tay estuary', *Proceedings of the Royal Society of Edinburgh*, 78: 1–13.

Allen J.R.L. (1981) 'Beach erosion as a function of variation in the sediment budget', Sandy Hook, New Jersey, USA, *Earth Surface Processes and Landforms*, 6: 139–50.

—— (1986) 'A short history of salt marsh reclamation at Slimbridge Warth and neighbouring areas, Gloucestershire', *Transactions of the Bristol & Gloucestershire Archaeological Society* 104 : 139–55.

—— (1990) 'The Severn Estuary in south-west Britain: its retreat under marine transgression and fine sediment regime', *Sedimentary Geology*, 66: 13–28.

—— (1993) 'Palaeowind: geological criteria for direction and strength', *Philosophical Transactions of the Royal Society of London*, B 341: 235–42.

—— (1997) 'Simulation models of salt marsh morphodynamics: some implications for high intertidal sediment couplets related to sea level change', *Sedimentary Geology* 113: 211–223.

Allen J.R.L. and Fulford M.G. (1990) 'Romano-British and later reclamations on the Severn salt marshes in the Elmore area, Gloucestershire', *Transactions of the Bristol & Gloucestershire Archaeological Society*. 108: 17–32.

Allison M.A., Nittrouer C.A., Faria L.E.C. and Mendes A.C. (1996) 'Sources and sinks of sediment to the Amazon margin: the Amapa coast', *Geo-Marine Letters*, 16(1): 36–40.

Allsop N.W.H. (1997) Contribution to 'Coastnet' internet discussion group on 'Tyres as sea defences'.

Al-Obaid E. and Al-Sarawi M. (1995) 'Artificial beaches along the Kuwait water front, Kuwait', *Coastal Change '95*, Bordeaux: 603–9.

Alverino Dias and Neal W.J. (1992) 'Sea cliff retreat in Southern Portugal: profiles, processes and problems', *Journal of Coastal Research*, 8: 641–65.

American Society of Civil Engineers (1982) *Failure of breakwater at Port Sines*, Portugal, ASCE, New York.

Anglin C.D., MacIntosh K.J., Baird W.F., and Warren D.J. (1987) 'The design of pocket beaches using a physical model', in Kraus N.C. (ed.) *Coastal Sediments '87*, Proceedings of a speciality conference on advances in understanding of coastal sediment processes, New Orleans, Louisiana, 12–14 May 1987: 1105–16.

Anthony E.J. (1994) 'Natural and artificial shores of the French Riviera: an analysis of their interrelationship', *Journal of Coastal Research*, 10(1): 48–58.

Anthony E.J. and Cohen O. (1995) 'Nourishment solutions to the problem of beach erosion in France: the case of the French Riviera', in Healy M.G. and Doody J.P. (eds) *Directions in European Coastal Management*, Samara Publishing, Cardigan: 199–206.

Arens B. (1996) 'Fore dunes and sand dykes in the Netherlands', *Coastline*, 1996(1): 33–5.

Armon J.W. (1979) 'Landward sediment transfers in a transgressive barrier island system', Canada, in Leatherman S.P. (ed.) *Barrier Islands: From the Gulf of St. Lawrence to the Gulf of Mexico*, Academic Press, New York.

Arthurton R. (1998) Resource, evaluation and net benefit, in Hooke J. (ed.) *Coastal Defence and Earth Science conservation*, The Geological Society, London.

Baba M. and Thomas K.V. (1987) 'Performance of a seawall with a frontal beach', in Kraus N.C. (ed.) *Coastal Sediments '87*, Proceedings of a speciality conference on advances in understanding of coastal sediment processes, New Orleans, Louisiana, 12–14 May 1987: 1051–61.

Baca B.J. and Lankford T.E. (1988) *Myrtle Beach nourishment project: biological monitoring*, Unpublished report, City of Myrtle Beach.

Bagley L.M. and Whitson D.H. (1982) 'Putting the beach back at Oceanside', *Shore and Beach*, 50(4): 24–32.

Bagnold R.A. (1940) 'Beach formation by waves: some model experiments in a wave tank', *Journal of the Institute of Civil Engineers*, 15: 27–54.

—— (1954) *The Physics of Wind-blown Sand and Desert Dunes*, Methuen, London.

Band W.T. (1979) 'Beach management for recreation in Scotland', in Guilcher A. (ed.) *Les côtes atlantiques de l'Europe, évolution, aménegement, protection*, Brest, France: 261–68.

Banks K. (1997) Contribution to 'Coastnet' internet discussion group on 'Tyres as sea defences'.

Barber P.C. (1984) 'Philosophy of coastline control – bay formation', unpublished paper presented at Cranfield, 1984, CEEMAID Services Ltd., Feltham.

Barcellos C., de Lacerda L.D. and Ceradini S. (1997) 'Sediment origin and budget in Septiba Bay (Brazil) – an approach based on multi-elemental analysis', *Environmental Geology*, 32(3): 203–9.

Barston R.P. (1994) 'International dimensions in coastal zone management', *Ocean and coastal Management*, 23: 93–116.

Bartberger C.E. (1976) 'Sediment sources and sedimentation rates, Chincoleague Bay, Maryland and Virginia', *Journal of Sedimentary Petrology*, 46(2): 326–36.

Basco D.R., Bellomo D.A., Hazelton J.M. and Jones B.N. (1997) 'The influence of seawalls on subaerial beach volumes with receding shorelines', *Coastal Engineering*, 30: 203–33.

Basinski T. (1994) 'Protection of Hel Peninsular', in Rotinicki K. (ed.), *Changes of the Polish Coastal Zone*, Quaternary Research Institute, Adam Mickiewicz University, Poznan: 53–6.

Baye P. (1990) 'Ecological history of an artificial foredune ridge of a north-eastern barrier spit', in Davidson-Arnott R.G.D. (ed.) *Proceedings of the symposium on coastal sand dunes*. National Research Council of Canada, Ottawa: 389–403.

Beatley T., Bower D.J. and Schwab A.K. (1994) *An introduction to coastal zone management*, Island Press, Washington DC.

Berg D.W. and Duane D.B. (1968) 'Effect of particle size and distribution on stability of artificially filled beach. Presque Isle Peninsular, Pennsylvania', *Proceedings of the 11th Conference on Great Lakes research*, 1968: 161–78.

Berenguer J.M. and Enriquez J. (1988) 'Design of pocket beaches: the Spanish case', *Proceedings 21st International Conference on coastal engineering*, Malaga/Costa del Sol, Spain.

Bird E.C.F. (1985) *Coastline Changes: a global review*, Wiley, Chichester.

—— (1990) 'Artificial beach nourishment on the shore of Port Philip Bay, Australia', *Journal of Coastal Research*, SI 6: 55–68.

—— (1993) *Submerging Coasts: the effects of rising sea level on coastal environments*, Wiley, Chichester.

—— (1996) *Beach Management*, Wiley, Chichester.

Birkemeier W.A. (1980) *The effect of structures and lake level on bluff and shore erosion in Berrigen County, Michigan 1970–74*. Misc. Report 80–2. Coastal Engineering Research Centre. USACE. 74p.

Blackpool Borough Council (1993) *Blackpool Shoreline Strategy Plan* (3 volumes), Unpublished report, Blackpool Borough Council.

Bocamazo L. (1991) 'Sea Bight to Manasques, New Jersey beach erosion coastal projects', *Shore and Beach*, 59: 37–42.

Boesch D.F., Josselyn M.N., Mehta A.J., Morris J.T., Nuttle W.K., Simenstad C.A. and Swift D.J.P. (1994), 'Scientific Assessment of coastal wetland loss, restoration and management in Louisiana', *Journal of Coastal Research*, Special Issue 20.

Boorman L.A. (1976) *Dune Management: a progress report*, unpublished report, Institute of Terrestrial Ecology, Cambridge.

—— (1989) 'The grazing of British sand dune vegetation', *Proceedings of the Royal Society of Edinburgh. Section B-Biological Sciences*, 96: 75–88.

Boorman L.A. and Fuller R.M. (1977) 'Studies on the impact of paths on the dune vegetation at Winterton, Norfolk, England'. *Biological Conservation* 12: 203–16.

Boorman L.A., Garbutt A. and Barratt D. (1998) 'The role of vegetation in determining patterns of the accretion of salt marsh sediment', in Black K.S., Paterson D.M. and Cramp A. (eds) *Sedimentary processes in the intertidal zone*, Geological Society Special Publications, 139: 389–99, Geological Society, London.

Boorman L.A., Goss-Custard J.D. and McGrorty S. (1989) 'Climate change, rising sea level and the British coast'. Institute of Terrestrial Ecology, Natural Environment Research Council. HMSO, London.

Boorman L. and Hazelden J. (1995) 'Saltmarsh creation and management for coastal defence', in Healy M.G. and Doody J.P. (eds) *Directions in European Coastal Management*, Samara Publishing, Wales: 175–83.

Bourman R.P. (1990) 'Artificial beach progradation by quarry waste disposal at Rapid Bay, South Australia', *Journal of Coastal Research*, SI 6: 69–76.

Bowen A.J. and Inman D.L. (1966) 'Budget of littoral sands at Point Arguello, California', *CERC Technical Memo No. 19*, US Army, Fort Belvoir, USA: 1–41.

Brampton A.H. (1992) 'Beaches – the natural way to coastal defence', in Barrett M.G. (ed.) *Coastal planning and management*, Institute of Civil Engineers, London.

—— (1998) 'Cliff conservation and protection: methods and practices to resolve conflicts', in Hooke J.M. (ed.) *Coastal defence and earth science conservation*, Geological Society of London: 21–31.

Brashears R.L. and Dartnell J.S. (1967) 'Development of the artificial seaweed concept', *Shore and Beach*, 35(2): 35–41.

Bray M.J., Carter D.J. and Hooke J.M. (1992) *Sea Level Rise and Global Warming: scenarios, physical impacts and policies*, Report to SCOPAC, Geography Department, Portsmouth University.

—— (1995) 'Littoral cell definition and budgets for central southern England', *Journal of Coastal Research*, 11(2): 381–400.

Bray M.J., and Hooke J.M. (1997) 'Prediction of soft-cliff retreat with accelerating sea level rise', *Journal of Coastal Research* 13(2): 453–67.

—— (1998a) 'Geomorphology and management of sites in Poole and Christchurch Bays', in Hooke J.M. (ed.) *Coastal Defence and Earth Science Conservation*, Geological Society of London: 233–66.

—— (1998b) 'Spatial perspectives in coastal defence and conservation strategies', in Hooke J.M. (ed.) *Coastal Defence and Earth Science Conservation*, Geological Society of London: 115–32.

Bray M.J., Hooke J.M. and Carter D. (1997) 'Planning for sea level rise on the south coast of England: advising the decision makers', *Transactions of the Institute of British Geographers*, NS 22: 13–20.

Bressolier-Bousquet C. (1991) 'Geomorphological effects of land reclamation in the 18th century at the mouth of the Leyre River, Arcachon Bay, France', *Journal of Coastal Research*, 7(1): 113–26.

Brickle C.I., Murphy B.J. and Pethick J.S. (1998) *Coastal Data Pack*, unpublished report, East Riding of Yorkshire Council.

Brooke J.S. (1991) 'Retreat, the best form of defence?', *Heritage Coast*, 6 (Dec): 4.

—— (1992) 'Coastal defence: The retreat option', *Journal of the Institute of Water and Environmental Management*, 6: 151–57.

Brooks A. (1979) *Coastlands*, British Trust for Conservation Volunteers, London.

Broome S.W. (1972) 'Stabilising dredge spoil by creating new salt marshes with *Spartina alterniflora*', Proceedings of the 15th Annual Meeting of the Soil Science Society of North Carolina. 15: 136–47.

—— (1990) 'Creation and restoration of tidal wetlands of the south-eastern United States', in Kusler J.A. and Kentula M.E. (eds) *Wetland Creation and Restoration: the status of the science*, Island Press, Washington: 37–72.

Brown S.L. (1998) 'Sedimentation on a Humber saltmarsh', in Black K.S., Paterson D.M. and Cramp A. (eds) *Sedimentary Processes in the Intertidal Zone* Geological Society Special Publications 139: 69–83, Geological Society, London.

Bruun P. (1952) 'Measures against erosion at groins and jetties', Proceedings of the third international conference on coastal engineering, Cambridge, MA, USA.

—— (1962) 'Sea level rise as a cause of shore erosion', *Journal of Waterways and Harbours Division ASCE*, 88: 117–30.

—— (1983) 'Beach scraping – is it damaging to beach stability?', *Coastal Engineering*, 7(2): 167–73.

—— (1985) *Design and Construction of Mounds for Breakwaters and Coastal Protection*, Developments in Geotechnical Engineering 37, Elsevier, New York.

Bull C.F.J, Davis A.M. and Jones R. (1998) 'The influence of fish-tail groynes (or breakwaters) on the characteristics of the adjacent beach at Llandudno, north Wales', *Journal of Coastal Research*, 14(1): 93–105.

Burby R.J. and Nelson A.C. (1991) 'Local government and public adaptation to sea level rise', *Journal of Urban Planning and Development*, 117(4): 140–53.

Burd F. (1995) *Managed Retreat: A Practical Guide,* English Nature, Peterborough.

Camfield F.E. (1993) 'Different views of beach fill performance', *Shore and Beach*, 61(4): 4–8.

Cammen L.M. (1976a) 'Macro-invertebrate colonisation of *Spartina* marshes artificially established on dredge spoil', *Estuarine and Coastal Marine Science*, 4: 357–72.

—— (1976b) 'Abundance and production of macro-invertebrates from naturally and artificially established salt marshes in North Carolina', *American Midland Naturalist*, 96: 487–93.

Campbell T.J. and Spadoni R.H. (1987) 'Beach restoration: an effective way to combat erosion at the south-east coast of Florida', *Shore and Beach*, 50: 11–12.

Carter R.W.G. (1980) 'Human activities and geomorphic processes: the example of recreation pressure on the Northern Ireland coast', *Zeitscrift für Geormophologie Suppl., bd.* 34: 155–64.

—— (1988) *Coastal Environments: an introduction to the physical, ecological and cultural systems of coastlines,* Academic Press, London.

Carter R.W.G., Nordstrom K.F. and Psuty N.P. (1990) 'The study of coastal dunes', in Nordstrom K.F., Psuty N.P. and Carter R.W.G. (eds) *Coastal Dunes: form and process,* Wiley, Chichester: 1–16.

Carter R.W.G. and Stone G.W. (1989) 'Mechanisms associated with the erosion of sand dune cliffs, Magilligan, northern Ireland', *Earth Surface Processes and Landforms,* 14(1): 1–10.

Castro-Ortiz C.A. (1994) 'Sea level rise and its impact on Bangladesh', *Ocean & Coastal Management,* 23: 249–70.

Chandler J.H. and Brunsden D. (1995) 'Steady state behaviour of the Black Ven mudslide complex: The application of archival analytical photogrammetry to studies of landform change', *Earth Science Processes and Landforms,* 20: 255–75.

Chappell J., Elliot I.G., Bradshaw M.P. and Lonsdale E. (1979) 'Experimental control of beach face dynamics by water table pumping', *Engineering Geology,* 14: 29–41.

Chapman D.M. (1981) 'Coastal erosion and the sediment budget, with special reference to the Gold Coast, Australia', *Coastal Engineering,* 4(3): 207–27.

Charlier R.H. and de Meyer C.P. (1989) 'Coastal defence and beach renovation', *Ocean and Shoreline Management,* 12: 525–43.

—— (1995a) 'New developments in coastal protection along the Belgian coast', *Journal of Coastal Research*, 11(4): 1287–93.

—— (1995b) 'Beach nourishment as efficient coastal protection', *Environmental Management and Health*, 6: 26–34.

—— (2000) 'Ask nature to protect and build-up beaches', *Journal of Coastal Research*, 16(2): 385–90.

Chasten M.A., Rosati J.D., McCormick J.W. and Randall R.E. (1993) *Engineering Design and Guidance for Detached Breakwaters as Shoreline Stabilisation Structures*, CERC technical report CERC-93-19. USACE.

Chen J. (1997) 'The impact of sea level rise on China's coastal areas and its disaster hazard evaluation', *Journal of Coastal Research*, 13(3): 925–30.

Child M. (1996) 'Soft engineering – a firm future?', *Proceedings of the Institution of Civil Engineers: Civil Engineering*, 114(3): 145–6.

Chill J., Butcher C. and Dyson W. (1989) 'Beach nourishment with fine sediment at Carlsbad, California', in American Society of Civil Engineers (ed.) *Coastal Zone '89* ASCE, New York: 2092–103.

Christiansen C. and Miller P.F. (1983) '*Spartina* in Mariager Fjord, Denmark: The effect on sediment parameters', *Earth Surface Processes and Landforms*, 8: 55–62.

Chung C.-H. (1993) 'Thirty years of ecological engineering with *Spartina* plantations in China', *Ecological Engineering*, 2: 261–89.

Cipriani L.E., Dreoni A.M. and Pranzini E. (1992) 'Nearshore morphological and sedimentological evolution induced by beach restoration: a case study', *Bollettino di Oceanologica Teorica ed Applicata* 10: 279–95.

Clark A.R. and Guest S. (1991) 'The Whitby cliff stabilisation and coast protection scheme, in Chandler R.J. (ed.) *Slope Stability Engineering*', Proceedings of International Conference on Slope Stability, Isle of Wight, 15–18 April 1991, Thomas Telford Press, London, Ch 62: 283–90.

Clayton K.M. (1980) 'Coastal protection along the East Anglian coast', *Zeitschrift für Geomorphologie*, Supp34: 163–72.

—— (1989a) 'Sediment input from the Norfolk Cliffs, Eastern England – a century of coast protection and its effect', *Journal of Coastal Research*, 5(3): 433–42.

—— (1989b) 'The implications of climatic change', in Institution of Civil Engineers (ed.) *Coastal Management*, Thomas Telford Press, London: 165–76.

—— (1993) *Coastal processes and coastal management*, Countryside Commission, London.

Clayton T.D. (1989c) 'Artificial beach nourishment on the Pacific shore: a brief overview', *Proceedings of Coastal Zone '89*: 2033–45.

Clements M. (1994) 'The Scarborough experience – Holbeck landslide, 3/4 June 1993', *Proceedings of the Institute of Civil Engineers: Municipal Engineers* 103(June): 63–70.

—— (1998) 'Design conditions for coastal works – practical experience of coastal risk assessment. *Proceedings of the Institute of Civil Engineers: Municipal Engineers*, 127(June): 49–55.

Cooper N.J. (1998) 'Assessment and prediction of Poole Bay (UK) sand replenishment schemes: application of data to Führböter and Verhagen models', *Journal of Coastal Research*, 14(1): 353–9.

Cooper N.J., King D.M. and Hooke J.M. (1996) 'Collaborative research studies at Elmer Beach, West Sussex, UK', in Taussik J. and Mitchell (eds) *Partnerships in Coastal Engineering*, Samara Press, Cardigan. 369–76.

Cooper W.S. (1958) 'Coastal sand dunes of Oregon and Washington', *Geological Society of America Bulletin*, 72.

Correia F., Dias J.A., Bosk T. and Ferraira O. (1996) 'The retreat of the eastern Quarteira cliffed coast (Portugal), and its possible causes', in Jones P.S., Healy M.G. and Williams A.T. (eds) *Studies in European Coastal Management*, Samara Publishing, Cardigan.

Cortright R. (1987) 'Fore-dune management on a developed shoreline: Nedonna Beach, Oregon', in American Society of Civil Engineers (ed.) *Coastal Zone '87*, New York: 1343–56.

Cozens-Hardy B. (1924) 'Cley-next-the-Sea and its marshes', *Transactions of the Norfolk & Norwich Naturalists' Society*, 12: 354–73.

Cruz-Colin M.E. and Cupul-Magana L.A. (1997) 'Erosion and sediment supply of sea cliffs of Todos Santos Bay, Baja California, from 1970 to 1991', *Ciencias Marinas*, 23(3): 303–15.

Cundy A.B. and Croudace I.W. (1996) 'Sediment accretion and recent sea level rise in the Solent, Southern England: Inferences from radiometric and geochemical studies', *Estuarine, Coastal and Shelf Science*, 43: 449–67.

Cunningham R.T. (1966) 'Evaluation of Bahaman oolitic aragonite sand for Florida beach replenishment', *Shore and Beach*, 34: 18–21.

Dahl B.E., Fall B.A. and Otteni L.C. (1975) 'Vegetation for creation and stabilisation of fore dunes, Texas coast', in Cronin L.E. (ed.) *Estuarine Research, Vol. II Geology and Engineering*, Academic Press: 457–70.

Dally R.G. and Fox J.E. (1979) 'Soil – landscape relationships of the tidal marshlands of Maryland', *Journal of the Soil Science Society of America*, 43: 534–541.

Dally W.R. and Pope J. (1986) *Detached breakwaters for shore protection*, Report CERC-86-1, US Army Engineer Waterways Experiment Station, Vicksberg, MS.

Darmody R.G. and Foss J.E. (1979) 'Soil-landscape realtionships of the tidal marshlands of Maryland,' *Journal of the Soil Science Society of America* 43: 534–41.

Davidson-Arnott R.G.D. and Keier H.I. (1982) 'Shore protection in the town of Stoney Creek, southwest Lake Ontario 1934–79. Historical changes and durability of structures', *Journal of Great Lakes Research*, 8: 635–47.

Davis J.E. and Landin M.C. (1998) *Geotextile tube structures for wetlands restoration and protection: an overview of information from the national Workshop on geotextile tube applications*, The CERCular, July 1998, US Army Corps of Engineers.

Davis R.A. (1996) *Coasts*, Prentice Hall, Englewood Cliffs, NJ.

Davis R.A., Fitzgerald M.V. and Terry J. (1999) 'Turtle nesting on adjacent nourished beaches with different construction styles: Pinellas County, Florida', *Journal of Coastal Research*, 15(1): 111–20.

Davison A.T., Nicholls R.J. and Leatherman S.P. (1992) 'Beach nourishment as a coastal management tool: an annotated bibliography on the developments associated with artificial nourishment of beaches', *Journal of Coastal Research*, 8(4): 984–1022.

Daws G. and Elson R.J. (1990) 'Cliff stabilisation works, East Cliff, Dover', in Burland J.B., Mortimore R.N., Roberts L.D., Jones D.L. and Corbett B.O. (eds) *Chalk*, Proceedings of International Chalk Symposium, Brighton 4–7 September 1990, Thomas Telford Press, London, Ch 81: 521–6.

Day J.W., Pont D., Hensel P.F. and Ibañez C. (1995) 'Impacts of sea level rise on deltas in the Gulf of Mexico ad the Mediterranean: The importance of pulsing events to sustainability', *Estuaries*, 18(4): 636–47.

de Jonge V.N., Essink K. and Boddeke R. (1993) 'The Dutch Wadden Sea: a changed ecosystem', *Hydrobiologia*, 265: 45–71.

de Lange W.P. and Healy T.R. (1990) 'Renourishment of a flood-tidal delta adjacent beach', Taurange Harbour, New Zealand. *Journal of Coastal Research*, 6(3): 627–40.

de Laune R.D., Pezeshki S.R., Pardue J.H., Whitcomb J.H. and Patrick W.H. (1990) 'Some influences of sediment accretion to a deteriorating salt marsh in the Mississippi river deltaic plain: A pilot study', *Journal of Coastal Research*, 6(1): 181–8.

de Meyer C.P. (1989) 'Case studies in coastal protection: Zeebrugge (Belgium) and Bali (Indonesia)', *Ocean and Shoreline Management*, 12: 517–24.

de Mulder E.F.J., van Bruchem A.J., Claessen F.A.M., Hannink G., Hulsbergen J.G. and Satijn H.M.C. (1994) 'Environmental impact assessment on land reclamation projects in the Netherlands: A case history', *Engineering Geology*, 37: 15–23.

de Ruig J.H.M. (1998) 'Seaward coastal defence: limitations and possibilities', *Journal of Coastal Conservation*, 4: 71–8.

de Ruig J.H.M. and Louisse C. J. (1991) 'Sand budget trends along the Holland coast', *Journal of Coastal Research*, 7(4): 1013–26.

Dean J.L. and Pope J. (1987) 'The Redington shores breakwater project: initial response', in Kraus N.C. (ed.) *Coastal Sediments '87*, Proceedings of a speciality conference on advances in understanding of coastal sediment processes, New Orleans, Louisiana, 12–14 May 1987: 1369–84.

Dean R.G. (1983) 'Principles of beach nourishment', in Komar P.D. (ed.) *C.R.C. Handbook of Coastal Processes and Erosion*, Boca Raton, Florida, CRC Press: 217–32.

—— (1991) 'Equilibrium beach profiles: characteristics and applications', *Journal of Coastal Research*, 7(1): 53–84.

Dean R.G., Chen R. and Browder A.E. (1997) 'Full scale monitoring study of a submerged breakwater, Palm Beach, Florida, USA', *Coastal Engineering*, 29: 291–315.

Den Elzen M.G.J. and Rotmans J. (1992) 'The socio-economic impact of sea level rise on the Netherlands: a study of possible scenarios', *Climatic Change*, 20: 169–95.

Dennis B., Conway, B.W., McCann D.M. and Grainger P. (1975) 'Investigation of a coastal landslip at Charmouth, Dorset', *Quarterly Journal of Engineering Geology*, 8: 119–40.

Dennis K.C., Niang-Diop, Nicholls R.J. (1995a) 'Sea level rise and Senegal: potential impacts and consequences', *Journal of Coastal Research*, SI14: 243–61.

Dennis K.C., Schnack E.J., Mouzo F.H. and Orona C.R. (1995b) 'Sea level rise and Argentina: potential impacts and consequences', *Journal of Coastal Research*, SI 14: 205–23.

Dette H.H. and Gartner J. (1987) 'Time history of a seawall on the Isle of Sjlt', in Kraus N.C. (ed.) *Coastal Sediments '87*, Proceedings of a speciality conference on advances in understanding of coastal sediment processes. New Orleans, Louisiana, 12–14 May 1987: 1006–22.

Dixon A.M., Leggett D.J. and Weight R.C. (1998) 'Habitat creation opportunities for landward coastal realignment: Essex case studies', *Journal of the Institute of Water and Environmental Management*, 12: 107–12.

Dixon K. and Pilkey O.H. (1991) 'Summary of beach replenishment on the US Gulf of Mexico shoreline', *Journal of Coastal Research* 7: 249–56.

DoE (1996) *Landslide investigation and management in great Britain: a guide for planners and developers*, Department of the Environment, HMSO, London.

Dolan R. and Godfrey P. (1973) 'Effects of Hurricane Ginger on the barrier islands of North Carolina', *Geological Society of America Bulletin*, 84: 1329–34.

Dolotov Y.S. (1992) 'Possible types of coastal evolution associated with the expected rise of the world's sea level caused by the greenhouse effect,' *Journal of Coastal Research* 8(3): 719–26.

Domurat G.W. (1987) 'Beach nourishment – A working solution', *Shore and Beach*, 55(3–4): 92–5.

Douglas B.C. (1997) 'Global sea level rise: a redetermination', *Surveys in Geophysics*, 18: 279–92.

Douglas S.L. (1987) 'Coastal responses to jetties at Murrel's Inlet, South Carolina', *Shore and Beach*, 55(2): 21–32.

Douglas S.L. and Weggel J.R. (1987) 'Performance of a perched beach – Slaughter Beach, Delaware', in Kraus N.C. (ed.) *Coastal Sediments '87*, Proceedings of a speciality conference on advances in understanding of coastal sediment processes, New Orleans, Louisiana, 12–14 May 1987: 1385–98.

Duncan J.R. (1964) 'The effects of water table and tide cycle on swash-backwash sediment distribution and beach profile development', *Marine Geology*, 2(3): 186–97.

Dunstan W.M., McIntire G.L. and Windom H.L. (1975) '*Spartina* revegetation on dredge spoil in SE marshes', *Journal of the Waterways Harbours and Coastal Engineering Division*, 101(WW3): 269–76.

Edelman T. (1968) 'Dune erosion during storm conditions', *Proceedings of the 11th Conference on Coastal Engineering*, American Society of Civil Engineers: 719–22.

—— (1972) 'Dune erosion during storm conditions', *Proceedings of the 13th Conference on Coastal Engineering*, American Society of Civil Engineers: 1305–12.

Eitner V. and Ragutzki G. (1994) 'Effects of artificial beach nourishment on nearshore sediment distribution (Island of Norderney, southern North Sea) *Journal of Coastal Research*, 10(3): 637–50.

El Raey M. (1997) 'Vulnerability assessment of the coastal zone of the Nile delta, to the impacts of sea level rise', *Ocean and Coastal Management*, 37(1): 29–40.

Eleuterius L.N. (1975) 'Submergent vegetation for bottom stabilisation', in Cronin L.E. (ed.) *Estuarine Research Vol. II: Geology and Engineering*, Academic Press: 439–56.

English Nature (1995) 'Taking the 'soft' option', *English Nature*, 19: 11.

Evangelista S., La Monica G.B. and Landini B. (1992) 'Artificial beach nourishment using fine crushed limestone gravel at Terracina (Latium, Italy)', *Bollettino di Oceanologica Teorica ed Applicata*, 10: 273–78.

Everts C.H. (1973) 'Particle overpassing on flat, granular boundaries', *Journal of Waterway, Harbour and Coastal Division, ASCE* 99: 425–38.

—— (1983) 'Shoreline changes down-drift of a littoral barrier', in Weggel J.R. (ed.) *Proceedings of 'Coastal Structures '83'* ASCE: 673–89.

—— (1985) Sea level rise effects on shoreline position. *Journal of waterways, Port, Coastal, and Ocean Engineering* 111(6): 985–99.

Fanos A.M., Khafagy A.A. and Dean R.G. (1995) 'Protective works on the Nile delta coast', *Journal of Coastal Research*, 11:516–28.

Fast D.E. and Pagan F.A. (1974) 'Comparative observations on an artificial tire reef and natural patch reefs off south-western Puerto Rico. Proceedings of an International conference on artificial Reefs', March 20–22 1974, Houston, Texas.

Favennec J. (1996) 'Coastal management by the French National Forestry Service in Aquitaine, France', in Jones P.S, Healy M.G. and Williams A.T. (ed.) *Studies in European Coastal Management*, Samara Publishing, Cardigan: 191–6.

Fenster M. and Dolan R. (1993) 'Historical shoreline trends along the outer bank, North Carolina – Processes and responses', *Journal of Coastal Research*, 9(1): 172–88.

—— (1994) 'Large-scale reversals in shoreline trends along the US Mid Atlantic coast', *Geology*, 22(6): 543–46.

Finkl C.W. (1996) 'What might happen to America's shoreline if artificial beach replenishment is curtailed: a prognosis for south eastern Florida and other sandy regions along regressive coasts', *Journal of Coastal Research*, 12(1): iii – ix.

Fitzhardinge R.C. and Bailey-Brock J.H. (1989) 'Colonisation of artificial reef materials by corals and other sessile organisms', *Bulletin of Marine Science*, 44(2): 567–79.

Fleming C.A. (1992) 'The development of coastal engineering', in Barrett M.G. (Ed) *Coastal planning and management*, Institute of Civil Engineers, London.

Fletcher C.H. (1992) 'Sea level trends and physical consequences: applications to the US shore', *Earth Science Reviews*, 33: 73–109.

Fletcher C.H., Mullane R.A. and Richmond B.M. (1997) 'Beach loss along armoured shorelines on Oahu, Hawaiian Islands', *Journal of Coastal Research*, 13(1): 209–15.

Fonseca M.S. and Cahalan J.A. (1992) 'A preliminary evaluation of wave attenuation by four species of sea grass', *Estuarine, Coastal and Shelf Science*, 35: 565–76.

Fonseca M.S., Fisher J.S., Ziema J.C. and Thayer G.W. (1982) 'Influence of the seagrass *Zostera marina* (L) on current flow', *Estuarine, Coastal and Shelf Science*, 15: 351–64.

Foster G.A., Healy T.R. and de Lange W.P. (1996) 'Presaging beach renourishment from a nearshore dredge dump mound, Mt. Maunganui Beach, New Zealand', *Journal of Coastal Research*, 12(2): 395–405.

Fox J. (1997) 'Has the Sidmouth sea defence scheme been successful in maintaining beach levels, restricting sediment movement, and protecting the town from storm action?' Unpublished BSc dissertation, Department of Geography, University of Lancaster.

Fraser R.J. (1993) 'Removing contaminated sediments from the coastal environment: The New Bedford Harbour project example', *Coastal Management*, 21: 155–62.

French J.R. (1993) 'Numerical simulation of vertical marsh growth and adjustment to accelerated sea level rise', North Norfolk, U.K. *Earth Surface Processes and Landforms*, 18: 63–81.

French J.R., Spencer T., Murray A.L. and Arnold N.S. (1995) 'Geostatistical analysis of sediment deposition in two small tidal wetlands, Norfolk, UK', *Journal of Coastal Research*, 11(2): 308–21.

French P.W. (1991) 'Natural set-back at Pagham Harbour', Unpublished document.

—— (1996) 'Long-term temporal variability of copper, lead, and zinc in salt marsh sediment of the Severn Estuary, UK', *Mangroves and Saltmarshes*, 1(1): 59–68.

—— (1997) *Coastal and Estuarine Management*, Routledge Environmental Management Series. Routledge, London.

—— (1999) 'Managed retreat: a natural analogue from the Medway estuary, UK', *Ocean & Coastal Management*, 42: 49–62.

French P.W. and Livesey J.S. (2000) 'The impacts of fish-tail groynes at Morecambe, north-west England', *Journal of Coastal Research*, 16 (forthcoming).

Führböter A. (1991) 'Eine theoretische betrachtung über Sandvorspülungen mit Wiederholungsintervallen', *Die Küste* 52.

Gagil R.L. and Ede F.J. (1998) 'Application of scientific principles to sand dune stabilisation in New Zealand: Past progress and future needs', *Land Degradation and Development*, 9(2): 131–42.

Galster R.W. and Schwartz M.L. (1990) 'Ediz Hook. A case history of coastal erosion and rehabilitation', *Journal of Coastal Research*, Special Issue, 6: 103–13.

Galvin C.J. (1968) 'Breaker type classification on three laboratory beaches', *Journal of Geophysical Research* 73: 3651–9.

Garbisch E.W., Bostian E.W. and McCallum R.J. (1975a) 'Biotic techniques for shore stabilisation', in Cronin L.E. (ed.) *Estuarine Research, Vol. II: Geology and engineering*, Academic Press: 405–26.

Garbisch E.W., Woller P.B. and McCallum R.J. (1975b) *Salt marsh establishment and development*, USACE Technical Memo No. 52, Coastal Engineering Research Centre, Fort Belvoir, Virginia. 110pp.

Gardner J. and Runcie R. (1995) 'Planning and construction of four offshore reefs in north Norfolk', *Proceedings of the 30th MAFF. Conference of River and Coastal Engineers*, Keele University, July, 1995.

Gares P.A. (1990) 'Eolian processes and dune changes at developed and undeveloped sites, Island Beach, New Jersey', in Nordstrom K.F, Psuty N. P. and Carter R.W.G. (eds) *Coastal Dunes: Form and Process*, Wiley, Chichester: 361–80.

Garford W. (1999) Planning policy and coastal erosion. Personal communication with author.

Gaughan M.K. and Komar P.D. (1977) 'Groin length and the generation of edge waves', Proceedings 15th Coastal Engineering Conference, American Society of Civil Engineers: 1459–76.

Geelan L.H.W., Cousin E.F.H. and Schoon C.F. (1995) 'Regeneration of dune slacks in the Amsterdam Waterwork dunes', in Healy M.G. and Doody J.P. (eds) *Directions in European Coastal Management*, Samara Publishing Ltd. Cardigan.

Ghiassian H., Gray D.H. and Hryciw R.D. (1997) 'Stabilisation of coastal slopes by anchored geosynthetic systems', *Journal of Geotechnical and Geoenvironmental Engineering*, 123(8): 736–43.

Giardino J.R., Bednarz R.S. and Bryant J.T. (1987) 'Nourishment of San Luis Beach, Texas: an assessment of impact, in American Society of Civil Engineers', (ed.) *Coastal Sediments '87*, vol. 2 ASCE New York: 1145–57.

Gibb J.G. and Adams J. (1982) 'A sediment budget for the east coast between Oamaru and Banks Peninsula, South Island, New Zealand', *New Zealand Journal of Geology and Geophysics*, 25(3): 335–52.

Gifford C.A. (1978) 'Use of floating tire breakwaters to induce growth of marsh and fore dune plants along a shoreline', in Cole D.P. (ed.) Proceedings of the 5th annual conference on restoration of coastal vegetation in Florida, Hillsborough Community College, Tampa, FL.

Ginsburg R.N. and Lowenstam H.A. (1958) 'The influence of marine bottom communities on the depositional environments of sediments', *Journal of Geology*, 66: 310–18.

Gleason M.C., Elmer D.A., Pien N.C. and Fisher J.S. (1979) 'Effects of stem density upon sediment retention by salt marsh cord grass', *Spartina alterniflora*, Louise, *Estuaries*, 2(4): 271–73.

Goldberg E.D. (1994) *Coastal zone space: Prelude to conflict?* UNESCO, Paris.

Good B. (1993) 'Louisiana's wetlands: combating erosion and revitalising native ecosystems', *Restoration and Management Notes*, 11(2): 125–33.

Goodhead T. and Johnson D. (1996) *Coastal Recreation and Management: the sustainable development of maritime leisure*, E. and F.N. Spon Publishing, London.

Goodwin P. and Williams P.B. (1992) 'Restoring coastal wetlands: the Californian experience', *Journal of Institute of Water and Environmental Management*, 6: 709–19.

Grainger P. and Kalaugher P.G. (1987) 'Intermittent surging movements of a coastal landslide', *Earth Surface Processes and Landforms*, 12: 597–603.

Granja H.M. (1995) 'Some examples of inappropriate coastal management practice in north-west Portugal', in Healy M.G. and Doody J.P. (eds) *Directions in European Coastal Management*, Samara Publishing, Wales.

Granja H.M. and de Carvalho G.S. (1995) 'Is the coastline 'protection' of Portugal by hard engineering structures effective?' *Journal of Coastal Research*, 11(4): 1229–41.

Grant U.S. (1948) 'Influence of the water table on beach aggradation and degradation', *Journal of Marine Research*, 7: 655–60.

Gray A.J. and Adam P. (1973) The reclamation history of Morecambe Bay. *Nature in Lancashire* 4: 13–17.

Green D.J. (1992) 'Coast protection works for Arun district Council – Elmer frontage', unpublished manuscript presented to seminar of IWE., September 1992.

Gregory D. (1997) Contribution to 'Coastnet' internet discussion group on 'Tyres as sea defences'.

Griggs G.B. (1990) 'Littoral drift impoundment and beach nourishment in northern Monterey Bay, California' *Journal of Coastal Research*, SI 6: 115–26.

—— (1995) 'Relocation or reconstruction of threatened coastal structures; a second look', *Shore and Beach*, 63(2): 31–6.

Griggs G.B. and Fulton-Bennett K.W. (1987) 'Failure of coastal protection at Seacliff State beach, Santa Cruz County, California', USA. *Environmental Management*, 11(2): 175–82.

Griggs G.B. and Tait J.F. (1988) 'The effects of coastal protection structures on beaches along northern Monterey Bay, California', *Journal of Coastal Research*, SI 4 93–111.

—— (1989) 'Observations on the end effects of sea walls', *Shore and Beach*, 57(1): 25–6.

Griggs G.B., Tait J.F. and Corona W. (1994) 'The interaction of seawalls and beaches: Seven years of monitoring, Monterey Bay, California', *Shore and Beach*, 62(3): 21–8.

Groenendijk A.M. (1986) 'Establishment of a *Spartina anglica* population on a tidal mudflat: a field experiment', *Journal of Environmental Management*, 22: 1–12.

Grove R.S., Sonu C.J. and Dykstra D.H. (1987) Fate of massive sediment injection on a smooth shoreline at San Onofre, California. *Proceedings of Coastal Sediments '87*: 531–38.

Guilcher A. and Hallégouët B. (1991) 'Coastal dunes in Brittany and their management', *Journal of Coastal Research*, 7(2): 517–33.

Gunbak A.R. (1985) 'Damage of Tripoli Harbour north-west breakwater', in Bruun P. (ed.) *Design and construction of mounds for breakwaters and coastal protection*, Developments in Geotechnical Engineering 37. Elsevier, New York: 676–95.

Gunbak A.R. and Ergin A. (1985) 'Damage and repair of Antalaya Harbour breakwater', in Bruun P. (ed.) *Design and construction of mounds for breakwaters and coastal protection*, Developments in Geotechnical Engineering 37. Elsevier, New York: 649–70.

Hackney C.T. and Cleary W.J. (1987) 'Saltmarsh loss in south eastern North Carolina lagoons: Importance of sea level rise and inlet dredging', *Journal of Coastal Research*, 3(1): 93–7.

Hails J.R. (1975) 'Submarine geology, sediment distribution and Quaternary history of Start Bay, Devon', *Journal of the Geological Society of London*, 131: 1–5.

Halcrow Ltd. (1997) 'Fairlight Cove scheme appraisal: performance review volume 1', unpublished report for Rother district Council, Sir William Halcrow, August 1997.

Hall J., Brampton A. and Terry S. (1995) 'The direction of littoral drift in Poole Bay, and the effectiveness of groynes', 30th Conference of MAFF, River and Coastal Engineering, Keele.

Hall M.J. and Pilkey O.H. (1991) 'Effects of hard stabilisation on dry beach width for New Jersey', *Journal of Coastal Research*, 7(3): 771–85.

Hallégouët B. and Guilcher A. (1990) 'Moulin Blanc artificial beach, Brest, western Brittany, France', *Journal of Coastal Research*, SI 6: 17–20.

Hands E.B. and Allison M.C. (1991) 'Mound migration in deeper water and method of categorising active and stable berms', *Proceedings of Coastal Sediments '91*, American Society of Civil Engineers: 1985–99.

Hansom J.D. (1988) *Coasts*, Cambridge University Press, Cambridge.

Hardisky M. (1978) 'Marsh restoration on dredged material, Buttermilk Sound, Georgia', in Cole D.P. (ed.) *Proceedings of the 5th Annual Conference of the Restoration and Creation of Wetlands*, Hillsborough Community college, Florida: 136–51.

Harris W.B. and Ralph K.J. (1980) 'Coastal engineering problems at Clacton-on-Sea, Essex', *Quarterly Journal of Engineering Geology*, 13: 97–104.

Harvey H.T., Williams P. and Haltiner J. (1983) *Guidelines for enhancement and restoration of diked historic baylands*, unpublished report, San Francisco Bay Conservation and Development Commission.

Haynes F.N. and Coulson M.G. (1982) 'The decline of *Spartina* in Langstone Harbour, Hampshire', *Hampshire Field Club and Archaeological Society*, 38: 5–18.

Healy T.R., Kirk R.M. and de Lange W.P. (1990) 'Beach renourishment in New Zealand', *Journal of Coastal Research*, SI 6: 77–90.

Helmer W., Vellinga P., Litjens G., Goosen H., Ruijgrok E. and Overmars W. (1986) *Growing with the Sea: creating a resilient coastlines*, World Wide Fund for Nature (Netherlands), Zeist, Netherlands.

Herron W.J. (1987) 'Sand replenishment in southern California', *Shore and Beach*, 55(3–4): 87–91.

Hesp P.A. and Thom B.G. (1990) 'Geomorphology and evolution of active transgressive dunefields', in Nordstrom K., Psuty N. and Carter B. (eds) *Coastal Dunes: Form and Process*, Wiley, Chichester.

Hill M.O. and Wallace H.L. (1989) 'Vegetation and environment in afforested sand dunes at Newborough, Anglesey', *Forestry* 62(3): 249–67.

Hillyer T.M., Stakhiv E.Z. and Sudar R.A. (1997) 'An evaluation of the economic performance of the U.S. army corps shore protection programme,' *Journal of Coastal Research* 13(1): 8–22.

Hinrichsen D. (1990) *Our Common Seas: coasts in crisis*, Earthscan, London.

Hjulström F. (1935) 'Studies of the morphological activity of rivers as illustrated by the River Fyris', *Bulletin of the Geological Institute of the University of Uppsala*, 25: 221–527.

Hobbs R.J., Gimingham C.H. and Band W.T (1983) 'The effects of planting technique on the growth of *Ammophila arenaria* (L)', and *Leymus arenarius* (L). *Journal of Applied Ecology*, 20: 659–72.

Holder C.L. and Burd F.H. (1990) *Overview of salt marsh restoration projects: an interim report*, Contract Surveys No. 83, English Nature, Peterborough.

Holland B. and Coughlan P. (1994) 'The Elmer coastal defence scheme', Proceedings of the 29th MAFF Conference of River and Coastal Engineers, Ministry of Agriculture, Fisheries and Food, London.

Hollis E., Thomas D. and Heard S. (1990) *The Effects of Sea Level Rise on Sites of Conservation Value in Britain and north-west Europe*, University College London and World Wildlife Fund.

Holman R.A. and Bowen A.J. (1982) 'Bars, bumps and holes: models for the generation of complex beach topography', *Journal of Geophysical Research*, 87: 457–68.

Hooke J. (ed.) (1998) Coastal Defence and Earth Science Conservation, The Geological Society, London.

Horn D.P. (1997) 'Beach research in the 1990s', *Progress in Physical Geography*, 21(3): 454–70.

Hotta S., Kraus N.C. and Horikawa K. (1987) 'Function of sand fences in controlling wind blown sand', in Kraus N.C. (ed.) *Coastal Sediments '87*, Proceedings of a speciality conference on advances in understanding of coastal sediment processes, New Orleans, Louisiana, 12–14 May 1987: 772–87.

Houston J.A. and Jones C.R. (1987) 'The Sefton coast management scheme: Project and process', *Coastal Management*, 15: 267–97.

Howard J.D., Kaufman W. and Pilkey O.H. (1985) 'Strategy for beach preservation proposed', *Geotimes*, 30 (Dec.): 15–19.

Hsu J.R.C. and Silvester R. (1990) 'Accretion behind single offshore breakwater', *Journal of Waterway, Port, Coastal and Ocean Engineering*, 116(3): 362–80.

Hubbard J.C.E. and Stebbings R.E. (1967) 'Distribution, dates of origin, and acreage of *Spartina townsendii* marshes in Great Britain', *Proceedings of the Botanical Society of the British Isles*, 7:1–7.

Hurme A.K. (1979) *Proceedings of Coastal Structures '79*, ASCE: 1042–51.

Hutchings C. (1994) 'Back to basics', *Geographical*, 1994(March): 20–21.

Hydraulics Research (1970) *Colliery Waste Dumping on the Durham Coast*, Report No. 521, Hydraulics Research Limited, Wallingford.

—— (1987) *The effectiveness of saltings*, Report SR 109. Hydraulics Research Limited, Wallingford. April, 1987.

—— (1992) 'Getting to the bottom of seawall scour', *Coastal Defence*, 1: 4 Ministry of Agriculture, Fisheries and Food, London.

Ibe A.C., Awosika L.F., Ibe C.E. and Inegbedion L.E. (1991) 'Monitoring of the 1985–86 beach nourishment project at Bar Beach, Victoria Island, Lagos, Nigeria', *Proceedings Coastal Zone '91*: 534–52.

Illenberger W.K. and Rust I.C. (1988) 'A sand budget for the Alexandria coastal dunefield, South Africa', *Sedimentology*, 35: 513–21.

Ince M. (1990) *The Rising Seas* Earthscan, London.

Inglis C.C. and Kestner J.T. (1958a) 'The long-term effect of training walls, reclamation and dredging on estuaries', *Proceedings of the Institute of Civil Engineers*, 9: 193–216.

—— (1958b) 'Changes in the Wash as affected by training walls and reclamation works', *Proceedings of the Institute of Civil Engineers*, 11: 435–66.

Inman D.L. and Dolan R. (1989) 'The outer banks of North Carolina: budget of sediment and inlet dynamics along a migrating barrier system', *Journal of Coastal Research*, 5(2): 193–237.

Inman L.D. and Frautschy J.D. (1966) 'Littoral processes and the development of shorelines', *Proceedings of the ASCE conference on Coastal Engineering*, Santa Barbara, CA: 511–36.

IECS (Institute of Estuarine and Coastal Sciences) (1993) *Sites of historical sea defence failure*, Phase II interim report, Unpublished Report, IECS, University of Hull.

ICE (Institution of Civil Engineering) (1989) *Coastal management*, Thomas Telford, London.

Jackson G.A. (1984) 'Internal wave attenuation by coastal kelp stands', *Journal of Physical Oceanography*, 14: 1300–06.

Jackson N.L. and Nordstrom K.F. (1998) 'Aeolian transport of sediment on a beach during and after rainfall, Wildwood, NJ, USA', *Geomorphology*, 22: 151–7.

Jagschitz J.A. and Wakefield R.C. (1971) 'How to build and save beaches and dunes: preserving the shoreline with fencing and beach grass', Bulletin 408, Rhode Island Agricultural Experimental Station, University of Rhode Island, USA.

Janin L.F. (1987) 'Simulation of sand accumulation around fences', in Kraus N.C. (ed.) *Coastal Sediments '87*, Proceedings of a speciality conference on advances in understanding of coastal sediment processes. New Orleans, Louisiana. 12–14 May 1987: 87–97.

Jennings S., Orford J.D., Canti M., Devoy R.J.N. and Straker V. (1998) 'The role of relative sea level rise and changing sediment supply on Holocene gravel barrier development: the example of Porlock, Somerset, UK', *The Holocene* 8(2): 165–81.

Jensen A. and Mallinson J. (1993) 'Notes on site visit to coastal defence rock islands on the beach at Elmer, 9th March 1993, with comments on techniques for biological enhancement', unpublished report, Department of Oceanography, University of Southampton.

Job D. (1993) 'Coastal management: Start Bay, Devon', *Geography Review*, 1993 (Nov.): 13–17.

Jokiel P.L., Hunter C.L., Taguchi S. and Watarai L. (1993) 'Ecological impact of a fresh water 'reef-kill' in Kanehoe Bay, Oahu, Hawaii', *Coral Reefs*, 12: 177–84.

Jones J.R., Cameron B. and Fisher J.J. (1993) Analysis of cliff retreat and shoreline erosion: Thompson Island, Massachusetts, USA, *Journal of Coastal Research*, 9(1): 87–96.

Jones P.S. (1996) Kenfig National Nature research: a profile of a British west coast dune system, in Jones P.S, Healy M.G. and Williams A.T. (eds) *Studies in European coastal management*, Samara Publishing, Cardigan: 255–67.

Jordan P. and Slaymaker O. (1991) 'Holocene sediment production in Lillooet River basin, British Columbia: a sediment budget approach', *Geographia Physique et Quaternaire*, 45(1): 45–57.

Kàdomatsu T., Uda T. and Fujiwara K. (1991) 'Beach nourishment and field observations of beach changes on the Toban coast facing Seto Inland Sea', *Marine Pollution Bulletin*, 23: 155–9.

Kana T.W. (1995) 'A mesoscale sediment budget for Long Island, New York', *Marine Geology*, 126 (1–4): 87–110.

Kay R. and Alder J. (1999) *Coastal planning and management*, E & FN Spon, London.

Kelletat D. (1992) 'Coastal erosion and protection measures at the German North Sea coast', *Journal of Coastal Research*, 8(3): 699–711.

Kellman M. and Kading M. (1992) 'Facilitation of tree seedling establishment in a sand dune succession', *Journal of Vegetation Science*, 3(5): 679–88.

Kerckaert P. *et al.* (1985) 'Artificial beach nourishment on the Belgian coast,' *Journal of Waterways, Ports, Coastal and Oceans Division* (ASCE), 112(5): 560–71.

Kiknadze A.G., Sakvarelidze V.V., Peshkov V.M. and Russo G.E. (1990) 'Beach forming process management of the Georgian Black Sea coast', *Journal of Coastal Research*, Special Issue 6: 33–44.

Kimura H. (1957) 'On the sand fence for beach stabilisation work', *Journal of the Japan Forestry Conservation Association*, 1(11): 1–5.

King D.M. (1996) 'Sediment transport studies at Elmer, West Sussex, 16th October/17th November: preliminary results and analysis', unpublished report, Department of Civil Engineering, University of Brighton.

King D.M., Cooper N.J., Morfett J.C. and Pope D.J. (2000) 'Application of offshore breakwaters to the UK: a case study at Elmer Beach', *Journal of Coastal Research*, 16(1): 172–87.

King S.E. and Lester J.N. (1995) 'The value of saltmarsh as a sea defence', *Marine Pollution Bulletin*, 30(3): 180–9.

Klein R.J.T. and Bateman I.J. (1998) 'The recreational value of Cley marshes nature reserve: an argument against managed retreat?', *Journal of the Institute of Water and Environmental Management*, 12: 280–5.

Knutson P.L. (1988) 'Role of coastal marshes in energy dissipation and shore protection', in Hook D.D. *et al.* (eds) *The ecology and management of wetlands*, Vol. 1: Ecology of wetlands. Croom Helm, London: 161–75.

Knutson P.L., Allen H.H. and Webb J.W. (1990) *Guidelines for vegetative erosion control on wave-impacted coastal dredged material sites*, unpublished report, US Army Corps of Engineers.

Knutson P.L., Brochu R.A., Seelig W.N. and Inskeep M. (1982) 'Wave damping in *Spartina alterniflora* marshes', *Wetlands*, 2: 87–104.

Knutson P.L., Ford J.C. and Inskeep M. (1981) 'National sowing of planted salt marsh (vegetation stabilisation and wave stress)', *Wetlands*, 1: 129–57.

Kochel R.C. and Dolan R. (1986) 'The role of overwash on a mid-Atlantic coast barrier island', *Journal of Geology*, 94(6): 902–6.

Koike K. (1990) 'Artificial beach construction on the shores of Tokyo Bay, Japan', *Journal of Coastal Research*, SI 6: 45–54.

Komar P.D. (1983) Coastal erosion in response to construction of jetties and breakwaters, in *Handbook of Coastal Protection and Erosion* CRC Press. Bocas Raton: 191–204.

—— (1998) *Beach Processes and Sedimentation*, 2nd ed, Prentice Hall, Englewood Cliffs, New Jersey.

Komar P.D. and McDougal W.G. (1988) 'Coastal erosion and engineering structures: The Oregon experience', *Journal of Coastal Research*, SI 4: 77–92.

Kowalski T. (1974) 'Scrap tire floating breakwaters', *1974 Floating Breakwaters Conference Proceedings*, April 23–25 1974. Newport, Rhode Island: 233–46.

Kraus N.C. (1988) 'The effects of sea walls on the beach: an extended literature review', *Journal of Coastal Research*, SI 4: 1–29.

Kriebel D.L. (1986) 'Verification study of a dune erosion model', *Shore and Beach*, 54: 13–21.

—— (1987) 'Beach recovery following hurricane Elena', *Coastal Sediments '87* American Society of Civil Engineers: 990–1005.

—— (1990) 'Advances in numerical modelling of dune erosion', *Proceedings of the 22nd Coastal Engineering Conference*, American Society of Civil Engineers: 2304–17.

Kriebel D.L. and Dean R.G. (1985) 'Numerical simulation of time-dependent beach and dune erosion', *Coastal Engineering*, 9: 221–45.

Kuhn G.G. and Shepard F.P. (1984) *Sea cliffs, beaches and coastal valleys of San Diego County*, University of California Press, Berkeley.

Kumbein W.C. and Slack H.A. (1956) 'The relative efficiency of beach sampling methods', *Technological Memo No. 90*, Beach Erosion Board, USA.

Kunz H. (1987) 'History of a sea wall revetments on the Isle of Nordeney', in Kraus N.C. (ed.) *Coastal Sediments '87*, Proceedings of a speciality conference on advances in understanding of coastal sediment processes, New Orleans, Louisiana, 12–14 May 1987: 974–89.

Lancaster N. and Baas A. (1998) 'Influence of vegetation cover on sand transport by wind: field studies at Owens Lake, California', *Earth Surface Processes and Landforms*, 23(1): 69–82.

Leafe R. (1992) 'Northey Island – an experimental sea back', *Earth Science Conservation*, 31: 21–2.

Leatherman S.P. (1986) 'Cliff stability along western Chesapeake Bay, Maryland', *Marine Technology Society Journal*, 20(3): 28–36.

Leidersdor C.B., Hollar R.C. and Woodwell G. (1994) 'Human intervention with the beaches of Santa Monica Bay, California', *Shore and Beach*, 62(3): 29–38.

Lelliott R.F.L. (1989) 'Evolution of the Bournemouth defences', in Institution of Civil Engineers (ed.) *Coastal Management*, Thomas Telford Press, London: 263–77.

Lee E.M. (1998) 'Problems associated with the prediction of cliff recession rates for coastal defence and conservation', in Hooke J.M. (ed.) *Coastal defence and earth science conservation*, Geological Society of London: 46–57.

Lee W.G. and Partridge T.R. (1983) 'Rates of spread of *Spartina anglica* and sediment accretion in the New River Estuary, Invercargill, New England', *New Zealand Journal of Botany*, 21: 231–6.

Leonard L.A., Clayton T. and Pilkey O. (1990a) 'An analysis of replenished beach design parameters on US east coast barrier islands', *Journal of coastal Research*, 6(1): 15–36.

Leonard L.A., Dixon K.L. and Pilkey O.H. (1990b) 'A comparison of beach replenishment on the US Atlantic, Pacific, and Gulf Coasts', *Journal of Coastal Research*, SI 6: 127–40.

Lewis R.R. (1990a) 'Creation and restoration of coastal plain wetlands in Florida', in Kusler J.A. & Kentula M.E. (eds) *Wetland creation and restoration: The status of the science*, Island Press, Washington: 73–101.

—— (1990b) 'Creation and restoration of coastal wetlands in Puerto Rico and the US Virgin Islands', in Kusler J.A. and Kentula M.E. (eds) *Wetland creation and restoration: The status of the science*, Island Press, Washington: 103–23.

Li L., Barry D.A. and Pattiaratchi C.B. (1996) 'Modelling coastal ground water response to beach dewatering', *Journal of Waterway, Port, Coastal, and Ocean Engineering*, 122(6): 273–80.

Lin J.C. (1996) 'Coastal modification due to human influence in south-western Taiwan', *Quaternary Science Reviews*, 15(8–9): 895–900.

Lisle L.D. (1986) 'United States sea level changes', in Sigbjarnarson (ed.) Proceedings of the Iceland Coastal and River Symposium, Reykjavik, Iceland, 1986: 277–286.

Louisse C.J. and van der Meulen F. (1991) 'Future coastal defence in the Netherlands: Strategies for protection and sustainable development', *Journal of Coastal Research*, 7: 1027–41.

LRDC International (1993) 'Saving the saltings – a strategy', LRDC International, Haselmere.

LSUCWR (1979) *Floating tire breakwaters (FTB'S)* 'Louisiana State University for Wetland Resources report LSU-TL-79-001, Natural Sea Grant College Program, Rhode Island.

Ly C.K. (1980) 'The role of the Akosombo Dam in the Volta river in causing erosion in central and eastern Ghana (west Africa)', *Marine Geology*, 37: 323–32.

McCave I.N. (1970) 'Deposition of fine-grained suspended sediments from tidal currents', *Journal of Geophysical Research*, 75: 4151–9.

MacDonald J. (1954) 'Tree planting on coastal sand dunes in Great Britain,' *Advancement of Science*, 11: 33–7.

MacDonald R.W., Solomon S.M., Cranston R.E., Welch H.E., Yunker M.B. and Gobeil C. (1998) 'A sediment and organic carbon budget for the Canadian Beaufort shelf', *Marine Geology* 144(4): 255–73.

McDougal W.G., Sturtevant M.A., and Komar P.D. (1987) 'Laboratory and field investigations of shoreline stabilization structures on adjacent properties', *Coastal Sediments '87*, American Society of Civil Engineers: 961–73.

McFarland S., Whitcombe L. and Collins M. (1994) 'Recent shingle beach renourishment schemes in the UK',: Some preliminary observations', *Ocean and Coastal Management*, 25: 143–9.

McGown A., Roberts A.G. and Woodrow L.K.R. (1987) 'Long-term pore pressure variation within coastal cliffs of North Kent, UK', Proceedings of the 9th European conference SMFE Dublin, 455–680.

—— (1988) 'Geotechnical and planning aspects of coastal landslides in the UK', in Bonnard C. (ed.) *Landslides, Volume 2*, Balkema, Rotterdam: 1201–6.

McInnes R.G. (1998) 'Practical experience in reconciling conservation and coastal defence strategies on the Isle of Wight', in Hooke J.M. (ed.) *Coastal Defence and Earth Science Conservation*, Geological Society of London: 67–86.

McLachlan A. (1990) 'The exchange of materials between dune and beach systems', in Nordstrom K., Psuty N. and Carter B. (eds) *Coastal Dunes: form and processes*, Wiley, Chichester.

McLusky D.S. (1989) *The estuarine ecosystem*, 2nd ed, Blackie, Glasgow.

McNinch J.E. and Wells J.T. (1992) 'Effectiveness of beach scraping as a method of erosion control, *Shore and Beach*, 60(1): 13–20.

Maddrell R.J. (1996) 'Managed coastal retreat, reducing the flood risks and protection costs, Dungeness nuclear power station, UK', *Coastal Engineering*, 28: 1–15.

Madge B. (1983) In *Shoreline Protection*, Thomas Telford, London 115–17.

MAFF (1993) 'Predicting long-term coastal morphology', *Flood and Coastal Defence*, 4(Nov): 1.

—— (1994) 'Full scale managed set back experiment starts', *Flood and Coastal Defence*, 5(Apr): 4.

—— (1995a) 'Studying the effects of control structures on shingle beaches', *Flood and Coastal Defence*, 7(July): 6–7.

—— (1995b) 'Artificial reefs and lobsters', *Flood and Coastal Defence*, 6 (Feb): 7.

—— (1995c) 'Managed set back experimental site', *Flood and Coastal Defence*, 6 (Feb): 3–4.

—— (1995d) 'Tollesbury breach experiment', *Flood and Coastal Defence*, 8 (Nov): 1–2.

—— (1995e) 'Managed set back experimental site', *Flood and Coastal Defence*, 6 (Feb): 3–4.

—— (1996) 'Design of submerged offshore breakwaters', *Flood and Coastal Defence*, 9 (June): 4.

—— (1997) 'Managed set back monitoring at Tollesbury', *Flood and Coastal Defence*, 10 (June): 7.

MAFF, Welsh Office, Association of District Councils, English Nature and NRA (1995) *Shoreline Management Plans: a guide for coastal authorities*, MAFF, London, May 1995.

Magoon O.T. (1976) 'Offshore breakwaters at Winthrop Beach, Massachusetts, *Shore and Beach*, 44(3): 34.

Magoon O.T., Calvarese V. and Clarke D. (1984), in Institution of Civil Engineering (eds) *Breakwaters, Design and Construction*, Thomas Telford, London.

Mailly D., N' Diaye P., Margolis H.A. and Pineau M. (1994) 'Stabilisation of shifting dunes and afforestation along the northern coast of Senegal, using the Filao (*Casuarina equisetifolia*)', *Forestry Chronicle*, 70(3): 282–90.

Marcus W.A., Nielsen C.C. and Cornwell J.C. (1993) 'Sediment budget-based estimates of trace metal inputs to a Chesapeake estuary', *Environmental Geology*, 22(1): 1–9.

Marino M.J. (1992) 'Implications of climatic change on the Ebro Delta', in Jeftic L., Milliman J.D. and Sestini G. (eds) *Climatic change and the Mediterranean*, Edward Arnold, London: 304–27.

Marsh G.A. and Turbeville D.B. (1981) 'The environmental impact of beach nourishment: Two studies in southeastern Florida', *Shore and Beach*, 49(3): 40–4.

Marsh W.M. (1990) 'Nourishment of perched sand dunes and the issue of erosion control in the Great Lakes', *Environmental Geology and Water Sciences*, 16(2): 155–64.

Marti J.L.J., Hernández C.G. and Montero G.G. (1995) 'Researches and measures for beach preservation: The case of Varadero Beach, Cuba', *Proceedings of the international conference: Coastal Change '95*, Bordomer – IOC, Bordeaux, 1995: 233–41.

Martin F.L., Losada M.A. and Medina R. (1999) 'Wave loads on a rubble mound breakwater crown walls,' *Coastal Engineering*, 37(2): 149–74.

Mason S.J. and Hansom J.B. (1988) 'Cliff erosion and its contribution to a sediment budget for part of the Holderness Coast, England', *Shore and Beach*, 56(4): 30–8.

May V. (1977) 'Earth cliffs', in Barnes R.S.K. (ed.) *The Coastline*, Wiley, London: 215–35.

—— (1990) 'Replenishment of resort beaches at Bournemouth and Christchurch, England', *Journal of Coastal Research*, SI 6: 11–15.

Meadows P.S., Meadows A., West F.J.C., Shand P.S. and Shaikh M.A. (1998) 'Mussels and mussel beds (*Mytilus edulis*) as stabilisers of sedimentary environments in the intertidal zone', in Black K.S., Paterson D.M. and Cramp A. (eds) *Sedimentary processes in the intertidal zone*, Geological Society Special Publications, 139: 331–47. Geological Society, London.

Metcalfe R.E. (1977) *The management of the Camber sand dunes, Sussex*, unpublished report, County Planning Department, Lewes, East Sussex County Council.

Mikkelson S.C. (1977) 'The effects of groins on beach erosion and channel stability at the Limfjord Barriers, Denmark', *Proceedings of Coastal Sediments '77*, American Society of Civil Engineers: 17–32.

Milton S.L., Achulman A.A. and Lutz P.L. (1997) 'The effect of beach nourishment with aragonite versus silicate sand on beach temperature and Loggerhead Sea Turtle nesting success', *Journal of Coastal Research*, 13(3): 904–15.

Misak R.F. and Draz M.Y. (1997) 'Sand drift control of selected coastal and desert dunes in Egypt: Case studies', *Journal of Arid Environments*, 35(1): 17–28.

Moeller I., Spencer T. and French J.R. (1996) 'Wind wave attenuation over salt marsh surfaces: Preliminary results from Norfolk, England', *Journal of Coastal Research*, 12(4): 1009–16.

Møller J.T. (1990) 'Artificial beach nourishment on the Danish North Sea coast', *Journal of Coastal Research*, 6: 1–9.

Moody S. (1997) 'The effects of offshore breakwaters upon beach sediment accretion the Elmer frontage, West Sussex',. unpublished BSc dissertation, Geography Department, Lancaster University.

Morton R.A. (1988) 'Interactions of storms, seawalls and beaches of the Texas coast', *Journal of Coastal Research*, SI 4: 113–34.

Munk W.H. and Taylor M.A. (1947) 'Refraction of ocean waves: a process linking underwater topography to beach erosion', *Journal of Geology*, 55: 1–26.

Muus B.J. (1967) *The Fauna of Danish Estuaries and Lagoons*, Copenhagen.

Myrick R.M. and Leopold L.M. (1963) *Hydraulic geometry of a small tidal estuary*, Professional paper No. 422–B. US Geological Survey.

Nagashima L.D., Moma J. and Dean J.L. (1987) 'Initial response of a segmented breakwater system', in Kraus N.C. (ed.) *Coastal Sediments '87*, Proceedings of a speciality conference on advances in understanding of coastal sediment processes, New Orleans, Louisiana, 12–14 May 1987: 1399–414.

National Research Council (1990) *Managing Coastal Erosion*, National Academy Press, USA.

Nelson S.M., Mueller G. and Hemphill D.C. (1994) 'Identification of tyre leachate toxicants and a risk assessment of water quality effects', *Bulletin of Environmental Contamination and Toxicology*, 52: 574–81.

Netto S.A. and Lana P.C. (1997) 'Influence of *Spartina alterniflora* on superficial sediment characteristics of tidal flats in Paranaguá Bay (south-eastern Brazil)', *Estuarine, Coastal and Shelf Science*, 44: 641–8.

Neumann A.C. and MacIntyre I. (1985) 'Reef response to a sea level rise: Keep up, catch-up, or give-up', *Proceedings of the 5th International Coral Reef congress*, Tahiti, 1985, 3: 105–10.

Newcombe C., Morris J.H., Knutson P.L. and Gorbics C.S. (1979) *Bank erosion control with vegetation, San Francisco Bay, California*, USACE Miscellaneous Report 79-2, Coastal Engineering Research Centre, Fort Belvoir, VA.

Newling C.J. and Landin M.C. (1985) 'Monitoring of habitat development at upland and wetland dredged material disposal sites 1974–1982', US. Army Engineer Waterways experimental station Technical Report D-85–5, Vicksburg, Mississippi.

Newman D.E. (1976) 'Beach replenishment: sea defences and a review or artificial beach replenishment', *Proceedings of the Institution of Civil Engineers*, 1(60): 445–60.

Nicholls R.J. (1998) 'Assessing erosion of sandy beaches due to sea level rise', in Maund J.G. and Eddleston M. (eds) *Geohazards in Engineering Geology*, Engineering Geology Special Publications, 15: 71–6, Geological Society, London.

Nicholls R.J., Leatherman S.P., Dennis K.C. and Volonte C.R. (1995) 'Impacts and responses to sea-level rise: qualitative and quantitative assessments, *Journal of Coastal Research*, SI 14: 26–43.

Nir Y. (1986) 'Offshore artificial structures and their influence on the Israel and Sinai Mediterranean beaches', *Proceedings 18th International conference on coastal engineering.*, ASCE: 1837–56.

Noble R.M. (1978) 'Coastal structures: effects on shorelines', *Proceedings 17th International Conference on Coastal Engineering*, ASCE, Sydney, Australia: 2069–85.

Nordstrom K.F. (1987) 'Dune grading along the Oregon coast, USA: a changing environmental policy', *Applied Geography* 8: 101–16.

Nordstrom K.F. and Allen J.R. (1980) 'Geomorphologically compatible solutions to beach erosion', *Zeitschrift für Geomorphologie*, S34: 142–54.

Nordstrom K.F, Allen J.R, Sherman D.J. and Psuty N.P. (1979) 'Management considerations for beach management at Sandy hook, New Jersey', *Coastal Engineering*, 2: 215–36.

Nordstrom K.F. and Psuty N.P. (1980) 'Dune district management: a framework for shorefront protection and land use control', *Coastal Zone Management Journal*, 7(1): 1–23.

Nordstrom K.F., Psuty N.P. and Carter B. (1990) *Coastal Dunes: form and process*, Wiley, Chichester.

North Holland Water Supply Company (1992) 'The North Holland Dune Reserve', *Coastline* 1992 (1–2): 17–32.

NRA (National Rivers Authority) (1991) 'Happisburgh to Winterton sea defences', *NRA Information Leaflet*, NRA Anglian Region, Peterborough.

NRA (undated) *Preserving shingle sea defences*, Information Sheet, National Rivers Authority, Southern Region, Worthing, Sussex.

OECD (1993) *Coastal Zone Management: integrated policies*, Organisation for Economic Co-operation and Development, Paris.

Oertel G. (1974) 'Review of the sedimentological role of dunes in shoreline stability', *Bulletin of Georgia Academy of Science*, 32: 48–56.

Open University (1989) *Waves, Tides and Shallow Water Processes*, Open University Press, Milton Keynes.

Orford J.D. (1988) 'Alternative interpretation of man-induced shoreline changes in Rosslare Bay, south-east Ireland', *Transactions of the Institute of British Geographers*, 13: 65–78.

Orford J.D. and Carter R.W.G. (1985) 'Storm generated rock armour on a sand-gravel ridge barrier system, south-eastern Ireland', *Sedimentary Geology*, 42(1–2): 65–82.

Ovington J.D. (1951) 'The afforestation of Tentsmuir sands', *Journal of Ecology*, 39: 363–75.

Owens J.S. and Case G.O. (1908) *Coast Erosion and Foreshore Protection*, St. Brides Press, London.

Packwood A.R. (1983) 'The influence of beach porosity on wave uprush and backwash', *Coastal Engineering*, 7(1): 29–40.

Page R.R., Davinha S.G. and Agnew A.D.Q. (1985) 'The reaction of some sand-dune plant species to experimentally imposed environmental change: a reduction-ist approach', *Vegetatio*, 61(1–3): 105–14.

Parker R. (1980) *Men of Dunwich*, Paladin Books, London.

Partridge T.R. (1992) 'Vegetation recovery following sand mining on coastal dunes at Kaitorete Spit, Canterbury, New Zealand', *Biological Conservation*, 61(1): 59–71.

Paskoff R. and Petiot R. (1990) 'Coastal progradation as a by-product of human activity: an example from Chañaral Bay, Atacama Desert, Chile', *Journal of Coastal Research*, Special Issue 6: 91–102.

Pearce F. (1992) 'Time to sound the retreat for sea defences', *New Scientist*, 136 (1843): 54.

—— (1993) 'When the tide comes in …', *New Scientist*, 137(1854): 22–6.

Penning-Rowsell E.C., Green C.H., Thompson P.M., Coker A.M., Tunstall S.M., Richards C. and Parker D.J. (1992) *The Economics of Coastal Management: a manual of benefit assessment techniques*, Belhaven Press, London.

Perry G. and D' Miel R. (1995) 'Urbanisation and sand dunes in Israel: direct and indirect effects', *Israel Journal of Zoology*, 41(1): 33–41.

Pethick J. (1984) *An Introduction to Coastal Geomorphology*, Edward Arnold, London.

—— (1992) 'Natural Change, in Barrett M.G. (ed.) *Coastal Zone Planning and Management*, Institute of Civil Engineers, Thomas Telford, London: 49–63.

—— (1993) *Sediment deposition under salt marsh vegetation*, unpublished report, Institute of Estuarine and Coastal Studies, University of Hull, May, 1993.

Pethick J. and Burd F. (1993) *Coastal Defence and the Environment: a guide to good practice*, MAFF, London.

Pethick J. and Reed D. (1987) 'Coastal protection in an area of salt marsh erosion', in Kraus N.C. (ed.) *Coastal Sediments '87*, Proceedings of a speciality conference on advances in understanding of coastal sediment processes, New Orleans, Louisiana, 12–14 May 1987: 1094–104.

Pierce J.W. (1969) 'Sediment budget along a barrier island chain. *Sedimentary Geology* 3: 5–16.

Pilkey O.H. (1990) 'A time to look back at beach replenishments,' *Journal of Coastal Research*, 6(1): iii–vii.

Pilkey O.H. and Clayton T.D. (1987) 'Beach replenishment: the national solution?', *Proceedings of Coastal Zone '87*, New York: 1408–19.

Pilkey O.H. and Wright H.L. (1988) 'Sea walls versus beaches', *Journal of Coastal Research*, SI 4: 41–64.

Pirazzoli P.A. (1991) 'Possible defences against a sea level rise in the Venice area, Italy', *Journal of Coastal Research*, 7: 231–48.

Pitts J. (1983) 'Geomorphological observations as aids to the design of coastal pro-tection works on a part of the Dee estuary', *Quarterly Journal of Engineering Geology*, 16: 291–300.

Plant N.G. and Griggs G.B. (1992) 'Interactions between nearshore processes and beach morphology near a seawall', *Journal of Coastal Research*, 8(1): 183–200.

Pope J. and Rowen D.D. (1983) 'Breakwaters for beach protection at Lorain, Ohio', *Proceedings of Coastal Structures '83*, American Society for Civil Engineers: 753–68.

Prestage M. (1991) 'The lost towns of Humberside', *Geology Today*, Jan–Feb 1991: 6.

Price W.A., Tomlinson K.W. and Hunt J.N. (1968) 'The effect of artificial seaweed in promoting the build-up of beaches', Proceedings of conference on Coastal Engineering, ASCE, New York, 36: 570–8.

Pringle A.W. (1995) 'Erosion of a cyclic salt marsh in Morecambe Bay north-west England', *Earth Surface Processes and Landforms*, 20: 387–405.

Psuty N.P. (1993) 'Foredune morphology and sediment budget, Perdito Key, Florida, USA', IN Pye K. (ed.) *The dynamics and environmental Context of aeolian sedimentary systems*, Geological Society Special Publication No. 72, Geological Society, London.

Psuty N.P. and Moreira M.E.S.A. (1990) 'Nourishment of a cliffed coastline, Praia de Rocha, The Algarve, Portugal, *Journal of Coastal Research*, SI 6: 21–32.

Psuty N.P. and Moreira M.E.S.A. (1992) 'Characteristics and longevity of beach nourishment at Praia da Rocha, Portugal, *Journal of Coastal Research*, 8(3): 660–76.

Pullen E.J. and Naqvi S.M. (1983) 'Biological impacts on beach replenishment and borrowing', *Shore and Beach*, 51(2): 27–31.

Pye K. (1990) 'Physical and human influences on coastal dune development between the Ribble and Mersey estuaries, north-west England', in Nordstrom K.F, Psuty N.P. and Carter R.W.G. (eds) *Coastal Dunes: form and process*, Wiley: Chichester: 339–59.

Pye K. and Bowman G.M. (1984) 'The Holocene marine transgression as a forcing function in episodic dune activity on the eastern Australian coast', in Thom B.G. (ed.) *Coastal Geomorphology in Australia*, Academic Press, Sydney.

Pye K. and French P.W. (1993a) *Erosion and Accretion Processes on British Salt-marshes. Volume 5: management of saltmarshes in the context of flood defence and coastal protection*. Report to MAFF, Cambridge Environmental Research Consultants Report No. ES23.

—— (1993b) *Targets for coastal habitat re-creation*, English Nature Science Series No. 13, English Nature, Peterborough.

Rakha K.A. and Kamphuis J.W. (1997) 'A morphology for an eroding beach backed by a sea wall', *Coastal Engineering*, 30: 53–75.

Ranwell D.S. and Boar R. (1986) *Coastal Dune Management Guide*. Institute of Terrestrial Ecology, NERC, Huntingdon.

Reckendorf F., Leach D., Baum R. and Carlson J. (1985) 'Stabilisation of sand dunes in Oregon', *Agricultural History*, 59(2): 260–8.

Reed D.J. (1995) 'The response of coastal marshes to sea level rise: survival or submergence?', *Earth Surface Processes and Landforms*, 20: 39–48.

Reilly F.J. and Bellis V.J. (1983) *The ecological importance of beach nourishment with dredged materials on the intertidal zone at Bogue Banks, North Carolina*, Miscellaneous Paper 83–3, USACE, Fort Bellvoir, Virginia.

Riby J. (1998) 'Design conditions for coastal works: practical experience of coastal risk assessments', *Proceedings of the Institute of Civil Engineers: Municipal Engineering*, 127(June): 49–55.

Riddle K.J. and Young S.W. (1992) 'The management and creation of beaches for coastal defence', *Journal of the Institute of Water and Environmental Management*, 6: 588–97.

Rijkwaterstaat (1979) *Zandwinning in de Waddenzee, resulttaten van een biologisch-ecologisch onderzoek*', Directie Friesland, Werkgroep II, Leeuwarden. (Referenced in Adriaanse & Coosen 1991).

Roberts A.G. and van Overeem J. (1993) 'Rebuilding of sea defences, Central

Parade, Herne Bay, Kent', *Proceedings of the Institution of civil Engineers: Municipal Engineers*, 98: 31–9.

Roberts T.H. (1989) 'Habitat value of man-made coastal marshes, Florida', in Webb F.J. (ed.) Proceedings of the 16th annual conference on wetlands restoration and creation, Hillsborough Community College, Florida.

Robinson A.H.W. (1980) 'Erosion and accretion along parts of the Suffolk coast of East Anglia, England', *Marine Geology*, 37: 133–46.

Rogers S.M. (1987) 'Artificial sea weed for erosion control,' *Shore and Beach*, 55(1): 19–29.

Rotnick K. (1994) *Changes of the Polish Coastal Zone*, Quaternary Research Institute. Adam Mickiewicz University, Poznan.

Rouch F. and Bellessort B. (1990) 'Man-made beaches more than 20 years on', *Proceedings of the 22nd Coastal Engineering Conference*: 2394–401.

Royal Commission on Coast Erosion (1907) *First report into coast erosion and the reclamation of tidal lands in the UK*, HMSO, London.

Rutin J. (1992) 'Geomorphic activity of rabbits on a coastal sand dune, Deblink dunes, the Netherlands', *Earth Surface Processes and Landforms*, 17(1): 85–94.

Sacco J.N., Booker S.L. and Seneca E.D. (1988) 'Comparison of the macrofaunal communities of a human-initiated salt marsh at two and fifteen years of age', Published abstract, Benthic Ecology Meeting, Portland, Maine, 1988.

Sayre W.O. and Komar P.D. (1988) 'The Jump-Off Joe landslide at Newport, Oregon: history of erosion, development and destruction', *Shore and Beach*, 56: 15–22.

Schwartz M.G., Juanes J., Foyo J. and Garcia G. (1991) 'Artificial nourishment at Varadero Beach, Cuba', in American Society of Civil Engineers (ed.) *Coastal Sediments '91*, ASCE, New York: 2081–8.

Schwartz R.K. and Musialawski F.R. (1977) 'Nearshore disposal: onshore sediment transport', *Proceedings of Coastal Sediments '77*: 85–101.

Scott-Anderson R., Borns H.W., Smith D.C. and Race C. (1992) 'Implications of rapid sediment accumulation in a small New England salt marsh', *Canadian Journal of Earth Science*, 29: 2013–17.

Seiji M., Uda T. and Tanaka S. (1987) 'Statistical study of the effect and stability of detached breakwaters', *Coastal Engineering in Japan*, 30(1): 131–41.

Seneca E.D. (1974) 'Stabilisation of coastal dredge spoil with S*partina alterniflora*', in Reimold R.J. and Queen W.H. (eds) *Ecology of halophytes*, Academic Press, New York: 525–9.

Seneca E.D., Woodhouse W.W. and Broome S.W. (1975) 'Salt water marsh creation', in Cronin L.E. (ed.) *Estuarine Research Vol. II: Geology & Engineering*, Academic Press.

Sexton W.J. and Moslow T.F. (1981) 'Effects of Hurricane David, 1979, on the beaches of Seabrook Island, Southern Carolina', *Northeastern Geology*, 3(3–4): 297–305.

Shaw J. and Ceman J. (1999) 'Salt marsh aggradation in response to late Holocene sea level rise at Amherst Point, Nova Scotia, Canada', *The Holocene*, 9(4): 439–51.

Shaw J., Taylor R.B., Solomon S, Christian H.A. and Forbes D.L. (1998) 'Potential impacts of global sea-level rise on Canadian coasts,' *The Canadian Geographer*, 42(4): 365–79.

Shepard M.J. (1987) 'Holocene alluviation and transgressive dune activity in the lower Mawnawatu Valley, New Zealand', *New Zealand Journal of Geology and Geophysics*, 39: 175–87.

Shennan I. (1993) 'Sea level changes and the threat of coastal inundation', *The Geographical Journal*, 159(2): 148–56.

Shennan I. and Woodworth P.L. (1992) 'A comparison of late Holocene and twentieth century sea level trends from the UK and North Sea region', *Geophysical Journal International*, 109: 96–105.

Shepard F.P. and Wanless H.R. (1971) *Our Changing Coastlines*, McGraw Hill, New York.

Shi Z., Pethick J.S. and Pye K. (1995) 'Flow structure in and above the various heights of a saltmarsh canopy. A laboratory flume study,' *Journal of Coastal Research*, 11(4): 1204–9.

Shisler J.K. (1990) 'Creation and restoration of coastal wetlands of the northeastern United States', in Kusler J.A. & Kentula M.E. (eds) *Wetland creation and restoration: The status of the science*, Island Press, Washington: 143–70.

Shisler J.K. and Charette D.J. (1984) *Evaluation of artificial salt marshes in New Jersey*, New Jersey Agriculture Experiment Station, Publication No. p–40502–01–84.

Shuisky Y.D. (1994) 'An experience of studying artificial ground terraces as a means of coast protection', *Ocean and Coastal Management*, 22: 127–39.

Shuisky Y.D. and Schwartz M.L. (1979) 'Natural laws in the development of artificial sandy beaches', *Shore and Beach*, 47: 33–6.

Simeonova G.A. (1992) 'Coastal protection against erosion along the Bulgarian Black Sea', *Journal of Coastal Research*, 8(3): 745–51.

Silvester R. (1974) *Coastal Engineering*, Elsevier Science, Amsterdam.

——(1979) 'What direction coastal engineering?', *Coastal Engineering*, 2(4): 327–49.

Snyder R.M. (1978) 'Revegetation on erosion prone estuarine beaches', in Cole D.P. (ed.) *Proceedings of the Fifth Annual Conference on Restoration of Coastal Vegetation in Florida*, Hillsborough Ccommunity College, Tampa, Florida.

Sonu C.J. and Warwar J.F. (1987) 'Evolution of the sediment budget in the lee of a detached breakwater', in Kraus N.C. (ed.) *Coastal Sediments '87*, Proceedings of a speciality conference on advances in understanding of coastal sediment processes. New Orleans, Louisiana, 12–14 May 1987: 1361–8.

Spătaru A.N. (1990) 'Breakwaters for the protection of Romanian beaches', *Coastal Engineering*, 14: 129–46.

Stanley D.J. and Warne A.G. (1983) 'Nile Delta: recent geological evolution and human impact', *Science*, 260: 628–34.

Stapor F.W. (1971) 'Sediment budgets on a compartmented low to moderate energy coast in north-west Florida', *Marine Geology*, 10: M1–M7.

Stauble D.K. and Hoel J. (1986) *Guidelines for beach restoration projects. Part 3: Engineering*, Report No. 77, Florida Sea Grant Institute, Gainsville, Florida.

Stevens T. (1995) Coastal protection works. Fairlight Cove. (Personal communication).

Stoddart D.R. (1990) 'Coral reefs and islands and predicted sea level rise', *Progress in Physical Geography*, 14(4): 147–71.

Stone G.W. and Stapor F.W. (1996) 'A nearshore sediment transport model for the north east Gulf of Mexico coast, USA', *Journal of Coastal Research* 12(3): 786–93.

Suanez S. and Provansal M. (1996) 'Morphosedimentary behaviour of the deltaic fringe in comparison to the relative sea level rise on the Rhone delta', *Quaternary Science Reviews*, 15: 811–18.

Suffolk Coastal District Council (1999) 'Planning Policy and Coastal Erosion', *Personal communication*.

Sunamura T. (1983) 'Processes of sea cliff and platform erosion', in Komar P.D.

(ed.) *CRC Handbook of coastal processes and erosion*, CRC Press, Baton Rough, Florida: 233–65.

—— (1991) 'Processes of cliff erosion', in *Geomorphology of Rocky Coasts*, Wiley, Chichester, 75–116.

Tainter S.P. (1982) *Bluff slumping and stability: a consumers guide*, Report MICHU–SG–82–902, Michigan Sea Grant Publications Office, University of Michigan, USA.

Tait J.F. and Griggs G.B. (1990) 'Beach response to the presence of a seawall', *Shore and Beach*, 58(2): 11–28.

Tanimoto K. and Goda Y. (1992) 'Historical development of breakwater structures in the world', in Institution of Civil Engineers (ed.) *Coastal Structures and Breakwaters*, Thomas Telford, London.

Taylor P.M. and Parker J.G. (1993) (eds) *The Coast of North Wales and North-west England: An Environmental Appraisal*, Hamilton Oil, London.

Terchunian A.V. (1988) 'Permitting coastal armouring structures: Can seawalls and beaches co-exist?', *Journal of Coastal Research*, SI 4: 65–75.

—— (1990) 'Performance of beachface dewatering: the Stabeach system at Sailfish Point (Stuart) Florida', *Proceedings of the National Conference on Beach Preservation Technology*. Florida shore and beach preservation association, Florida, 1990: 185–201.

Thom B.G., Bowman G.M. and Roy P.S. (1981) 'Late quaternary evolution of coastal sand barriers, Port Stephens – Myall Lakes area, Central New South Wales, Australia, *Quaternary Research*, 15: 345–64.

Thyme F. (1990) 'Beach nourishment on the west coast of Jutland', *Journal of Coastal Research*, Special Issue 6: 201–9.

Titus J.G. (1986) 'Greenhouse effect, sea level rise, and coastal zone management. *Coastal Zone Management Journal* 14(3): 147–71.

—— (1990) 'Greenhouse effect, sea level rise and barrier islands: case study of Long Beach Island, New Jersey', *Coastal Management*, 18: 65–97.

—— (1991) 'Greenhouse effect and coastal wetland policy: how Americans could abandon an area the size of Massachusetts at minimum cost', *Environmental Management*, 15(1): 39–58.

Toft A.R. and Townend I.H. (1991) *Saltings as a sea defence*, NRA, R&D Note 29, National Rivers Authority, Bristol.

Townend I.H. and Fleming C.A. (1991) 'Beach nourishment and socio-economic aspects', *Coastal Engineering*, 16: 115–27.

Toyoshima O. (1972) 'Coastal engineering for the practising engineer: Erosion', *Gemba no tame no kaigan kogaku*, Japan (Translation).

Trent S.A, Sellery K. and Gordinier T. (1983) 'The re-vegetation potential of California coastal sand dunes using containerised native plant species', *Hortscience*, 18(4): 622.

Turner I.L. and Leatherman S.P. (1997) 'Beach dewatering as a 'soft' engineering solution to coastal erosion: A historical and critical review', *Journal of Coastal Research*, 13(4): 1050–63.

Turner N. (1994) 'Recycling of capital dredging arisings: the Bournemouth experience', in SCOPAC (eds) *Proceedings of conference on Inshore dredging for beach replenishment*, SCOPAC, 28th October 1994, Lymington, Hampshire.

Turner N. (1999) Beach nourishment at Bournemouth, (*personal communication*), 9th September 1999.

Underwood G.J.C. (1997) 'Microalgal colonisation in a saltmarsh restoration scheme', *Estuarine, Coastal and Shelf Science*, 44: 471–81.

USACE (1984) *Shoreline Protection Manual*, 4th Edn. US Army Corps of Engineers, Washington DC.

—— (1986) *Designing breakwaters and jetties*, US Army Corps of Engineers, Report EM 1110-2-2904, Washington DC.

——(1991) *Sand bypassing system selection*, US Army Corps of Engineers, Report EM 1110-2-1616, Washington DC.

—— (1992) *Coastal groynes and nearshore breakwaters*, US. Army Corps of Engineers, Report EM 1110-2-1617, Washington DC.

van Aarde R.J., Ferreira S.M., Kritzinger J.J., van Dyk P.J., Vogt M. and Wassenaar T.D. (1996) 'An evaluation of habitat rehabilitation on coastal dune forests in northern Kwa-Zulu Natal, South Africa', *Restoration Ecology*, 4(4): 334–45.

van Aarde R.J., Smit A.M. and Claassens A.S. (1998) 'Soil characteristics of rehabilitating and unmined coastal dunes at Richard's Bay, Kwa-Zulu Natal, South Africa', *Restoration Ecology*, 6(1): 102–10.

van Dijk H.W.J. (1989) 'Ecological impact of drinking water production in Dutch coastal dunes', in Meulen F, van der Jungerius P.D. and Visser J. (eds) *Perspectives in Coastal Dune Management*, SBP publishers, The Hague: 183–182.

van Dolah R.F., Knott D.M. and Calder D.R. (1984) *Ecological Effects of Rubble Weir Jetty Construction at Murrell's Inlet, South Carolina*, CERC, US Army Technical Report EL84-4, Vicksberg, Missouri.

van Malde J. (1991) 'Relative rise of mean sea level in the Netherlands in recent times', in Tooley M.J. and Jelgersma S. (eds) *Impacts of Sea Level Rise on European Coastal Lowlands*, Blackwell, Oxford.

van Oorschot J.H. and van Raalte G.H. (1991) 'Beach nourishment: execution methods and developments in technology', *Coastal Engineering*, 16: 23–42.

van Rijn L.C. (1997) 'Sediment transport and budget of the central coastal zone of Holland', *Coastal Engineering*, 32(1): 61–90.

van de Graaff J., Niemeyer H.D. and van Overeem J. (1991) 'Beach nourishment, philosophy and coastal protection policy', *Coastal Engineering*, 16: 3–22.

van der Linden M. (1985) 'Golfdempende constructies, evaluatie van drijvende golfdempende constructies in het grevelingenmear, Report to TU-Delft, The Netherlands.

van der Maarel E. (1979) 'Environmental management of coastal dunes in the Netherlands', in Jefferies R.L. and Davy A.J. (eds) *Ecological Processes in Coastal Environments*, Blackwell Scientific, London: 543–70.

van der Meulen F. and Salman A.H.P.M. (1996) 'Management of Mediterranean coastal dunes', *Ocean and Coastal Management*, 30(2–3): 177–95.

Vera-Cruz D. (1972) 'Artificial nourishment of Copacabana Beach', *Proceedings of the 13th Coastal Engineering Conference*. American Society of Civil Engineers 1451–63.

Verhagen H.J. (1990) 'Coastal protection and dune management in the Netherlands', *Journal of Coastal Research*, 6: 169–79.

—— (1992) 'Model for artificial beach nourishment', *Proceedings of the 23rd International Conference on Coastal Engineering*, ICCE, Venice, 1992: 2474–85.

—— (1996) 'Analysis of beach nourishment schemes', *Journal of Coastal Research*, 12(1): 179–85.

Viles H. and Spencer T. (1995) *Coastal Problems: geomorphology, ecology and society at the coast*, Edward Arnold, London.

Vincent C.E. (1979) 'Longshore sand transport rates: a simple model for the East Anglian coastline', *Coastal Engineering*, 3: 113–36.

Viner-Brady N.E.V. (1955) 'Folkestone Warren landslips: investigations 1948–1954', *Proceedings of the Institution of Civil Engineers, Railway Paper*, 57: 429–41.

Walker H.J. and Mossa J. (1986) 'Human modification of the shoreline of Japan', *Physical Geography*, 7: 116–39.

Walker J. R. (1987) 'Santa Barbara breakwater: an update', *Shore and Beach*, 55(3–4): 56–60.

Walker J.R., Clark D. and Pope J. (1981) 'A detached breakwater system for beach protection', *Proceedings 17th International Conference on Coastal Engineering*, Sydney, Australia: 1968–87.

Walton T.L. and Purpura J.S. (1977) 'Beach nourishment along the south-eastern Atlantic and Gulf coasts', *Shore and Beach*, 45: 10–18.

Walton T.L. and Sensabaugh W. (1978) 'Seawall design on sandy beaches', *University of Florida Sea Grant Report. No.29*, University of Florida, USA.

Wang X. (1989) 'Trend of sea level toward rise become apparent', *People's Daily, Overseas Edition* 29/3/89.

Warner M.F. and Barley A.D. (1997) 'Cliff stabilisation by soil nailing, Bouley Bay, Jersey, CI', in Davies M.C.R. and Schlosser F. (eds) *Ground improvement Geosystems*, Proceedings of 3rd International Conference on Ground Improvement Geosystems, Institution of Civil Engineers, London, 3–5 January 1997, Ch. 62: 468–76.

Warren R.S. and Niering W.A. (1993) 'Vegetation change on a north east tidal marsh: Interaction of sea level rise and marsh accretion', *Ecology*, 74(1): 96–103.

Watson I. and Finkl C.W. (1990) 'State of the art in storm surge protection', *Journal of Coastal Research*, 6: 739–64.

Watts P. (1998) 'Cliff recession: Friend or Foe?', The Naze Cliffs, Essex, unpublished BSc dissertation, Department of Geography, Lancaster University.

Webb J.W. and Dodd J.D. (1976) 'Vegetation establishment and shoreline stabilisation: Galveston Bay, Texas', *US Army Corps of Engineers Technical Paper No. 76-13*, Coastal Engineering Research Centre, Forth Belvoir, VA.

—— (1978) 'Shoreline plant establishment and use of a wave-stilling device', *US. Army Corps of Engineers Misc. Report No. 78-1*, Coastal Engineering Research Centre, Forth Belvoir, VA.

Weggel J.R. (1988) 'Seawalls: the need for research, dimensional considerations, and a suggested classification', *Journal of Coastal Research*, SI 4: 29–39.

Weggel J.R. and Sorenson R.M. (1991) 'Performance of the 1986 Atlanta City, New Jersey beach nourishment project', *Shore and Beach*, 59: 29–36.

Whatmough J.A. (1995) 'Grazing on sand dunes: the re-introduction of the rabbit *Oryctolagus cuniculus* L. to Murlough NNR, Co. Down', *Biological Journal of the Linnean Society*, 56(SA): 29–43.

Whitcombe L.J. (1996) 'Behaviour of an artificially replenished shingle beach at Hayling Island, UK', *Quarterly Journal of Engineering Geology*, 29: 265–71.

Wiegel R.L. (1964) *Oceanographic engineering*, Prentice Hall, Englewood Cliffs, New Jersey.

Wiegel R.L. (1993) 'Artificial beach construction with sand/gravel made by crushing rock', *Shore and Beach*, 61(4): 28–9.

Wigley T.M.L. and Raper S.C.B. (1990) 'Future changes in global mean temperature and thermal expansion-related sea level rise', in Warrick R.A. and Wigley T.M.L. (eds) *Climate and Sea Level Change: observations, projections, and implications*, Cambridge University Press, Cambridge.

Williams A.T. and Davies P. (1980) 'Man as a geological agent: the sea cliffs of Llantwit Major, Wales, UK', *Zeitschrift für Geomorphologie*, Supplementband 34: 129–41.

Willmington R.H. (1983) 'The nourishment of Bournemouth beaches 1974–1975', in Institution of Civil Engineers (ed.) *Shoreline Protection*, Thomas Telford Press, London: 157–62.

Wilson R.L. and Smith A.K.C. (1983) 'The construction of a trial embankment on the foreshore at Llandulas', Institution of Civil Engineers (eds) *Shoreline protection*, Thomas Telford Press, London: 223–33.

Wong P.P. (1981) 'Beach evolution between headland breakwaters', *Shore and Beach*, 49(3): 3–12.

Wood A. (1978) 'Coast erosion at Aberystwyth: the geological and human factors involved', *Geological Journal*, 13: 61–72.

Wood W.L. (1988) 'Effects of seawalls on profile adjustment along Great Lakes coastlines', *Journal of Coastal Research*, SI 4: 135–46.

Woodhouse W.W., Seneca E.D. and Broome S.W. (1974) 'Propogation of *Spartina alterniflora* for substrate stabilisation and salt marsh development,' USACE technical memo 46, Coastal Engineering Research Centre, Fort Belvoir, VA.

—— (1976a) 'Ten years of development of man-initiated coastal barrier dunes in North Carolina', *Bulletin of the North Carolina Department of Agriculture*, No. 453.

—— (1976b) 'Propogation and use of *Spartina alterniflora* for shoreline erosion abatement', USACE Report TR76-2, Coastal Engineering Research Centre, Fort Bellvoir, VA.

Woodward R. (1998) *'To what extent can natural retreat sites in Essex be used as an analogue for managed retreat?'*, unpublished BSc dissertation, Department of Geography, Lancaster University.

Woodworth P.L. (1987) 'Trends in UK mean sea level', *Marine Geology*, 11: 57–87.

Work P.A. and Rogers W.E. (1997) 'Wave transformation for beach nourishment projects', *Coastal Engineering*, 32: 1–18.

Worth H.R. (1909) 'Hallsands and Start Bay, Part 2', *Report and Transactions of the Devonshire Association for the Advancement of Science*, 41: 301–8.

Wrigley A. (1991) *Morecambe Coastal Defences*, unpublished report to the Institute of Water and Environmental Management, North-west and north Wales Branch.

Yang S.L. (1998) 'The role of *Scirpus* marsh in attenuation of hydrodynamics and retention of fine sediment in the Yangtze Estuary, *Estuarine, Coastal and Shelf Science*, 47: 227–33.

Zedler J.B. (1984) *Salt marsh restoration: A guidebook for southern California*, Report No. 7–CSGCP–009, California Sea grant Institute, USA.

Zenkovich V.P. and Schwartz M.L. (1987) 'Protecting the Black Sea – Georgian SSR gravel coast', *Journal of Coastal Research*, 3: 201–9.

Zwamborn J.A., Fromme G.A.W. and Fitzpatrick J.B. (1970) 'Underwater mound for protection of Durban's beaches', *Proceedings of the 12th Coastal Engineering Conference*, American Society of Civil Engineers: 975–94.

Subject index

(Italic entries signify non-text references)

Geographical index